Solid State Chemistry and Physics

VOLUME 1

Solid State Chemistry and Physics

AN INTRODUCTION

VOLUME 1

Edited by Paul F. Weller

Department of Chemistry
State University College
Fredonia, New York

MARCEL DEKKER, INC.   New York   1973

CHEMISTRY

MARCEL DEKKER, INC.
95 Madison Avenue, New York, New York 10016

LIBRARY OF CONGRESS CATALOG CARD NUMBER: 73-85302

ISBN: 0-8247-1776-7

Printed in the United States of America

PREFACE

   During the last decade the field of solid state science – or materials
science – has become increasingly interdisciplinary. Physicists and
metallurgists have been joined by chemists in solving many problems
relating to computer and communications technology, microcircuitry,
lasers, photography, and so on. And now the importance of electrical
and magnetic effects in biological processes has involved many biologists in the study and application of solid state principles.

   This text is intended for students of these interdisciplinary fields.
We have made an effort to present the fundamental principles and prac-
tices of the solid state in a language and a form that is understandable
not only to those acquainted with solid state physics, but also to chemists and biologists – or anyone sincerely interested in the properties of
solids. The concepts and topics, along with the necessary mathematical
background, covered in a typical undergraduate physical chemistry course
are assumed requirements for the entire book. In a few specific cases
somewhat more sophisticated mathematical treatments are used and
described. We have attempted to prepare the reader for these more
detailed sections by dividing the text into four major parts, with Part I
serving as a relatively nonmathematical introduction to the concepts
used throughout the rest of the text.

   The introductory chapter of Part I treats many of the concepts and
properties covered in more detail in later sections of the book. Many
intuitive, nonmathematical physical pictures and analogies are used to

introduce necessary principles. Crystal structures are covered in Chapter 2 without the use of sophisticated mathematics, and the very important topic of chemical bonds and their roles in solids is developed in Chapter 3. Then the effects and interactions of various imperfections in solids are treated in considerable detail. Rather strong emphasis has been placed on imperfections throughout the text. Their presence can have pronounced effects on the properties of materials, properties that are of critical importance to present and future technology and possibly even to fundamental life processes.

Part II builds on the concepts introduced in Part I and covers many of the properties of solids – electrical, optical, magnetic, etc. – in more rigorous detail. Since the study and correct understanding of many of the properties described in Part II require highly pure single crystals or crystals with carefully controlled impurity contents, Part III considers many aspects of sample preparation, purification, and single crystal growth. In Part IV the principles developed in Parts I through III are applied to the cases of polymers and biological processes.

While this text is certainly not an exhaustive treatment of all topics important to the solid state scientist, it is hoped that the presentation will whet the appetite and stir the interest for further investigation by individuals with various backgrounds and training. We believe that the coverage is sufficient in scope and depth to establish a firm foundation in the principles of the solid state, from which a more detailed study of the entire field or of a given specialized area can be launched.

Paul F. Weller

LIST OF CONTRIBUTORS

JOHN D. AXE, Department of Physics, Brookhaven National Laboratory, Upton, New York

BILLY L. CROWDER, IBM Thomas J. Watson Research Center, Yorktown Heights, New York

PAUL H. KASAI, Union Carbide Research Institute, Tarrytown, New York

JEROME H. PERLSTEIN,[*] Department of Chemistry, The Johns Hopkins University, Baltimore, Maryland

J. J. STEGER, Department of Chemistry, Cornell University, Ithaca, New York

LAWRENCE SUCHOW, Chemistry Division, Department of Chemical Engineering and Chemistry, Newark College of Engineering, Newark, New Jersey

PAUL F. WELLER, Department of Chemistry, State University College, Fredonia, New York

---

[*]Present address: Research Laboratories, Eastman Kodak Company, Rochester, New York

CONTENTS

PART I
CONCEPTS AND PROPERTIES

3   BONDING MODELS OF SOLIDS   139
          Billy L. Crowder

PART II

PHYSICAL PROPERTIES AND IMPERFECTIONS

4   ELECTRICAL PROPERTIES OF SOLIDS   189
          Jerome H. Perlstein

CONTENTS OF VOLUME 2

Solid State Chemistry and Physics

VOLUME 1

PART I

CONCEPTS AND PROPERTIES

Chapter 1

AN INTRODUCTION TO PRINCIPLES OF THE SOLID STATE

Paul F. Weller

Department of Chemistry
State University College
Fredonia, New York

I. INTRODUCTION

Practical use of materials, or solids, depends to a large extent on our ability to change or modify the properties of a certain specific material or group of materials. Economically, technologically, or scientifically, needs for solids with certain characteristics continually arise. These needs are for materials with new mechanical properties, or with different magnetic properties, or with special optical characteristics, or with higher thermal conductivities, or with an unlimited assortment of combined characteristics such as high chemical stability, high solid state symmetry, strong optical absorption bands at particular wavelengths, and sharp visible fluorescent transitions.

In all cases it is the task of the materials scientist – physicist, chemist, metallurgist, biologist, or whoever – to devise ways in which desired properties can be obtained. This requires, of course, a basic understanding of the concepts and principles of materials. Physicists and metallurgists have dealt with problems of the solid state for many years; chemists have more recently become involved. It is clear that the concepts of solids will become increasingly important to the study and understanding of surface properties of materials, of polymers and large organic molecules, and of biochemical and life processes. Increasingly, scientists interested in materials, the solid state and properties thereof, will be drawn from disciplines other than solid state physics. Throughout this text, therefore, we have attempted to present and develop the concepts and principles of solids in a way that will permit – and encourage – the understanding of the solid state by all students and practitioners in the sciences.

II. MATERIALS, THEIR CLASSIFICATIONS

There are many ways in which we can broadly classify different types of solid materials. They can, for example, be grouped into two major categories: (1) crystalline solids, such as NaCl and GaAs, in which the atomic building blocks occur in more or less regular patterns, and (2) amorphous solids, such as glasses and polymers, in which there is no definite macroscopic pattern to the atom arrays. We have, for the most part, considered crystalline materials in this text, although glasses are mentioned in a few places and Chapter 13 is devoted entirely to polymers and their solid state properties.

While it is helpful to separate solids into these two categories, it is clear that the categories are so very broad that other classification schemes are necessary. We can also group solids into categories according to the chemical bond types that are thought to exist in the solids. For example, while NaCl and Al metal are both crystalline solids, there are vast differences between the two solids (see Table 1). Often we say that NaCl is an ionic solid, i.e., it contains ionic bonds in which positive and negative ions attract one another [Fig. 1(a)], and that Al is a metal in which metallic bonding is present, i.e., metal cations are surrounded by an electron gas or embedded in an electron sea [Fig. 1(b)].

There are two other bond types [Fig. 1(c) and (d)] that characterize crystalline materials: the covalent solids, diamond (C) and InSb, and the molecular solids, H_2 and Ar (see Table 1). Diamond and InSb have properties that we often attribute to solids that contain covalent chemical

TABLE 1

Classification of Solids According to Bond Type

Bond type	Examples	Attractive force	Binding energy	Properties
Van der Waals (molecular solids)	Ar H_2 CO_2	Dipole-dipole	Weak, about 0.1 eV/molecule or 2 kcal/mole	Low melting and boiling points; compressible; soft; electrical insulators
Ionic	NaCl CaF_2 Al_2O_3	Electrostatic	High, about 10 eV/molecule or 200 kcal/mole	High melting and boiling points; hard and brittle; electrical insulator at low temperatures and ionic conductor at high temperatures
Covalent	Si InSb PbTe	Electron pair sharing	Moderate to high, about 1 eV/molecule or 20 kcal/mole	Moderate to high melting and boiling points; moderate to very hard; electrical insulator or semiconductor
Metallic	Na Cu Fe	Free electron gas	Moderate to high, about 1 eV/molecule or 20 kcal/mole	Moderate to high melting and boiling points; soft or hard; electrical conductors

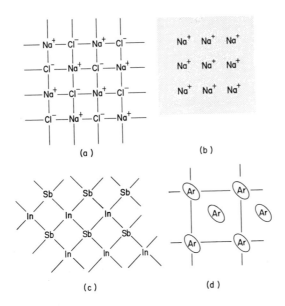

FIG. 1. Four different bond types: (a) ionic, (b) metallic,
(c) covalent, and (d) molecular.

bonds, i.e., the atoms are attracted by sharing electrons. In molecular
solids, van der Waals interactions serve as the effective intermolecular
attractive forces. These are much weaker forces, as indicated by the
properties listed in Table 1. (Sometimes solids containing hydrogen
bonds, such as ice, are considered as a separate bond type.)

Solids characterized by the above bond types, sometimes referred
to as different lattice networks, can also be categorized by the particular way in which their lattice units are arrayed in space, i.e., by their
crystal structures. The ionic solid NaCl, for example, crystallizes
with the sodium chloride structure in which Na^+ and Cl^- ions are both
cubic close packed (ccp), forming interpenetrating face centered cubic
(fcc) space lattices (see page 80). Or, alternatively, we can view
the NaCl crystal structure as an array of ccp Na^+ ions with the Cl^- ions
occupying all of the octahedral holes formed by the Na^+ array (see
page 115). An extended description of crystal structures [1] is given
in Chapter 2.

We can also group solids into three categories according to their conductivity properties: (1) metals, (2) semiconductors, and (3) insulators. Such groupings permit a discussion of various types of materials and the differences among them, using the band theory of solids. Since the ideas of energy bands in solids will be used throughout the text, an introduction to these important concepts will be given here.

III. ELECTRICAL CONDUCTIVITY

Some materials conduct electricity quite well, copper, silver, aluminum (all metals), while other materials are very poor conductors, diamond, sodium chloride, aluminum oxide (all insulators). This difference in behavior between metals and insulators is quite interesting and is considered in more detail later. First, let us consider the conductivity property of metals. We know that electrical charge can be transported from, say, one end of a metal wire to the other through the movement of electrons. One question that we might ask is, "What determines the conductivity of the metal, i.e., how easily is charge transported from one end of the wire to the other?" A simple model for this process was developed early in the twentieth century.

A. The Free Electron Theory

The free electron theory of metals was first proposed by several scientists shortly after the discovery of the electron; among these were P. Drude in 1902 and H.A. Lorentz in 1909 [2]. Very briefly, the ideas of Drude and Lorentz were as follows (see Chapter 4) [3]. They assumed that a metal was composed of positive ion cores embedded in a perfect gas of freely moving electrons. When an electric field was applied to the metal, the electrons were free to flow along the potential gradients giving rise to an electric current. After some mean time of motion, the electrons were envisioned as colliding with one another or with the ion cores, which destroyed the component of their motion in the direction of the field. Then using classical equations of motion for a particle in a field F and defining the conductivity σ as $\sigma = j/F$ where j is the current obtained, they were able to derive the relation

$$\sigma = ne\mu$$

where n is the number of free electrons, e is the electron charge, and μ is the electron mobility (free electron velocity per unit field strength). Common units are: $(ohm\text{-}cm)^{-1}$ for σ, e^-/cm^3 for n, coulomb for e^-, and $cm^2/V\text{-}sec$ for μ.

While the free electron theory of Drude and Lorentz succeeds in explaining some aspects of the electrical conductivity of metals, it fails when applied to other metallic properties, such as specific heats and paramagnetic properties of the free electrons. Furthermore, since it treats the valence electrons in conductors essentially as gaseous particles, it does not provide a very good working model for solids in general. For example, the differences among metals, insulators, and semiconductors are unexplained. We know that metals are good conductors and insulators very poor conductors, while semiconductors show intermediate conductivities. There is also another difference in the electrical properties of these three materials. The electrical conductivity of a metal decreases as its temperature is increased, while the conductivities of semiconductors and insulators are higher at increased temperatures.

However, the properties of solids in general as well as many of the observed differences between various materials can be explained by application of the band theory.

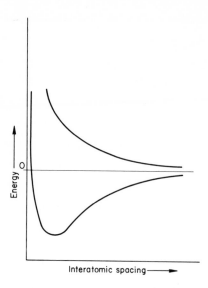

FIG. 2.   Potential energy vs internuclear separation for two hydrogen atoms (1s atomic orbitals) forming a hydrogen molecule, H_2.

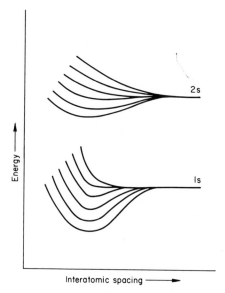

FIG. 3.   A potential energy diagram, similar to Fig. 2, for a row of six equally spaced atoms (after Shockley [7]).

IV. BAND THEORY, FORMULATIONS

A. Energy Band Formation in Solids [4, 5]

The formation of energy bands or relatively broad electron energy levels in solids can be considered using two different approaches: (1) the interaction of localized electrons and their energy levels (often called the tight binding approximation) and (2) a collective or essentially free electron approximation. Only a brief and rather qualitative discussion of these two approaches to band formation is given here. They are treated in more detail in later chapters, particularly in Chapters 3 and 4.

1. Localized Electrons

When two hydrogen atoms interact to form a diatomic hydrogen molecule, their 1s wave functions overlap producing two energy levels (each of which can contain two electrons) appropriate to the H_2 molecule [6], as diagrammed in Fig. 2. If a group of six atoms are placed in a row with equal separations between the atoms, a treatment similar to that for two hydrogen atoms yields a set of six allowed energy levels split from the 1s atomic levels in a fashion similar to that for the two levels in H_2 [7]. As illustrated in Fig. 3 the 2s wave functions overlap at larger atomic spacings, but the same general pattern of energy level splittings is preserved.

Solids, of course, contain not six but about 10^{22} atoms/cm^3. When these huge numbers of atomic wave functions combine, the maximum amount of splitting of the atomic energy levels is not changed appreciably, but the number of energy levels split from any given atomic level is vast. Consequently, there is a very small energy difference between the levels split from any one atomic level; these energy levels essentially form an energy continuum, i.e., an energy band. There remains an energy separation between the 1s and 2s (the 2s and 2p,

2p and 3s, etc.) energy bands, but within a given band the discrete energy levels have an extremely small separation in energy [8].

2. Free Electrons

A similar description of energy bands in solids is obtained by considering an assembly of essentially free electrons and the manner in which they interact with, and are affected by, the atoms in a crystal lattice.

An approximate solution to this problem, first given by Sommerfeld [9] in 1928, is obtained by requiring that the electrons behave according to the principles of quantum mechanics and that the potential energy well (created by the atoms at the lattice sites) in which the electrons move have infinitely high walls, i.e., $V = 0$ within the solid (metal) but $V = \infty$ outside the solid, so that electrons can move freely within but cannot escape the solid. This is the "particle in a box" problem [10], and in the one-dimensional case the Schrödinger equation is

$$\frac{d^2\psi}{dx^2} + \frac{8\pi^2 m}{h^2}(E - V)\psi = 0 .$$

Here E is the total energy, V is the potential energy of the electron of mass m, h is Planck's constant, and ψ is the wave function associated with the electron. Since $V = 0$ within the solid, the equation reduces to

$$\frac{d^2\psi}{dx^2} + \frac{8\pi^2 mE}{h^2}\psi = 0 .$$

The solutions for this equation are standing waves of the form

$$\psi = e^{\underline{i}kx}$$

where k is called a wave number and

$$\underline{k} = \frac{2\pi p}{h} = \frac{2\pi}{\lambda}$$

where ρ and λ are the electron momentum and wavelength, respectively. The corresponding energy levels are given by

$$E = \frac{k^2 h^2}{8\pi^2 m} = \frac{n^2 h^2}{8ma^2} \quad,$$

where n is the principal quantum number and a is the "box" dimension.

This approximation, then, yields essentially one continuous band of energy levels for different values of k (or n). An improvement is needed, and a major flaw in the original set of assumptions is rather apparent. Since the atoms in a crystal lattice occur periodically, the potential energy of the electrons within a solid will not be zero but will also vary periodically, because of the charges at the lattice sites. This periodicity must be included in the solutions to the wave equation and in the associated expressions for the allowed energy levels.

The problem for a single electron moving in a linear array of square potential wells (corresponding to the atomic cores at the lattice sites) was first solved by Kronig and Penney [11]. The wave function for an electron in this case, ψ_B, is composed of the wave function of a free electron, $\psi(x)$, essentially the same as the one already given, combined with a function which has the periodicity of the lattice, u(x), or

$$\psi_B = \psi(x)\,u(x) \quad.$$

Functions of this form are called Bloch functions and have solutions composed of running waves modulated by the periodic function and given by

$$\psi_B = e^{ikx}\,u(x) \quad.$$

At this point one might be tempted to say "so what," but what appears to be only a slight modification in results actually turns out to be quite significant. The periodic function permits only specific values for electron energies within the solid; other energy values are not allowed. Just as a periodic array of lattice atoms reflects certain x-ray wavelengths while other wavelengths pass through a solid, this same atomic array permits only certain electron energies (or wavelengths) and disallows others. Instead of one (essentially) continuous band of energy levels, as in the case of the "electron in a box," the allowed electron energy levels fall in a series of energy bands, with the bands separated by regions of forbidden energy or energy gaps. The energy bands and gaps correspond to certain values of electron energies E (or wave numbers $\underline{k}$ or wavelengths λ since $\underline{k} = 2\pi/\lambda$).

An expression for E similar to that given above for the "electron in a box" can be used if a factor called the effective electron mass is substituted for the free electron mass,

$$E = \frac{k^2 h^2}{8\pi^2 m^*} \quad .$$

The effective mass m^* depends on $\underline{k}$

$$m^* = \frac{h^2}{d^2 E/d\underline{k}^2}$$

and, consequently, can be used to account for the energy band structure.

3. Energy Band Diagrams

A very common method of describing allowed electron energy bands and their corresponding forbidden energy gaps is shown in Fig. 4. The band of highest energy that is completely filled by electrons is generally called the valence band, i.e., electrons associated with this band are involved in chemical bonding and are, consequently, rather localized and not free to move throughout the solid. Bands of lower energy are usually not considered (nor shown in a diagram such as the one in Fig. 4) since they involve, primarily, core electrons and are even more localized to the individual atoms. The lowest lying

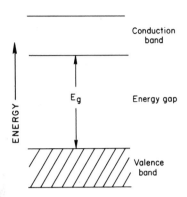

FIG. 4. A typical energy band diagram for solids showing a conduction band containing no electrons separated by a forbidden energy region, the energy gap E_g, from the valence band which has energy states completely occupied by electrons.

energy band that is not completely occupied by electrons (it can be, and often is, empty) is generally called the conduction band. Electrons located in this band are free (approximately) to move throughout the solid with essentially zero activation energy. There are, of course, bands of higher energy; these are usually not shown in typical diagrams.

There is a forbidden energy region between the valence band and the conduction band; this energy separation is generally called the energy gap or band gap (E_g in Fig. 4).

B. Energy Bands: Metals, Semiconductors, and Insulators

1. Band Diagrams

The differences among metals, semiconductors, and insulators can be described in terms of energy band diagrams, as shown in Fig. 5. In metals, Fig. 5(a), the conduction band is partially occupied by electrons at all temperatures including 0 °K. In insulators, Fig. 5(c), the conduction band is essentially empty at all temperatures. In semiconductors, Fig. 5(b), the conduction band is partially populated at "higher"

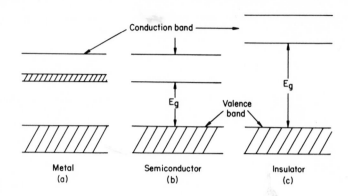

FIG. 5. Typical band diagrams illustrating the differences between: (a) metals, the conduction band is partially occupied by electrons; (b) semiconductors, the conduction band contains essentially no
electrons if $E_g > kT$; when $E_g \approx kT$, electrons are excited across the
band gap, from the valence band into the conduction band; (c) insulators,
the conduction band is essentially free of electrons at all normal temperatures since $E_g \gg kT$.

temperatures but is completely empty at $0\,°K$ (or, more generally, at
"low" temperatures).

While metals differ from the other two types of materials in the
occupancy of the conduction band states, the difference between semiconductors and insulators is much less definite. Conduction band electrons in intrinsic semiconductors and insulators are produced by excitation of electrons from the valence band (out of the chemical bonds).
Thermal excitation of these electrons, then, depends on the band gap,
i.e., the energy difference between the valence and conduction bands.
In insulators the band gap energy is large with respect to available
thermal energies, at temperatures below the melting point. In semiconductors (intrinsic) thermal production of electrons, say at or above
room temperature, becomes possible since band gaps are smaller than
they are in insulators. At lower temperatures, less thermal energy is
available and fewer electrons are excited across the forbidden energy
gap; consequently, fewer conduction electrons are present at low temperatures in semiconductors.

2. Conductivity: Temperature Dependence

The differences shown in Fig. 5 and described in the previous section indicate the reasons for different conductivity variations with changing temperature between metals and semiconductors (or insulators). Using the free electron relationship for the specific conductivity, $\sigma = ne\mu$, we see that σ depends on both n and μ. Then the temperature dependence of σ stems from the variations of n or μ, or a combination of the two, with temperature.

In the case of metals, the free electron concentration n (the number of conduction electrons per cm^3) remains essentially constant at all temperatures below the melting point [electrons occupying conduction band states in Fig. 3(a)]. Then the temperature dependence of the conductivity σ is caused by the variation in electron mobility μ. An increase in the temperature increases the thermal vibrations of the metal ions at the lattice sites. Conduction electrons interact more strongly with the more vigorously vibrating positive ions and are scattered more often from their straight-line motion and pathway. These scattering processes [12] decrease the electron mobility and, consequently, the conductivity as the temperature is increased. At most temperatures (above a certain temperature, the Θ or Debye temperature, characteristic of each metal) the mobility and conductivity are proportional to $1/T$. Hence, a plot of σ v T is a straight line.

For semiconductors the situation is quite different since n can change as the temperature is varied. In fact, the production of charge carriers follows an exponential dependence on temperature. The number of conduction electrons n is given by

$$n = n_0 e^{-E_g/kT} \quad ,$$

where n_0 is the concentration of atoms at the lattice sites, E_g is the band gap energy, and k is the Boltzmann constant (assuming $E_g > kT$). Then the conductivity relationship becomes

$$\sigma = (n_0 e^{-E_g/kT}) e\mu \quad .$$

Since μ varies approximately as $1/T$ (a very weak temperature dependence), the major temperature variation in the conductivity of a semiconductor is caused by the dependence of the carrier concentration n. Very often n_0, e, and μ are combined into a preexponential constant, assumed to be temperature independent, to give

$$\sigma = A e^{-E_g/kT} \quad ,$$

where

$$A = n_0 e\mu$$

and

$$\log \sigma = \log A - \frac{E_g}{2.303\,kT} \quad .$$

Then the plot of $\log \sigma$ v $1/T$ should give a straight line with a slope of $-E_g/2.303\,k$. The measurement of σ as a function of temperature will, therefore, permit calculation (approximate) of the band gap energy E_g of a semiconducting material. (A similar treatment can be used for insulators, but energy gaps are generally so large that thermal production of free carriers is negligible.)

3. Energy Gaps and Ionicity

A relationship exists between the band gap energy E_g and the amount of ionic character present in a solid (or the electronegativity difference in a binary compound). As the percent ionic character of a chemical bond increases, the electrons become more tightly bound to the cores of the atoms involved. Since there is a greater degree of localization of the possible charge carriers with strong ionic bonding, we would predict that highly ionic compounds would have large band gap energies. At least qualitatively, this trend is observed. A highly ionic substance, such as NaF, has a band gap (at 25 °C) of about 12 eV, while a compound with a low percent ionic character, such as InSb, has a band gap of about 0.17 eV. Further examples are given in Table 2.

V. EXTRINSIC SEMICONDUCTORS

So far, our treatment has covered only the case of <u>intrinsic semiconductors</u> in which charge carriers arise from the chemical bonds of the semiconductor itself. Most common semiconductors that are important to technology, however, are not of the intrinsic type but are <u>extrinsic semiconductors,</u> in which charge carrier production is determined by lattice imperfections or by trace amounts of impurities.

In the important semiconductor gallium arsenide, for example, the room temperature band gap energy is about 1.4 eV. This energy corresponds to a temperature of about 16,000 °K and, therefore, thermal production of significant numbers of electron charge carriers would require temperatures near or exceeding the GaAs melting point of about 1511 °K. At room temperature, therefore, a sample of GaAs showing only intrinsic conductivity would be a very poor conductor and, in fact, could be classed as an insulator. But unless very special circumstances are present, GaAs samples usually exhibit semiconductor behavior at room temperature with either electron or hole charge carriers present at relatively high concentrations, say 10^{17} carriers/cm^3 or more, yielding sample conductivities of about 100 reciprocal ohm-cm or higher.

The carrier type, carrier concentrations, and sample conductivities can be selectively controlled for GaAs, and a few other important semiconductors, via careful addition of particular impurities (generally present at very low concentrations, i.e., between about 0.1% and less than 1 ppm). These extrinsic semiconductors permit many technological applications.

The presence of <u>imperfections</u> or <u>defects</u> [13-16]— foreign impurities dissolved in the host material or irregularities in the orderly

TABLE 2

Dependence of E_g on Percent Ionic Character

Substance	E_g[a] (eV)	Approximate % ionic character[b]
NaCl	7.7	67
NaI	5.8	47
AlN	5.1	43
AgCl	4.1	26
AgBr	3.6	19
ZnS	3.6	19
ZnSe	2.6	15
AlP	2.8	9
AgI	2.7	9
ZnTe	2.3	6
AlAs	2.2	6
AlSb	1.6	4
InSb	0.16	1
GaN	3.3	39
GaP	2.2	6
GaAs	1.4	4
GaSb	0.67	2
ZnTe	2.3	6
CdTe	1.5	4
HgTe	0.15	1

[a]Values for E_g are for 25 °C; 1 eV = 23.05 kcal/mole and corresponds to a temperature of 1.16×10^4 °K, a wave number of 8066 cm^{-1}, and a wavelength of 1.24 μ.

[b]Calculated using Pauling's electronegativity differences.

geometric array of the host crystal lattice – can create isolated centers in a solid which can ionize and thereby contribute electrons to the conduction band of the material. Such defects are called <u>donor centers</u> or <u>donors</u> since they donate conduction electrons to the semiconductor. When these negatively charged electrons are the primary charge carriers, the material is termed an <u>n-type</u> semiconductor. Other crystal defect types create <u>acceptor centers</u> or <u>acceptors</u>, which remove (or accept) electrons from the chemical bonds of the semiconductor, producing electron vacancies. These vacancies, or <u>holes</u> as they are called, behave as positive charge carriers located in the valence band. A <u>p-type</u> semiconductor results when holes are the predominant charge carrier.

Imperfections, or defects which are found in all real materials and usually at rather high concentrations, fall into three broad categories: (1) <u>point defects</u>, (2) <u>line defects</u>, usually referred to as dislocations, and (3) <u>plane defects</u>, which are two-dimensional disruptions of a perfect crystal such as grain boundaries. All three imperfections or defect types affect crystal properties and are covered in Chapter 2, but point defects are considerably more important to the effects that we are now considering.

A.   Point Defects

Point defects can be classified in two ways: those that are <u>native</u> to the material itself and those that involve the presence of a <u>foreign</u> impurity.

1.   Native Defects

There are several different kinds of native defects. One, illustrated in Fig. 6, is called an <u>interstitial</u>. There is an extra atom in the crystal, located somewhere between, but not on, lattice sites. This defect is illustrated in Fig. 6 by the Cd atom that does not occupy a regular geometric site in the CdS space lattice. Interstitials are often

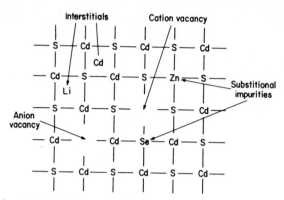

FIG. 6. Point defects in CdS.

limited by available-space considerations in the crystal lattice. Consequently, the cation-forming atom is a much more likely interstitial candidate (Cd rather than S in CdS).

A second type of native defect is the <u>vacancy</u>. When a vacancy is present, an atom is absent from its normal lattice site. Any atom may be omitted from a solid as it forms, giving rise to cation and anion vacancies. In Fig. 6 both a Cd cation vacancy and a S anion vacancy are shown.

Native vacancies and interstitials usually occur together in any real solid. Two combinations of defects have specific designations. The Schottky defect consists of a cation and anion vacancy combination in the crystal lattice, while a Frenkel defect is an interstitial-vacancy combination.

2. Foreign Defects

An impurity or foreign atom can be located either at regular lattice sites, a <u>substitutional impurity</u>, or at positions not normally occupied in the lattice, an <u>interstitial impurity</u>. These defects are shown in Fig. 6, with Li occurring as an interstitial and Zn and Se as substitutional cation and anion impurities, respectively.

3. Defect Charges

There are two things that we should note about the above defects.

(1) Charge balance must always be maintained. This can occur in different ways, many of which are often present in the same material. If an interstitial with a +1 charge is present, one more negative or one less positive charge is required somewhere else in the crystal. In CdS with a Li^+ interstitial, for example, charge compensation could occur via a free (conduction) electron, or a Na^+ ion substituting for a Cd^{2+} ion, or a P^{3-} ion substituting for a S^{2-} ion, and so on. Naturally, the various charge compensation mechanisms affect the material in different ways.

(2) We must be conscious of, and careful of, two different methods of describing the charges that exist on the lattice atoms and any imperfections that are present. Chemists usually discuss "ionic-type" materials, such as CdF_2, ZnS, InP, and PbTe, in terms of the ions that exist at the lattice sites. In solid ZnS, for example, we often talk about Zn^{2+} and S^{2-} ions. That is a perfectly good approach, which physicists also often use, but notice that when a Na^+ ion substitutes for a Zn^{2+} ion in ZnS, a chemist might refer to a substitutional atom with a +1 charge while a physicist might refer to the same substitutional sodium atom as having a -1 charge. Actually it is the same defect. Remember that the substitutional Na^+ ion must be referred to the rest of the lattice. There is a +1 charge deficiency at the substitutional site, which appears to be a -1 charge to the surrounding lattice. There are many problems similar to this one; keep the two different approaches in mind (see Chapter 7 and Ref. [15]).

B. Imperfections and Properties

The presence of imperfections affects the properties of materials in many different ways. The structural, mechanical, magnetic, optical, and electrical properties – essentially all properties – of solids depend,

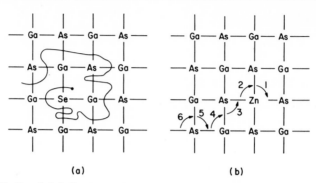

(a)   (b)

FIG. 7. (a) A Se impurity in GaAs provides an "extra" electron, relatively weakly bound to the Se; (b) a Zn impurity in GaAs provides one too few electrons for bonding. Motion of this hole is from position 1 to position 6, as successive electron "jumps" into the "dangling bond" are indicated by the arrows labeled 1, 2, etc.

and sometimes dramatically, on the type of defects or impurities present in the material. To illustrate these impurity effects let us first consider the creation of donor and acceptor centers by the presence of point defects. Later, effects on the electrical conductivity and impurity solubilities will be introduced.

1. Native Defects

A donor level is formed by an anion vacancy in a semiconductor. We can justify this in the following way. You can think of the formation of an anion (or nonmetal) vacancy, a S^{2-} vacancy in CdS as diagrammed in Fig. 6, for example, as the removal of a neutral S atom from the lattice – which leaves two electrons behind and loosely bound to the vacant site. The electrons are weakly attracted to the vacant site, since the surrounding Cd^{2+} cations "relax" into the vacancy creating a small positive charge, and can be rather easily excited into the conduction band. An n-type semiconductor results.

Similar reasoning indicates that cation (or metal) vacancies produce acceptor centers. Removal of a neutral Cd atom from CdS leaves two holes (electron vacancies) loosely bound to the cation vacancy. Each hole can trap – or accept – an electron from the valence band, or

alternatively, holes can be excited ("down") into the valence band.
This creates positive charge carriers and a p-type semiconductor.

2. Foreign Defects

Impurities dissolved in a semiconducting material can also produce donor and acceptor centers. Donors can be formed by small metallic interstitials, such as Li in GaAs. Upon ionization, Li^+ is formed
along with an electron located in the conduction band of GaAs.

Donor centers can also be formed with substitutional impurities.
If a lattice atom is replaced by an atom that contains more valence
electrons, the substituted atom is generally a donor. For example, the
substitution of Ge at a Ga site or Se at an As site in GaAs will produce
donor centers and n-type GaAs. This is diagrammed in Fig. 7(a). Se
has six valence electrons compared to five for As. Since only five
electrons are required from either Se or As to form the required bonds
in GaAs, Se contributes one "extra" electron, i.e., an electron not
used in bond formation. This electron is attracted to the substituted
Se atom but is not localized or held as strongly as the bonding electrons. Hence, it can be excited, say thermally, into the GaAs conduction band where it is free to move throughout the sample under the
influence of an electric field.

Figure 7(b) shows p-type GaAs containing a substitutional Zn at
a Ga site. There is a deficiency of one electron for complete bond formation. This electron deficiency, or hole, is attracted to the Zn atom.
(It might be easier to visualize this attraction if you think of a Zn^{2+}
substituting for a Ga^{3+}, with the lower positive charge appearing to the
rest of the crystal as a negatively charged region.) The hole can be
excited down into the valence band and travel through the GaAs sample,
which corresponds to the schematic movements in Fig. 7(b). Note that
the hole "movement" is actually the result of electrons moving in the
opposite direction (see Chapter 4).

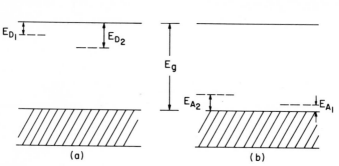

FIG. 8. (a) An energy band diagram for an n-type semiconductor containing two donor centers with different activation energies, E_{D_2} and E_{D_1}; (b) a p-type semiconductor with two different acceptor centers.

3. Energy Band Diagrams

The presence of donors and acceptors in a particular semiconducting material is conveniently indicated using energy band diagrams, as shown in Fig. 8. Donor centers are located – in energy terms – near the conduction band [Fig. 8(a)]. Ionization of an electron from the donor level into the conduction band requires the energy E_D; various types of donors have different activation energies, E_{D_1}, E_{D_2}, etc. A similar description applies to the acceptors. The activation energy E_A for the ionization of a hole from the acceptor level into the valence band (or, alternatively, of a valence band electron into the acceptor) depends on the type of acceptor present, hence, E_{A_1} and E_{A_2} in Fig. 8(b).

It is worthwhile to emphasize that donor (or acceptor) centers are effectively isolated from one another in their solid host material. There are no easy interaction pathways directly connecting the donors (or acceptors). Interesting changes occur in the semiconductor electrical properties when donor (or acceptor) centers are present in high enough concentrations to establish direct interactions (energy level overlap and impurity banding, for example; see Chapter 4).

We can qualitatively understand why the donor levels should be located somewhere near the conduction band (or the acceptor levels

near the valence band) using the following argument. Let us consider,
for example, an n-type GaAs sample which contains Se substituting for
some of the As atoms. Selenium needs to use only 5 of its 6 valence
electrons to complete its 4 bonds to the surrounding (tetrahedrally) Ga
atoms. The sixth electron of Se is not used in bond formation and is,
therefore, not held as strongly to the Se atom – or better, between the
Se and Ga atoms. Then the energy level associated with the sixth
electron lies <u>above</u> the valence band states. And yet the electron is
still attracted by the substituting Se atom nuclear charge and is, therefore, not free to roam throughout the entire sample. Hence, the energy
level associated with this sixth electron must lie somewhere below the
conduction band edge. This places the Se donor energy level somewhere within the forbidden energy gap (E_g in Fig. 8) of the host semiconductor GaAs.

A similar argument can be used for acceptor centers, but it is
(for most people) less easily visualized since the behaviors of holes
often seem to violate our intuitive feelings of how things operate, i.e.,
holes prefer high energies, they are "excited" to lower energies, etc.

While we can qualitatively locate a donor (or acceptor) level
somewhere within the band gap of the host material, it is, in general,
not possible to calculate the activation energy for a particular donor
center in a given semiconductor. In a few cases, for example, Si or
Ge containing As or Ga, the experimentally observed donor or acceptor
activation energies correspond closely to those calculated using a
hydrogen-like atom approach [17].

4. Conductivity Temperature Dependence

From the previous discussions we would expect increasing conductivity for an extrinsic semiconductor as its temperature is increased.
Indeed this is just what is observed experimentally. An extrinsic semiconductor's conductivity increases exponentially with temperature. This
dramatic rise in conduction is caused by increased numbers of charge

carriers and is, to a first approximation, unaffected by the mobility of the carriers. The considerations are similar to those used in discussing the temperature dependence of intrinsic semiconductors. An extended treatment is given in Chapter 4.

C.   Presence of Donors Plus Acceptors

In the previous discussion we treated n-type semiconductors as if donor centers were present by themselves and p-type cases as if only acceptors were present. That is , we assumed that donor concentrations, and consequently electron charge carrier concentrations, were much greater than acceptor, or hole, concentrations in n-type materials (and vice versa for p-type conductors). In many cases this is a valid and useful assumption, but there are important effects that arise when both acceptor and donor centers are present in the same semiconductor. This is, of course, the usual situation since native or foreign defects which serve as donors and as acceptors are present in all real solids, even though we might try hard to eliminate them [15, 16, 18].

Let us consider just two of the effects that are caused by varying relative concentrations of donors and acceptors in the same host material. One is concerned with changes in the solubility of a given impurity and the second with changes in the electrical conductivity of the material. It has been found experimentally, and subsequently explained theoretically, that, for example, the conductivity of an n-type semiconductor can be varied over a wide range (several orders of magnitude if the acceptor concentration in the material can be selectively and accurately controlled). The case is similar for solubilities, say, of donors in a given solid. If the acceptor concentration is changed, the solubility of the donor is affected.

1. Solubility Effects: Equilibrium Treatments

The interactions of donor and acceptor centers in solids can be considered in several different ways, and in quantitative fashion. One can, for example, write down all appropriate reactions involving the donors and acceptors and solve the corresponding mass action expressions simultaneously. Or a kinetic approach can be used in which the rates of electron excitation into, and trapping out of, the conduction band are used. Both approaches afford quantitative solutions to the interaction characteristics of donors and acceptors, but considerable tedium is involved in definitions and algebraic manipulations (see Chapter 7 and Refs. [15] and [18]).

We can get an insight, however, into the methods involved, and some qualitative answers, using the following arguments. Consider an n-type semiconductor containing a donor D. Conduction band electrons are produced by excitation from the donor, yielding an ionized donor and the free charge carrier. This process is represented by the equation

$$D \rightarrow D^+ + e^- \ .$$

At a given temperature, equilibrium concentrations of the three species are present. According to LeChatelier's principle, this equilibrium condition can be changed by applying a stress, such as the removal of the electrons from the conduction band, to the system. Conduction electrons can be readily "trapped" by acceptor centers, which can be represented by the equation

$$e^- + A \rightarrow A^- \ .$$

Then if acceptors A are added to an n-type semiconductor containing donors, the un-ionized – ionized donor equilibrium is upset and new D and D^+ concentrations are established:

$$D \rightarrow D^+ + e^-$$
$$+$$
$$A \rightarrow A^- \ .$$

This entire process favors a higher solubility of D in the host semi-conductor.

We can develop a similar argument using appropriate equilibria and the corresponding mass action expressions. Silicon containing an As donor is an n-type semiconductor. Conduction electron production is governed by the donor center ionization and its corresponding equilibrium constant:

$$As \rightarrow As^+ + e^- \; ,$$

$$\frac{[e^-][As^+]}{[As]} = K_{As} \approx e^{-E_D/kT} \; .$$

At a given temperature, a certain [As] is established.

If an acceptor such as boron is added to the Si host along with As, the solubility of As is increased. This is true since B accepts a valence band electron forming an ionized acceptor B^- and an electron vacancy, or a hole h^+, in the valence band:

$$B + e^-_{vb} \rightarrow B^- + V_{e^-}$$

or

$$B \rightarrow B^- + h^+ \; .$$

The equilibrium constant for the recombination of electrons and holes is very high. Consequently, when we consider the simultaneous donor and the acceptor equilibria

$$As \rightarrow As^+ + e^- \; ,$$

$$B \rightarrow B^- + h^+$$

$$\downarrow$$

$$h^+ e^-$$

we see that the conduction band electrons (essentially free charge carriers) and the valence band holes (also free charge carriers) combine and effectively annihilate one another. That is, they are no longer charge carriers; the electron is now localized in a chemical bond between atoms (recall Fig. 7). This combination of electrons and holes (h^+e^-) is very highly favored; hence, very few free e^- and h^+ exist in the presence of one another.

Then for the equilibrium constant expression $[e^-][As^+]/[As]$ to be equal to K_{As}, the $[As]$ must decrease (and the $[As^+]$ increase) to compensate for the decrease in $[e^-]$. Thus the $[As]$ is lowered below the solubility limit and more As can be dissolved in the Si sample containing the acceptor boron.

2. Compensation Effects

It is apparent from what we have just said about the effect of acceptors on free electron charge carrier concentrations (or donors on $[h^+]$) that the conductivity of a semiconductor can be changed markedly by adjusting donor and acceptor concentrations. These are called compensation effects since the presence of an acceptor effectively negates the effect of a donor by trapping its "free" electron.

From an equilibrium point of view, we can consider the case of the n-type CdTe semiconductor containing Li interstitials, which act as donors. Essentially free electron charge carriers in the CdTe conduction band are produced by thermal ionization of the interstitial Li atoms,

$$Li \rightarrow Li^+ + e^- \ .$$

At a given temperature and $[Li]$, the conduction electron concentration $[e^-]$ is controlled by the equilibrium constant for the reaction. Let us assume for simplicity that the Li ionization reaction goes essentially to completion, i.e., $[e^-] \approx [Li]$. Then the conductivity value for the CdTe sample is determined by $[Li]$.

Now if acceptor centers, such as Sb atoms substituting for Te atoms, are also present in the CdTe sample, which still contains

interstitial Li at the same concentration, the conductivity of the CdTe
will be decreased. This occurs because the acceptors reduce the conduction electron concentration. The effect of an acceptor can be considered in the following way. You can think of an un-ionized acceptor,
the substitutional Sb atom with its one incomplete bond, strongly
attracting one of the free conduction electrons. This process completes
the local bonding at the expense of a conduction band electron and forms
an ionized acceptor, i.e., an Sb atom that has lost its hole and now is
associated with one electron more than the number of protons in the Sb
nucleus,

$$Sb + e^- \rightarrow Sb^- \; .$$

Then for each Sb acceptor center present in the CdTe one electron
is removed from the conduction band. For the constant Li donor concentration and assumed complete ionization, the decrease in conductivity
will depend directly on the Sb acceptor concentration,

$$Li \rightarrow Li^+ + e^-$$
$$+$$
$$Sb \rightarrow Sb^- \; .$$

Since $\sigma = ne\mu$ and $n = [e^-]$, σ decreases as n decreases, and n
decreases as [Sb] increases.

VI.   THE WATER ANALOGY

There is an analogy for semiconductors that is especially appropriate for chemists and biologists, water and its dissociation equilibrium. We will consider briefly here some of the solid state concepts
covered, using the water analogy [19].

Dissociation of pure water is analogous to the simultaneous formation of conduction band electrons and valence band holes, via excitation from the valence band, in intrinsic semiconductors. These two equilibria can be represented by

$$H_2O \rightleftarrows H^+ + OH^- \; ,$$

$$h^+e^- \rightleftarrows h^+ + e^- \; .$$

Note that both are weakly dissociated systems. Ionization of H_2O molecules is difficult as is excitation of valence band electrons. It is true that $[H^+] = [OH^-]$ and that these concentrations are low; it is true that $[h^+] = [e^-]$ and that these concentrations are low.

We know that the H^+ and OH^- concentrations in aqueous solution can be drastically changed by the addition of acids or bases to the solution. Acids ionize to form H^+ ions and bases produce OH^- ions. Similarly in semiconductors: the h^+ and e^- concentrations are affected greatly by the presence of impurities. Acceptors ionize to form h^+, holes, and donors produce e^-, conduction electrons. Extrinsic semiconductors are formed. These relationships are shown by the equilibria

$$NaOH_{(aq)} \rightarrow Na^+_{(aq)} + OH^-_{(aq)} \; ,$$

$$D_{(s)} \rightarrow D^+_{(s)} + e^-_{(cb)}$$

and

$$HCl_{(aq)} \rightarrow Cl^-_{(aq)} + H^+_{(aq)} \; ,$$

$$A_{(s)} \rightarrow A^-_{(s)} + h^+_{(vb)} \; ,$$

where the subscripts imply aqueous, solid, conduction band, and valence band.

The activation energy for an acceptor (or donor) in a semiconductor is analogous to the ease of (or percent) dissociation of an acid or base:

$$HClO_4 \xrightarrow{\sim 100\%} H^+ + ClO_4^-$$

$$A_1 \rightarrow h^+ + A_1^-, \qquad \text{low } E_{A_1}$$

but

$$HCN \xrightleftharpoons{\text{low \%}} H^+ + CN^-$$

$$A_2 \rightleftharpoons h^+ + A_2^-, \quad \text{higher } E_{A_2}$$

The solubility of a donor is increased by the presence of an acceptor, and vice versa. In H_2O the solubility of $NH_{3(g)}$ can be increased by the addition of an acid to the solution:

$$NH_{3(g)} + H_2O \rightleftharpoons NH_4OH \rightleftharpoons NH_4^+ + OH^-$$
$$+$$
$$HCl \rightarrow Cl^- + H^+$$
$$\downarrow$$
$$H_2O$$

The analogous equilibria in a solid are:

$$D \rightleftharpoons D^+ + e^-$$
$$+$$
$$A \rightarrow A^- + h^+$$
$$\downarrow$$
$$h^+ e^-$$

The solubility of NH_4OH and the donor D are increased by the addition of HCl and the acceptor A.

We also know that the $[H^+]$ is affected by the presence, or addition, of OH^-:

$$HCl \rightarrow Cl^- + H^+$$

$$+$$

$$NaOH \rightarrow Na^+ + OH^-$$

$$\downarrow$$

$$H_2O$$

The $[H^+]$ is rapidly decreased when OH^- is added to the solution. Similarly in the extrinsic semiconductors: hole concentrations, $[h^+]$, in a p-type semiconductor are lowered by the addition of a donor, which produces free electrons, e^- :

$$A \rightarrow A^- + h^+$$

$$+$$

$$D \rightarrow D^+ + e^-$$

$$\downarrow$$

$$h^+ e^- \quad .$$

The $[h^+]$ is rapidly decreased when electrons are added to the semiconductor via the donor D.

VII.   THE FERMI ENERGY

We have seen previously that electron energy levels in solids are given by

$$E = \frac{k^2 h^2}{8\pi^2 m^*} = \frac{n^2 h^2}{8m^* a^2} \quad .$$

However, when electrons are assigned to these energy levels, it is necessary that the Pauli exclusion principle be obeyed. Only two electrons of opposed spin can occupy any one energy level. The level

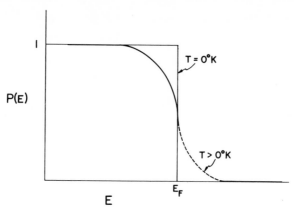

FIG. 9. Fermi-Dirac distribution function.

is then full and any other electrons must enter higher energy levels. If
we plot this energy level occupancy scheme as a distribution function,
i. e. , the probability of filling an energy level P(E) as a function of
energy E, its shape is evident. Electrons occupy the lowest energy
levels – the probability of filling the level is 1– up to some maximum
energy value, which we can designate E_F. At energies greater than E_F
the filling probability is 0.

This distribution function, called the Fermi-Dirac distribution
function [20], is plotted in Fig. 9. Note that P(E) = 1 until the energy
E_F is reached when P(E) = 0. This situation is true only at 0 °K. At
any higher temperature some electrons with energies near E_F can be
thermally excited into levels with energies greater than E_F. The electron distribution at temperatures above absolute zero will therefore vary
more smoothly, as shown in Fig. 9. An important aspect of the distribution at any temperature above 0 °K is that P(E) = $\frac{1}{2}$ when an energy
level with energy E_F is considered. Then at E_F the probability that
an energy level is occupied by an electron is one-half.

The equation that describes this distribution function as a function of temperature and energy is

$$P(E) = \frac{1}{e^{(E-E_F)/kT} + 1} \quad .$$

As we have just noted, the equation indicates that there is a fifty-fifty probability, $P(E) = \frac{1}{2}$, that an energy level is occupied when $E = E_F$. This energy E_F is called the <u>Fermi energy</u> (and the associated energy level the <u>Fermi level</u>). It plays a very important role in the theory of solids and is used throughout the book; see the discussions in Chapters 3 and 4. As is mentioned in these chapters, it can be shown that the Fermi energy is the electrochemical potential of the electron [4]. This identification of E_F is quite helpful, especially for people who are more familiar with thermodynamic concepts than with the concepts of solids.

VIII. THERMAL PROPERTIES

Much of what we have considered thus far clearly concerns the electrical characteristics of solids. In order to introduce some other properties of solids and their related theoretical concepts as well as broaden the application of the ideas already presented, we will cover briefly a few elementary thermal and optical characteristics of solids.

As a solid absorbs heat energy its temperature rises and its internal energy E increases. The internal energy in most solids is composed primarily of the vibrational energy of the atoms about their lattice sites and of the kinetic energy of the free electrons. Then the important thermal properties of solids, such as heat capacity, thermal conductivity, and thermal expansion, depend on energy changes in the lattice atoms and the free electrons [18, 21]. A similar situation applies to the related thermoelectric properties discussed in Chapter 4. Thermal behaviors associated with phase equilibria are covered in Chapter 11.

One principal way of studying lattice vibrations is through an investigation of the specific heat (heat capacity per mole) of the solid [18]. Two thermodynamic functions, the specific heat at constant pressure C_p and the specific heat at constant volume C_v, can be defined by

$$C_v = \left(\frac{\partial E}{\partial T}\right)_v \quad \text{and} \quad C_p = \left(\frac{\partial H}{\partial T}\right)_p$$

where E is the internal energy and H the enthalpy. Both C_v and C_p have the units calories per mole- °K.

Because measurements are more easily made at constant pressure, C_p is of most interest experimentally. C_v on the other hand is more easily treated theoretically. Fortunately C_p and C_v are related by the equation

$$C_p = C_v + \frac{\alpha_v^2 VT}{\beta}$$

where α_v, the coefficient of volume expansion, and β, the compressibility, can be determined experimentally, and V is the volume per mole of the solid.

We can see that C_v is more easily handled theoretically by considering the absorption of heat by a substance. As the temperature is increased, the internal energy of the substance increases. For an ideal monatomic gas, for example, the atoms have three translational degrees of freedom each of which, according to equipartition of energy, contributes an average of $\frac{1}{2}kT$ per atom. Then for one mole of ideal monatomic gas the average energy is

$$N \times 3 \times \tfrac{1}{2}kT = \tfrac{3}{2}RT$$

where N is Avogadro's number and R is the ideal gas constant. The specific heat for this gas is

$$C_v = \left(\frac{\partial E}{\partial T}\right)_v = \frac{d(\tfrac{3}{2}RT)}{dT} = \tfrac{3}{2}R \ .$$

The case of a solid is more complicated, however, since atoms are bound together. These bonded atoms can be approximated by linear harmonic oscillators, where energy is shared between kinetic and potential energies. That is, energy is absorbed by atoms that vibrate about their mean lattice positions. Since each atom has three vibrational degrees of freedom, three mutually perpendicular oscillators are required to represent each atom. The average kinetic energy of a single classic harmonic oscillator is $\frac{1}{2}kT$; the average potential energy is also $\frac{1}{2}kT$. Thus the total average energy that can be absorbed by any one solid atom is $1\,kT$ per degree of freedom or $3\,kT$ per atom. For one mole of solid atoms this average energy is $N \times 3\,kT = 3\,RT$, and the specific heat is

$$C_v = \frac{d(3\,RT)}{dT} = 3R = 5.96 \text{ cal/mole-}°K \ .$$

This theoretical treatment of specific heats served as a justification and explanation for the law of Dulong and Petit. In 1819 they made the observation that the specific heat of most solid elements at reasonably high temperatures was approximately 6 cal/mole-°K. Experimentally it was found that some elements, particularly those of atomic weight greater than 40, do have $C_v \approx 3R$ at room temperature, but many others, such as B, Be, C, Si, Na, Cs, Mg, and Ca, were found to deviate considerably from the law. Moreover, specific heats were found to be temperature dependent, decreasing to zero at low temperatures. Clearly the law of Dulong and Petit and the explanation for the law presented here serve only as crude initial attempts to understand specific heats of solids.

A. Einstein Model

Einstein was the first (1906) to offer an explanation of specific heats that accounted for their low temperature behavior and also permitted differences between various solids at higher temperatures. He approximated the solid by replacing the atoms with harmonic oscillators, all with the same frequency ν. He then applied the quantum hypothesis to this assembly of oscillators, requiring that they take on only discrete energies which were integer multiples of $h\nu$.

Using these assumptions, each atom has the set of energy levels
of a harmonic oscillator

$$E = nh\nu + \tfrac{1}{2}h\nu$$

(including the zero point energy), where $n = 0, 1, 2, 3, \ldots$ and h is
Planck's constant. Transitions between energy levels are given by,
for example,

$$E_{n_2} - E_{n_1} = (n_2 - n_1)h\nu \quad .$$

Since n_2 and n_1 differ by unity, the transition requires the absorption
of one quantum of energy $h\nu$. By analogy to the absorption (or emission) of quanta of electromagnetic radiation, i.e., photons, the term
phonon has been adopted to describe quanta of thermal energy. Just as
photons are often considered as particles rather than electromagnetic
waves, it is often helpful to consider phonons as thermal particles in
their interactions in solids rather than as a propagating elastic wave.
Phonons are covered in more detail in Chapter 7.

The number of harmonic oscillators in each quantum state N_n
relative to the number in the zero energy state is given by the Boltzmann
function

$$N_n = N_0 e^{-(E_n/kT)} = N_0 e^{-(nh\nu + \tfrac{1}{2}h\nu)/kT} \quad .$$

Assuming free, uncoupled oscillators the average energy of an oscillator
is

$$\bar{E} = \frac{E}{N} = \sum_{n=0}^{\infty} \frac{E_n N_n}{N} \quad ,$$

which can be shown to be

$$\bar{E} = \frac{h\nu}{e^{h\nu/kT} - 1} + \frac{h\nu}{2} \quad .$$

This equation, which is of the form of Planck's radiation law equation,
also indicates that the harmonic oscillator behaves like a set of quanta —
or phonons — with energy $h\nu$.

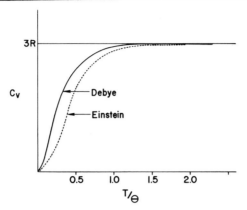

FIG. 10. The Einstein and Debye equations for specific heat.
At very low temperatures the Debye equation agrees more closely with
experimental data than does the Einstein equation.

For three degrees of freedom and one mole of solid, 3N harmonic
oscillators are required which yields an average vibrational energy
(internal energy) per mole of 3NE. The specific heat is then obtained
by differentiating with respect to temperature:

$$C_V = 3R \left(\frac{h\nu}{kT}\right)^2 \frac{e^{h\nu/kT}}{(e^{h\nu/kT} - 1)^2}$$

or

$$C_V = 3R \left(\frac{\theta}{T}\right)^2 \frac{e^{\theta/T}}{(e^{\theta/T} - 1)^2}$$

where $h\nu/k$ has been replaced by defining a characteristic temperature θ.

We can see that this Einstein equation for specific heat, diagrammed in Fig. 10, is a vast improvement over the previous model.
At low temperatures when $T \ll \theta$,

$$C_V \approx 3R(\theta/T)^2 e^{-\theta/T} \quad,$$

and C_V approaches zero exponentially as $-1/T$, as is observed experimentally. At high temperatures when $T \gg \theta$, C_V approaches the
Dulong and Petit value of 3R. The Einstein temperature θ is

characteristic for various solid materials and is the temperature at which C_v departs significantly from the value 3R.

B.   Debye Model

Although the Einstein model for specific heats fits experimental data quite well, some deficiencies were present. It was not obvious how to choose the value of θ for any particular substance, and the temperature dependence of the specific heat was not quite right at very low temperatures.

One evident difficulty in Einstein's model was his assumption of uncoupled harmonic oscillators. In 1912 Debye developed a more sophisticated model for a solid. He assumed that solids were composed of a continuous vibrating medium and was able to derive an expression for the specific heat of the form

$$C_v = 9Nk \frac{1}{x_0^3} \int_0^{x_0} \frac{x^4 e^x}{(e^x - 1)^2} \, dx \ ,$$

where $x = h\nu/kT$ and $x_0 = h\nu_D/kT$, with the Debye characteristic frequency ν_D giving the maximum vibration frequency for the continuum. Making the conversion $R = Nk$ and defining the Debye temperature $\theta_D = x_0 T$ (similar but not equal to the Einstein θ), the equation becomes

$$C_v = 9R \left(\frac{T}{\theta_D}\right)^3 \int_0^{x_0} \frac{x^4 e^x}{(e^x - 1)^2} \, dx \ .$$

At low temperatures, $T \ll \theta_D$, C_v approaches zero as T^3, which is often called the third-power law of Debye and agrees more closely with known specific heat data at very low temperatures than does the Einstein model. When $T \gg \theta_D$, C_v approaches 3R. The Debye equation is diagrammed in Fig. 10 and compared to the Einstein equation.

The Debye temperature θ_D is characteristic of a particular substance and is used in several later chapters, particularly in relation to thermoelectricity and charge carrier scattering processes. Typical values of the Debye temperature are:

	C (diamond)	Na	Ca	Cu	Mn	NaCl
θ_D; °K	1860	150	230	315	400	308

C. Electronic Specific Heat

As indicated earlier, a solid can absorb heat via atom vibrations and also by free electrons. But not every conduction band electron is able to absorb thermal energy [22]. We can see this by considering the Fermi-Dirac distribution plotted in Fig. 9. Only those electrons within about kT of the Fermi energy E_F can absorb thermal energy and be excited into vacant, higher energy levels. Electrons that have energies lower than about $(E_F - kT)$ cannot be excited by absorbing heat energy since most (or all) energy levels within kT above them are full. It is, therefore, only those electrons near E_F that contribute to the specific heat – a fraction of the order of $2kT/E_F$. This fraction is quite small for most metals, less than 1%, since E_F is about 5 eV and kT about 0.025 eV at room temperature.

The specific heat of free electrons can be estimated as was done above for an ideal monatomic gas. That is, $C_v = \frac{3}{2}R$ for one mole of the solid, assuming that all of the free electrons absorb thermal energy. Since we have just estimated that only about $2kT/E_F$ of the free electrons can absorb heat, the total contribution to the specific heat is about

$$C_{v(electronic)} = \frac{3}{2}R\left(\frac{2kT}{E_F}\right) = 3R\frac{kT}{E_F} .$$

This shows the linear dependence of the electronic specific heat on the temperature.

The total specific heat of a metal is made up of the atomic specific heat and the electronic specific heat:

$$C_v = C_{v(\text{atomic})} + C_{v(\text{electronic})} \;.$$

At most temperatures $C_{v(\text{atomic})}$ is the dominant term. At low temperatures, $C_{v(\text{atomic})}$ is given by the Debye third-power law and $C_{v(\text{electronic})}$ by the equation derived above,

$$C_v = \alpha T^3 + \gamma T \;,$$

where α and γ are constants and include the appropriate coefficients in the Debye and $C_{v(\text{electronic})}$ equations. From this last expression we can see that the electronic contribution to the specific heat becomes increasingly important at very low temperatures, below about 2 °K.

If we rearrange the last equation to give

$$\frac{C_v}{T} = \alpha T^2 + \gamma \;,$$

it is clear that a plot of C_v/T v T^2 should give a straight line whose slope is α and intercept γ. Such a graph permits calculation of the electronic specific heat using the value of γ and also determines the Debye temperature θ_D from the slope α.

D.   Thermal Conductivity

At normal temperatures heat is transferred through a solid by (primarily) phonons, i. e., lattice vibrations, and by free charge carriers (electrons or holes). The thermal conductivity σ_T is a measure of the ease of heat transfer in a particular solid, just as the electrical conductivity indicates the ease of transport of electric charge, when the solid is placed in a thermal gradient [23]. If the heat flux Q is defined as the calorie flow per second across a unit cross section, then σ_T is given by

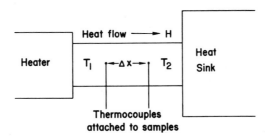

FIG. 11. Schematic for measuring the thermal conductivity σ_T of a solid. If the cross-sectional area of the sample is A, then $\sigma_T = -\dfrac{Q}{(dT/dx)} = -\dfrac{(H/A)}{(T_2 - T_1)/\Delta x}$. Values of σ_T generally differ for different crystallographic directions in the solid.

$$Q = -\sigma_T \frac{dT}{dx}$$

or

$$\sigma_T = -\frac{Q}{(dT/dx)} \quad .$$

The units of σ_T are usually cal/cm-sec-°K or W/cm-°K. A schematic for measuring σ_T is shown in Fig. 11, and some representative values are given in Table 3.

TABLE 3

Representative Thermal Conductivity and

Wiedemann-Franz Values at 300°K

Material	σ_T cal/(cm-sec °K)	L $(V/°K)^2 \times 10^8$
C (diamond)	1.5	-
Ag	1.0	2.31
Al	0.53	2.2
Cu	0.94	2.23
Cu-Zn (70:30)	0.24	-
Fe	0.18	2.47
Steel	0.12	-
NaCl	0.017	-
AgCl	0.0026	-

The relative importance of phonons and free electrons in determining the thermal conductivity of any particular solid depends on the presence of free carriers and also on the mean free path of the moving particles, i.e., either phonons or electrons. In general, the mean free path of free electrons in metals is larger than that of phonons. Metals, consequently, usually have high thermal conductivities, as shown in Table 3.

The processes of electron or phonon scattering within solids are quite complex, however, and it is possible for highly pure nonmetallic solids to have a phonon thermal conductivity that is comparable to the thermal conductivities of metals in certain temperature ranges. Note, for example, the high σ_T of diamond in Table 3. These scattering processes also play a prominent role in the temperature dependence of σ_T. For pure materials the average phonon and electron energies and velocities increase with increasing temperature while their mean free paths decrease. These two competing effects tend to cancel one another, at least in pure metals, and the thermal conductivity shows little temperature dependence above about 1000 °K.

Foreign impurities and lattice imperfections affect not only the value of σ_T but also its temperature dependence. In general, crystal defects lower σ_T by controlling the mean free paths of the phonons and electrons. In these cases σ_T often increases with increasing temperature. Some comparisons between Cu and Cu-Zn, Fe and steel are given in Table 3.

Since the thermal and electrical conduction processes in pure metals both occur via free electrons, it is not surprising that the thermal conductivity σ_T and the electrical conductivity σ should be related. This relationship, called the Wiedemann-Franz law (1853), is

$$\frac{\sigma_T}{\sigma T} = L$$

where T is the absolute temperature and L is the Wiedemann-Franz constant [21]. As seen from some of the values for L in Table 3, the theoretically constant value of $L = 2.45 \times 10^{-8}$ $(V/°K)^2$ is only approximately found for most metals.

IX. OPTICAL PROPERTIES

Electromagnetic radiation such as visible light interacts with solid materials in many different ways. Light is refracted, reflected, scattered, absorbed, and emitted in and by solids. We will, in this very brief introduction, consider only a few cases of absorption and emission of electromagnetic radiation, primarily in the ultraviolet, visible, and near-infrared portions of the electromagnetic spectrum.

In many instances absorption and emission in solids are caused by impurities (often called activators) which have energy levels lying within the energy gap of the host material. Ruby, for example, is aluminum oxide with a small amount of Cr^{3+}, $Al_2O_3 : Cr^{3+}$. The Cr^{3+} serves as a substitutional impurity replacing Al^{3+} $(Cr^{3+}/Al^{3+} \approx 1/2000$ in the solid). While pure Al_2O_3 is colorless, ruby is deep red. The red color is caused by the absorption of visible light by the Cr^{3+} impurity ions; its energy levels are thought to lie within the band gap of Al_2O_3 .

If a ruby crystal is exposed to ultraviolet radiation, say the 3660Å radiation of a mercury discharge lamp, it emits intense red light. This luminescent behavior occurs in many solids that we call phosphors. These materials give off light (visible to near-visible) when they are exposed to a source of relatively high energy, such as ultraviolet light, x-rays, and electrons. If emission occurs within about 10^{-8} sec after excitation, a luminescent material is said to fluoresce; if emission takes longer than 10^{-8} sec, the material is termed phosphorescent. Television screens and luminous watch dials make use of these properties.

Actually, ruby, along with a few other solid materials, has been fabricated as a laser. In a laser (an acronym for light amplification by stimulated emission of radiation) the fluorescence that ordinarily occurs spontaneously and is emitted in all possible directions is stimulated so that almost all luminescent centers emit simultaneously and in the same direction. The laser thereby produces highly monochromatic and spatially coherent radiation (see p. 62).

A. Absorption

The absorption of light by metals, semiconductors, and insulators can be understood qualitatively (a more extensive treatment is given in Chapter 7) using the band theory of solids introduced previously. Electrons absorbing light energy can be excited between bands of the solid, or from donor levels into the conduction band, or within the energy levels of an impurity present in a host solid, and so on. These transitions do, of course, occur in different spectral regions since quite different energies are involved.

1. Metals

The free electrons present in the conduction band in metals absorb incident radiation by making transitions to bands that are higher in energy. (The free electron theory of the optical properties of metals is considered in Chapter 7.) This requires a characteristic energy for any given metal since the energy band scheme is different for every metal. The absorption coefficients associated with these transitions are very large.

After excitation into higher energy bands, electrons can return to the lower energy levels. The energy radiated by these transitions causes the characteristic metallic luster displayed by most metals.

2. Semiconductors

The band theory appears to give a somewhat more satisfactory picture for the absorption of radiation in semiconductors than it does

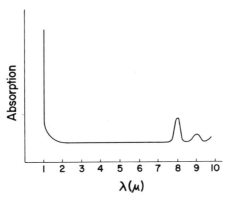

FIG. 12. Diagram of an absorption spectrum of p-type silicon showing the absorption edge near 1.1 μ and valence band to acceptor center absorption bands at longer wavelengths.

for metals [24]. Most semiconductors, Ge, Si, InSb, etc., do not absorb infrared radiation too strongly at room temperature or below. At some critical incident wavelength, however, absorption begins, often rather abruptly, and the absorption coefficient then increases very rapidly until the solid is ensentially opaque to radiation of any shorter wavelength (higher energy).

This abrupt and characteristic absorption of light in semiconductors can be attributed to excitation of electrons from the valence to the conduction bands, across the forbidden energy gap E_g, in the semiconductor. Since the E_g values are different for, and characteristic of, any given semiconducting material, these absorption edges, as they are often called, are also characteristic of the particular solid. Germanium, for example, begins absorbing strongly around 1.7 μ, which corresponds to its energy gap of about 0.7 eV. Silicon, on the other hand, is transparent in the infrared until approximately 1.1 μ, which corresponds to E_g = 1.1 eV.

A schematic absorption spectrum is given in Fig. 12 for p-type Si showing the characteristic absorption edge near 1.1 μ as well as some absorption bands occurring at longer wavelengths (lower energies) in the infrared. These latter absorptions can often be associated with electron transitions from donor levels to the conduction band or from valence band states to localized acceptor levels. The donor or acceptor ionization energies can be determined from the positions of the absorption bands. Generally, experimental studies must be done at low temperatures (usually at 4.2°K, liquid He temperature) to prevent thermal ionization of these shallow donor or acceptor states as well as electron interactions with lattice vibrations (phonons) which increase the spectral bandwidths above about 20°K. Because interactions between the impurity atoms themselves also broaden bands, impurity concentrations should be about 10^{16} atoms/cm^3 or less. As indicated previously these shallow donor or acceptor states can, to a first approximation, be treated using a hydrogen atom model.

3. Insulators

The absorption of relatively high energy radiation by pure insulators is similar to that described for semiconductors [14]. Electrons are excited across the forbidden energy gap. This transition from the valence to the conduction band occurs at higher energy in insulators, however, since their band gaps are larger. While semiconductors often have intrinsic absorption edges in the near-infrared spectral region, insulators generally absorb band gap energy in the ultraviolet. For example, the absorption edge of CdF_2 is near 2000Å, or about 6 eV, that of Al_2O_3 about 7 eV, and near 8 eV for NaCl.

Detailed spectral observations on thin films of insulating materials, such as the alkali halides, have revealed a series of very strong absorption peaks that occur at slightly lower energies than the band gap energy. These peaks are associated with electron transitions from the valence band to energy levels just below the conduction band, as shown in Fig. 13. These energy levels are envisioned as excited state levels of an electron-hole pair, called an exciton, formed by the absorption of energy. The electron is bound to the hole, in a hydrogen-like orbit, and the pair can move distances of about 1 μ through the solid before recombining. Excitons are considered in more detail in Chapter 7.

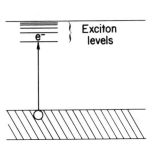

FIG. 13. Possible exciton levels in an insulator. The excited
state exciton levels lie below the conduction band edge.

Since the absorption coefficients for the observed exciton transitions are so large, care must be taken when determining the forbidden
energy gap of a material via optical absorption measurements. However,
if absorption data are combined with electrical conductivity measurements, the energy gap (optical) can be determined. Exciton states are
quite localized; the excited electron is not free to participate in electrical conduction processes. Consequently, light of greater than exciton energies is required to produce free conduction band electrons from
the valence band, which significantly increases the electrical conductivity of the solid. This process of creating higher electrical conduction by exposure to incident radiation is called photoconductivity [25].
Photoconduction can occur via valence-to-conduction band transitions,
intrinsic photoconductivity, or it can be caused by excitation of free
charge carriers from donor (electrons) or acceptor (holes) centers.
Radiant energies required for photoconductivity are, therefore, dependent on donor or acceptor activation energies. Hence, photoconductors
can be used as quantum counters for light of various wavelengths.

Absorption of light in insulators is often not caused by the pure
material itself but by impurities that are present in it. Transition
metals, for example, are prime candidates for impurities that cause
color in normally colorless solids. Color is produced by the absorption of light by the transition metal impurity and not by the host. Pure
aluminum oxide compared to ruby, $Al_2O_3 : Cr^{3+}$, is an example. The

ruby red color is imparted to the colorless Al_2O_3 (white powder) by the substitutional Cr^{3+} ions. The visible absorption bands that occur for Cr^{3+}, a d^3 ion, in octahedral symmetry can be understood using crystal field theory (see Chapter 3).

4. Color Centers

Other types of imperfections can also cause color in certain insulators [26]. If alkali halides are heated in their alkali metal atom vapor (often termed additive coloration), they turn various colors, LiF pink, KCl blue, NaCl yellow, etc. This coloration is attributed to defects; the most common is known as the F center. These defects cause the crystal to become very slightly nonstoichiometric (metal rich).

A simplified model for the formation of an F center in NaCl can be envisioned as follows. A sodium metal vapor atom strikes the NaCl surface and sticks. It then forms more halide by ionizing and attracting a chloride ion from the existing surface. This forms an anion vacancy on the surface and a free electron. Both migrate into the crystal bulk. The electron is attracted to, and trapped at, an anion vacancy (which is positively charged) within the crystal. This defect is called an F center.

We can think of the electron in an F center as being shared by the six surrounding Na^+ (or M^+) ions in a hydrogen atom-like orbit. Alternatively, the anion vacancy can be replaced by a box of dimension d and the "electron-in-a-box" approach used to calculate the energy levels available to the trapped electron (see Chapter 7). The observed absorption bands of the F center are then attributed to the promotion of the trapped electron into excited states, calculated according to the above models. For example, the F band corresponds to the transition from an approximate 1s ground state to a 2p excited state.

The proposed models agree, at least qualitatively, with the experimental observations that the F band depends systematically on the lattice constant d of the alkali halides, as indicated in Table 4,

TABLE 4

F Band Absorption in Several Alkali Halides at 300°K

Halide	Lattice constant (Å)	Peak (Å)	Position (eV)
NaCl	5.64	4650	2.67
KCl	6.29	5630	2.20
KBr	6.60	6890	1.80
KI	7.07	6850	1.81
RbCl	6.58	6300	1.97
RbBr	6.85	7000	1.77
RbI	7.34	7450	1.66

and that the band half-width as well as peak position depend on the temperature. At lower temperatures, in a given additively colored alkali halide, the absorption bands become sharper and the absorption peaks shift toward higher energies. These defects are caused by the temperature dependence of the lattice ion vibrations (see Chapter 7).

Many other defect types have been observed in the alkali halides as well as other defect preparation methods used. For example, two electrons can be trapped at an anion vacancy giving rise to an F' center and an absorption band, occurring at lower energies than the F-center band. Defects termed V centers can be formed by heating a crystal in halide vapor $X_{2(g)}$, at least for KBr and KI. Exposure of a pure alkali halide to x-rays produces F, F', and V centers, along with their related absorption bands, as well as several other complex defects (see Chapter 7).

Alkali halides containing F centers also show photoconductivity. Light absorbed in the F band produces free electrons which become trapped at F centers, producing F' centers, and leaving other empty F centers, called alpha (α) centers. Then light in the F' band releases the electrons and F centers are regenerated from the α centers, as the free electrons are trapped.

5. Photographic Process

When silver halides are exposed to electromagnetic radiation
with energies in the visible portion of the spectrum or higher, color
centers are not produced as they are in alkali halides [14]. Instead
neutral Ag atoms precipitate and rapidly form colloidal sized particles
of Ag metal, which appear as dark spots within the silver halide solid.
These dark regions are opaque to visible light and form the basis of the
photographic process.

A model for the photographic process was proposed by Gurney and
Mott in 1938. Considerable experimental and theoretical work since
that time has failed to yield a better model, even though their explanation has deficiencies. In the Gurney-Mott model an incident photon
strikes the silver halide, AgBr, for example, producing an electron-hole
pair. The free electron migrates via the conduction band (AgBr is
photoconducting in visible light) to the AgBr surface where it is trapped,
possibly by an impurity such as Ag_2S. This trapped negative charge
attracts an interstitial Ag^+ ion, present as a Frenkel defect in the AgBr;
they combine to form a neutral Ag Atom. This neutral Ag atom can, in
turn, trap another electron generated by a second photon, which produces an Ag^- ion. This negatively charged species can attract another
Ag^+ ion to begin the formation of the Ag colloidal particle. The entire
process is repeated until the colloidal dimensions are attained, when
we say a latent image has formed.

The proposed model requires a relatively high mobility for the
Ag^+ ions in the AgBr, which has been observed experimentally, and the
destruction of positively charged holes to preserve charge neutrality.
This last condition is envisioned to occur by diffusion of holes to the
surface, combination of the holes with Br^- ions to form neutral Br atoms,
and loss of Br atoms or Br_2 molecules by evaporation from the surface.
The latent image can be rendered opaque in a photographic emulsion by
a chemical development process that aids in depositing Ag at the initially formed center in the exposed AgBr.

B. Luminescence

Emission of light by solids is generally attributed to the presence of impurities or activators within the solids [27]. The fluorescence and phosphorescence processes are therefore many, varied, and often quite complex. In some way electrons must be excited from their ground states to upper energy levels and then return, either directly or via intermediates, releasing light in the process. The luminescence observed depends on the types of excitation and emission processes involved. There are, for example, several common methods of excitation: (1) photoluminescence, in which electrons are excited by absorption of photons. This process is used in fluorescent lamps; phosphors absorb ultraviolet energy produced by a mercury vapor discharge in the lamp; (2) cathodoluminescence, in which high energy electrons strike phosphors as in television screens; (3) electroluminescence, in which an electric field is used to excite electrons. Small indicating lamps of various colors using GaP with varying dopants as activators are examples of this process.

The types of emission processes possible within any luminescent material are very numerous and can be quite complex. They can, or do, depend on the activator identity, its concentration, the other kinds and amounts of impurities present, the presence of a coactivator (second impurity that needs to be present to obtain luminescence), and so on. A few specific cases are considered here.

1. Thermoluminescence

When electrons are excited in a solid by some means, such as the absorption of light or x-rays, at low temperatures and the sample is then heated, say from 77 to 300 °K or from 300 to 600 °K, etc., visible light is sometimes observed as the sample warms. This process is called thermoluminescence.

We can envision the following mechanism for thermoluminescence. At low temperatures photon absorption produces a defect in the crystal;

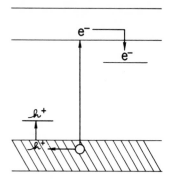

FIG. 14. Trapping of electrons and holes at low temperatures
after excitation of an electron from the valence to the conduction band.

for example, an electron is excited from the valence to the conduction
band, as shown in Fig. 14. The excited electron is then trapped at
another defect center (the electron-hole recombination probability is
low) which has an activation energy too large for significant thermal
ionization, for example, $E_D \gg kT$ at 77 °K. As the sample temperature
increases, kT approaches E_D. When kT is approximately equal to E_D,
electrons are rapidly ionized from the traps into the conduction band.
From there they can recombine with a hole via several different mechanisms, several of which are diagrammed in Fig. 15.

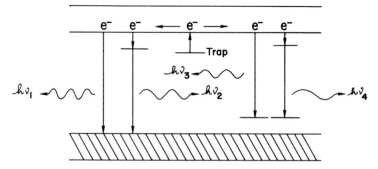

FIG. 15. Various thermoluminescence pathways, producing light
of different wavelengths. Electrons are first excited from traps thermally, at characteristic temperatures.

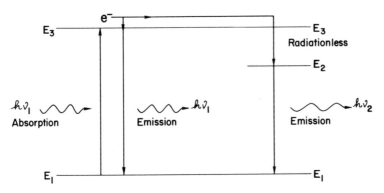

FIG. 16. Absorption and emission processes. Luminescence from the E_3 and E_2 energy levels occurs spontaneously.

There are two obvious sets of data that are important in thermo-luminescence experiments. The temperature at which the luminescence occurs is directly related to the activation energy of the trap. And the wavelength of the luminescence indicates the energy separation between the upper and terminal fluorescent (or phosphorescent) levels. The particular levels involved are, of course, not directly identified.

2. Lasers

Solid state lasers, such as ruby, Nd^{3+}, doped $CaWO_4$, and Sm^{2+} doped CaF_2, illustrate some of the principles of absorption and emission mentioned and serve to introduce several other related optical properties [28].

In general we can think of electron transitions corresponding to absorption and emission processes, as shown in Fig. 16. Electrons are excited from the ground state into a higher energy level by incoming radiation with energy $E = h\nu_1 = E_3 - E_1$. Photons of energy $h\nu_1$ can then be emitted randomly by electron transitions from E_3 to E_1. This process is known as spontaneous emission.

Another mechanism for emission is also given in Fig. 16. Radiation of energy $E = h\nu_1$ excites ground-state electrons to level E_3. These electrons then undergo radiationless (no near-visible light

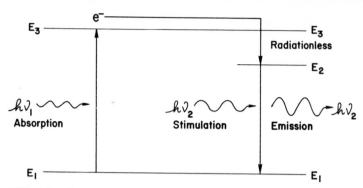

FIG. 17. Stimulated emission of light with energy $E = h\nu_2 = E_2 - E_1$. The emitted light at frequency ν_2 has a greater intensity than the stimulating ray.

emitted) transitions to level E_2. The following transitions back to the ground-state level E_1 emit photons with energy $E = h\nu_2 = E_2 - E_1$. This latter emission process can occur spontaneously, i.e., the transition of excited electrons to a lower energy level is a random process, or the electrons in excited states can be caused or stimulated to return to lower energy levels if they are placed in the proper radiation field. This is illustrated in Fig. 17. If the system is exposed to radiation with $E = h\nu_2$, electrons at level E_2 will be stimulated to emit photons with energy $E = h\nu_2$. Hence, the process is called stimulated emission. This is, of course, just the reverse of the E_1 to E_2 absorption process (it occurs with the same probability) and is one of the required conditions for laser operation.

The stimulated emission process also indicates another prerequisite for lasing to occur. In order for the emission process to dominate the absorption process, under stimulated conditions, a population inversion is necessary. For example, in Fig. 17 the number of excited atoms with energy E_2 must be greater than the number of atoms with energy E_1 for the $E_2 \rightarrow E_1$ stimulated emission process to dominate the $E_1 \rightarrow E_2$ absorption. This is true since the transition probabilities for both processes are identical. Needless to say, an inverted population situation is very unusual. Under normal conditions the populations of these states are governed by the Boltzmann relation. Since E_2 is generally much greater than E_1, essentially all atoms have energy E_1 at normal temperatures. Even at very high temperatures, only equal electron populations can be attained at E_2 and E_1; populations at E_2 cannot exceed those at E_1 under normal thermal equilibrium conditions.

One possible way of obtaining the necessary population inversion makes use of transition probability differences between various energy levels. In Fig. 17, for example, suppose that the probability of an electron transition $E_3 \rightarrow E_2$ is much greater than the $E_3 \rightarrow E_1$ transition. If electrons are excited into E_3, most will funnel into level E_2 rather than returning to the E_1 level. Then if the transition probability for the $E_3 \rightarrow E_2$ process is greater than that for the $E_2 \rightarrow E_1$ transition, electrons will pile up at E_2. If enough power is pumped into the $E_1 \rightarrow E_3$ absorption band, a population inversion can be obtained with more atoms in the excited E_2 state than in the lower energy state E_1.

Often E_1 is the ground state. However, an even more favorable population inversion situation can be present if the lower fluorescent level, E_1 as described here, lies somewhat above the ground state level. Then E_1 can be rapidly depopulated thermally, with the return of the electrons to empty positions in the ground state. Such a situation is often referred to as a four-level laser, with the two fluorescent levels lying between the absorption and ground-state levels.

We can consider the case of ruby to illustrate these requirements for lasers. In $Al_2O_3 : Cr^{3+}$ the substitutional Cr^{3+} ions are located at lattice sites with octahedral symmetry. From crystal field theory (see Chapter 3) we find that the ^4F ground state of the Cr^{3+} free ion (a d^3 electron configuration) is split into three energy levels in the crystal, a $^4A_2$ ground state along with $^4T_2$ and $^4T_1$ excited states. As shown in Fig. 18, the ^2G free ion term yields ^2E, $^2T_1$, $^2T_2$, and $^2A_1$ crystal field split levels; higher energy levels are not shown. The two principal absorption bands in ruby correspond to the $^4A_2 \rightarrow {}^4T_2$ and the $^4A_2 \rightarrow {}^4T_1$ transitions and the principal emission line (red at 6943 Å) to a $^2E \rightarrow {}^4A_2$ transition.

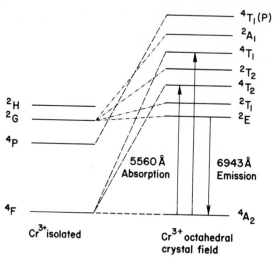

FIG. 18. Octahedral crystal field splitting of some free ion Cr^{3+} (d^3) energy levels.

While both absorption bands are important in a ruby laser, we will consider only the $^4A_2 \rightarrow \, ^4T_2$ transition centered around 5560 Å (yellow) for simplicity. Since the energy difference between these two states is large,

$$E = h\nu = \frac{hc}{\lambda} = \frac{(6.63 \times 10^{-27} \text{ erg-sec})(3.0 \times 10^{10} \text{ cm/sec})}{5.560 \times 10^{-5} \text{ cm}}$$

$$E = 3.58 \times 10^{-12} \text{ erg,}$$

essentially all of the Cr^{3+} ions in ruby are in the ground state at room temperature. This Boltzmann distribution of Cr^{3+} ions between the $^4A_2$ and $^4T_2$ states at 300 °K is given by

$$\frac{N_{T_2}}{N_{A_2}} = \exp\left[-\frac{E}{kT}\right] = \exp\left[\frac{-3.58 \times 10^{-12}}{(1.38 \times 10^{-16})(300)}\right]$$

$$\frac{N_{T_2}}{N_{A_2}} = \exp(-87) \ .$$

Absorption of 5560 Å light produces $^4T_2$ excited Cr^{3+} ions at a rate that is proportional to the radiation density (quanta/cm^2-sec) of the incident light ρ_{A-T} and the number of Cr^{3+} ions in the $^4A_2$ ground state N_A,

$$-\frac{dN_A}{dt} = B_{A-T}\,\rho_{A-T}\,N_A \quad ,$$

where B_{A-T} is a constant called the <u>Einstein coefficient,</u> in this case for absorption.

The $^4T_2$ excited state can be depopulated via two processes, spontaneous and stimulated emission. Stimulated emission is the inverse of the absorption process; its rate is given by

$$-\frac{dN_T}{dt} = B_{T-A}\,\rho_{A-T}\,N_T \quad .$$

Quantum mechanically it is true that $B_{A-T} = B_{T-A}$. The spontaneous emission rate is

$$-\frac{dN_T}{dt} = A_{T-A}\,N_T \quad ,$$

where A_{T-A} is the Einstein coefficient for spontaneous emission.

For the stimulated emission process to dominate absorption, as is the case during lasing,

$$\left(-\frac{dN_T}{dt}\right)_{stim} > \left(-\frac{dN_A}{dt}\right)_{abs} \quad .$$

This can only be achieved if $N_T > N_A$, which is a very unusual circumstance and is contrary to the Boltzmann distribution law. This inverted population is one requirement for laser operation.

In ruby the population inversion is the result of a set of happy circumstances that correspond to the general description given for Fig. 17. We can identify the $^4A_2 \rightarrow {}^4T_2$ absorption at 5560 Å with the $E_1 \rightarrow E_3$ transition in Fig. 17. A rapid, radiationless transition then occurs from the $^4T_2$ to the ^2E state. The rate of this phonon-assisted transition (release of energy via lattice vibrations) is about 2×10^7 sec^{-1} compared to the slower rate of about 3×10^5 sec^{-1} for the $^4T_2 \rightarrow {}^4A_2$ emission. Consequently, energy quickly funnels from the $^4T_2$ into the ^2E excited state ($E_3 \rightarrow E_2$ in Fig. 17). Energy piles up in the ^2E level because of its slow emission rate, about 2×10^2 sec^{-1}. (The $^2E \rightarrow {}^4A_2$ transition is multiplicity forbidden.)

It is possible, then, to pump sufficient energy into the 5560 Å absorption band of ruby, depopulating the $^4A_2$ ground state rather rapidly, so that the required population inversion is attained. The number of Cr^{3+} ions in the ^2E upper fluorescent, excited state is larger than the number of $^4A_2$ ground state Cr^{3+} ions.

After the required population inversion is attained, another condition must be met before a laser such as ruby will actually lase. The stimulated emission of light must be greater than all of the losses caused by spontaneous emission, scattering, absorption, and any other possible mechanism. This is accomplished by fashioning a proper resonating cavity. The quality of such an arrangement is often expressed in terms of a value Q for the oscillation process involved. Q is defined as 2π times the number of periods T_0 that the oscillation lasts before its total energy is reduced to $1/e$ of its initial value,

$$Q = 2\pi \left(\frac{\tau}{T_0} \right) = 2\pi \nu_0 \tau \quad .$$

Here τ is the decay time to $1/e$ of the initial energy and ν_0 the frequency of the emitted light. If the Q of the cavity is high, the losses in the cavity are low and stimulated emission of light is favored.

In solid lasers the oscillation cavity is obtained by machining a single crystal (usually into cylindrical form), cutting and grinding the ends parallel to optical tolerances, and plating the highly polished ends with a reflecting coating. Often silver is used to plate the ends, one end silvered opaque and the other very slightly transparent to the emitted light.

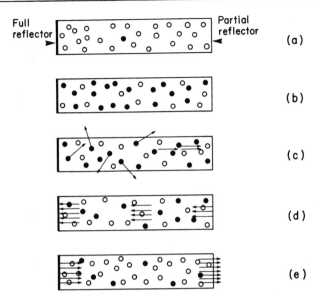

FIG. 19. Schematic showing stimulated and laser emission in ruby. (a) Thermal equilibrium has essentially all Cr^{3+} ions in ground state ○. (b) Excitation by flash lamp produces the inverted population of Cr^{3+} excited state ions ●. (c) A few Cr^{3+} ions emit spontaneously. Emission along the crystal axis stimulates other Cr^{3+} excited state ions to emit in the same direction. (d) Stimulated emission increases as the building wave reflects back and forth between the reflecting ends of the crystal. (e) The beam intensity becomes great enough to penetrate the partially reflecting end.

The laser rod is then placed within an arrangement of high intensity excitation lamps, such as a helical xenon flash lamp which is often used with ruby, that produce very intense pumping radiation of the wavelength(s) required by the laser (5560 Å for ruby, $^4A_2$ to $^4T_2$ transition).

Figure 19 is a schematic showing the stimulated emission and lasing processes for a ruby laser rod. Prior to flashing the xenon lamp, essentially all Cr^{3+} impurity ions are in the $^4A_2$ ground state. Flash lamp light is then absorbed in the pumping bands of the ruby, particularly the $^4A_2 \rightarrow {}^4T_2$ 5560 Å transition. The rapid, phonon-assisted transfer to the ^2E state produces the necessary inverted population. A few excited Cr^{3+} ions spontaneously emit 6943 Å photons directly

along the cylindrical cavity axis (spontaneous emission in other directions is lost from the crystal). These photons induce other ^2E Cr^{3+}
ions to emit. As the light bounces back and forth between the silvered
ends of the crystal, more and more stimulated emission occurs, providing the Q of the device is high enough so that losses are overcome.
Finally the light intensity is sufficient so that a beam passes through
the partially silvered crystal end.

A laser emits a beam with unusual properties. It is very highly
monochromatic, has a very small line width, is very intense, and is
coherent. The coherence of the radiation may be its most important
characteristic. Light is coherent if all sections of the beam are in
phase. The sinusoidal wave propagating down the crystal rod increases
in amplitude as each excited ion emits its energy, and the wavefronts
of the stimulated photons are exactly in phase with the propagating
beam. Spontaneous emission, on the other hand, is a random process.
The light waves produced by successive emitted quanta are not in
phase.

The above properties of laser light have led to applications in
photography, photographic and spectrographic research, precise measurement, special types of surgery, metallurgical purposes, welding,
and potential communications applications. The most important applications lie, no doubt, in the future.

Solid state lasers can also be prepared from semiconductor p-n
junction diodes. They are called injection lasers because they operate
by injecting electrons and holes into the n and p regions of the laser
structure. Emission then occurs at the p-n junction. Injection lasers
have the advantages of small size and of being electrically operated.
Since population inversion is attained electrically, it is easily controlled. Output wavelengths can be varied by using different dopants
in a given host semiconductor such as GaAs.

REFERENCES

1. E. A. Wood, Crystals and Light, Van Nostrand, Princeton, New Jersey, 1964; M. C. Day and J. Selbin, Theoretical Inorganic Chemistry, 2nd ed., Chap. 6, Reinhold, New York, 1969; C. Kittel, Introduction to Solid State Physics, 2nd ed., Chap. 1, Wiley, New York, 1956.

2. P. Drude, Ann. Physik, 7, 687 (1902); H. A. Lorentz, Proc. Acad. Sci. Amst., 1, 438, 585, 684 (1904).

3. N. F. Mott and H. Jones, The Theory of the Properties of Metals and Alloys, Dover, New York, 1936, pp. 240-242.

4. N. B. Hannay (ed.), Semiconductors, Reinhold, New York, 1959; N. B. Hannay, Solid-State Chemistry, Prentice-Hall, Englewood Cliffs, New Jersey, 1967; W. J. Moore, Seven Solid States, Benjamin, New York, 1967.

5. F. C. Brown, The Physics of Solids, Benjamin, New York, 1967; C. Kittel, Introduction to Solid State Physics, 3rd ed., Wiley, New York, 1966; A. J. Dekker, Solid State Physics, Prentice-Hall, Englewood Cliffs, New Jersey, 1957; F. Seitz, D. Turnbull, and H. Ehrenreich (eds.), Solid State Physics, Academic Press, New York, a series.

6. See, for example, M. C. Day and J. Selbin, Theoretical Inorganic Chemistry, 2nd ed., Reinhold, New York, 1969, pp. 176-184; L. Pauling and E. B. Wilson, Introduction to Quantum Mechanics, McGraw-Hill, New York, 1935, pp. 340-345.

7. W. Shockley, Electrons and Holes in Semiconductors, Van Nostrand, Princeton, New Jersey, 1950, pp. 129-134.

8. A.J. Dekker, Solid State Physics, Prentice-Hall, Englewood Cliffs, New Jersey, 1957, pp. 257-262.

9. A. Sommerfeld, Z. Physik, 47, 1 (1928).

10. See, for example, G.W. Castellan, Physical Chemistry, Addison-Wesley, Reading, Massachusetts, 1964, pp. 406-412.

11. R. de L. Kronig and W.G. Penney, Proc. Roy. Soc. (London), A130, 499 (1930).

12. E.H. Putley, The Hall Effect and Related Phenomena, Butterworths, London, 1960, pp. 138-153.

13. H.G. Van Bueren, Imperfections in Crystals, 2nd ed., North-Holland Publ., Amsterdam, 1961.

14. N.F. Mott and R.W. Gurney, Electronic Processes in Ionic Crystals, 2nd ed., Dover, New York, 1964.

15. F.A. Kröger, The Chemistry of Imperfect Crystals, North-Holland Publ., Amsterdam, 1964, particularly Chap. 7.

16. F.A. Kröger and H.J. Vink, in Solid State Physics (F. Seitz and D. Turnbull, eds.), Vol. 3, Academic Press, New York, 1956.

17. See, for example, R.A. Levy, Principles of Solid State Physics, Chap. 9, Academic Press, New York, 1968.

18. R.A. Swalin, Thermodynamics of Solids, Wiley, New York, 1962.

19. C.S. Fuller, in Semiconductors (N.B. Hannay, ed.), Reinhold, New York, 1959.

20. See, for example, J.S. Blakemore, Solid State Physics, Chap. 3, Saunders, Philadelphia, Pennsylvania, 1969.

21. J. M. Ziman, Principles of the Theory of Solids, Chaps. 2 and 7, Cambridge Univ. Press, London and New York, 1964; E. W. Montroll, in Handbook of Physics (E. U. Condon and H. Odishaw, eds.), 2nd ed., P. 5, Chap. 10, McGraw-Hill, New York, 1967.

22. A. H. Wilson, The Theory of Metals, 2nd ed., Chap. 6, Cambridge Univ. Press, London and New York, 1965.

23. D. Grieg, in Progress in Solid State Chemistry (H. Reiss, ed.), Vol. 1, Chap. 4, Macmillan, New York, 1964.

24. H. J. Hrostowski, in Semiconductors (N. B. Hannay, ed.), Chap. 10, Reinhold, New York, 1959.

25. R. H. Bube, Photoconductivity of Solids, Wiley, New York, 1960.

26. J. H. Schulman and W. D. Compton, Color Centers in Solids, Macmillan, New York, 1962.

27. D. Curie, Luminescence in Crystals, Methuen, London, 1963; D. W. G. Ballentyne, in Progress in Solid State Chemistry (H. Reiss, ed.), Vol. 1, Chap. 5, Macmillan, New York, 1964.

28. W. V. Smith and P. P. Sorokin, The Laser, McGraw-Hill, New York, 1966.

Chapter 2

CRYSTALLOGRAPHY

Lawrence Suchow

Chemistry Division
Department of Chemical Engineering and Chemistry
Newark College of Engineering
Newark, New Jersey

The ways in which atoms and ions pack together in the solid state to yield what we call crystal structures are fascinating in themselves. In the fields of solid state chemistry and solid state physics, however, the fascination is perhaps only secondary, for most of the interesting and useful properties of solids are directly related to their structures. Few predictions or interpretations in depth would be possible without structural information.

Although atoms are, for many purposes, considered to be uncharged hard spheres, they are in truth not such, so that it seems desirable first to consider the forces that bind them together.

I.   CHEMICAL BONDING

Volumes can and have been written on chemical bonding; it appears necessary here, however, only to summarize the types of bonds and to emphasize those factors that are of special importance in crystals.

A.   The Ionic Bond

The ionic bond is one in which valence electrons from one atom are transferred to the outer shell of another atom. This leaves both atoms charged, one positively and the other negatively, and the attraction between the charged atoms, called ions, is then electrostatic.

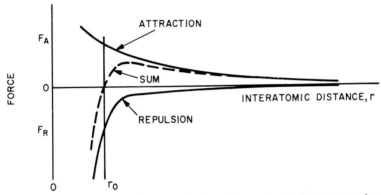

FIG. 1. Ionic bond: forces of attraction and repulsion as a function of interatomic distance.

The force of attraction between the ions, F_A, is given by Coulomb's law,

$$F_A = -\frac{(+Z_1 e)(-Z_2 e)}{r^2} \quad ,$$

where Z_1 and Z_2 are the ionic charges,[*] e is the value of the electronic charge, and r is the interatomic distance. Since the force of attraction must be positive for ions of opposite charge, a minus sign precedes the expression.

The force of repulsion, F_R, between the two ions is given by Born's expression,

$$F_R = \frac{-nb}{r^{n+1}} \quad ,$$

where r is again the interatomic distance, and n and b are constants.

Plots of these expressions are given in Fig. 1, in which it is seen that the repulsive force becomes very large at short separations. This is due to the increasing resistance to compression which results from mutual repulsion of electrons in the two atoms as the atoms are forced

---

[*]As used here, Z_1 and Z_2 represent signless numerical values of the ionic charges; the actual signs precede the Z's.

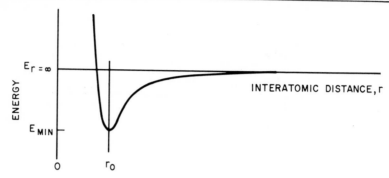

FIG. 2.   Ionic bond: bond energy as a function of interatomic distance.

together.   The situation is analogous to squeezing two rubber balls together – the more one squeezes them together, the greater becomes the force required to squeeze them further together. Figure 1 also shows that the attractive and repulsive forces balance each other exactly at one point, r_0, which is called the equilibrium interatomic distance.

The expression for the bond energy may be derived by integration of the sum of the forces as follows:

$$E = \int_{\infty}^{r} (F_A + F_R)\, dr$$

$$E = \int_{\infty}^{r} \left[\frac{-(+Z_1)(-Z_2)\, e^2}{r^2} + \left(\frac{-nb}{r^{n+1}} \right) \right] dr$$

$$E = \frac{(+Z_1)(-Z_2)\, e^2}{r} + \frac{b}{r^n} \quad .$$

Plotting this expression yields Fig. 2, in which it is seen that the equilibrium interatomic distance is that at which the bond energy is at a minimum value. The sharp rise in energy at small separation distances is, for the reason already discussed, due to the repulsion term. Note that the bond energy reaches the value found at infinite separation long before infinity; once the atoms are sufficiently far apart that their fields no longer interact, there is, in effect, infinite separation.

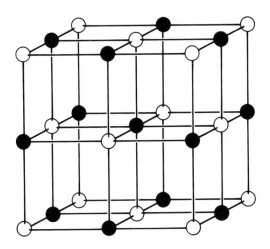

FIG. 3. The fcc unit cell of NaCl (the rocksalt structure). Either the open or filled circles may be taken to represent Cl^-; the remaining circles then represent Na^+.

The expressions above are for the case of a single ion pair. In an ionic solid, however, all the ions are attracted and repelled by all the other ions in the crystal, though the forces of attraction and repulsion become less and less significant as the distance between ions increases. Let us consider, for instance, what the situation is in sodium chloride, NaCl, the structure of which is depicted in Fig. 3. Each Na^+ ion is surrounded by 6 nearest neighbor Cl^- ions at a distance r, 12 next nearest neighbor Na^+ ions at a distance $\sqrt{2}r$, 8 more distant Cl^- ions at $\sqrt{3}r$, 6 Na^+ ions at 2r, 24 Cl^- ions at $\sqrt{5}r$, etc. The electrostatic interaction energy of the Na^+ ions with each of the surrounding ions is as given above for the single ion pairs, so that:

$$E_{attraction} = \frac{6e^2}{r}(+Z_1)(-Z_2) + \frac{12e^2}{\sqrt{2}r}(+Z_1)^2 + \frac{8e^2}{\sqrt{3}r}(+Z_1)(-Z_2) +$$

$$\frac{6e^2}{2r}(+Z_1)^2 + \frac{24e^2}{\sqrt{5}r}(+Z_1)(-Z_2) + \ldots$$

In NaCl and other compounds with its structure, both ions are equally

TABLE 1

Some Madelung Constants (A)

Structure	A
Rocksalt (NaCl)	1.747558
CsCl	1.76267
Zincblende (Cubic ZnS)	1.63806
Wurtzite (Hexagonal ZnS)	1.641
Fluorite (CaF_2)	5.03878
Corundum (Hexagonal (α-) Al_2O_3)	25.0312

charged but with opposite sign $[+Z_1 = -(-Z_2)$ or $Z_1 = Z_2 = |Z|]$ so that

$$E_{attraction} = - \frac{e^2 |Z|^2}{r} (6 - \frac{12}{\sqrt{2}} + \frac{8}{\sqrt{3}} - \frac{6}{2} + \frac{24}{\sqrt{5}} \pm \ldots) \quad .$$

The sum of the terms in parentheses is called the Madelung constant (denoted by A), which is characteristic only of the geometrical arrangement of the ions and independent of their identities or charges. The above series converges to 1.747558... ; Table 1 lists Madelung constants for several common structures.

For a solid, the b in the repulsion term is also modified by the crystal geometry and is then denoted B (known as the Born repulsion constant), so that

$$E_{repulsion} = \frac{B}{r^n} \quad .$$

Both the attractive and repulsive energy expressions must be multiplied by Avogadro's number to give the energy per mole, but only the interactions of a mole of negative or positive ions are considered because using both is equivalent to counting each bond twice. The inherently negative lattice energy [the energy of a reaction such as $Na^+(g) + Cl^-(g) \rightarrow NaCl(s)$] is

$$U = \frac{N(+Z_1)(-Z_2)Ae^2}{r} + \frac{NB}{r^n} .$$

Values of n can be derived from measurements of the compressibility of the solid and may also be estimated theoretically. If n is known, B can be determined because, at the experimentally determinable equilibrium interatomic distance r_0, the forces of attraction and repulsion (the derivatives of the energy terms) are equal but opposite in sign, so that

$$\frac{-(+Z_1)(-Z_2)ANe^2}{r_0^2} = -\left[\frac{-nNB}{r_0^{n+1}}\right] ,$$

and therefore

$$B = \frac{-A(+Z_1)(-Z_2)e^2 r_0^{n-1}}{n} .$$

For calculations of greatest accuracy, the lattice energy equation should also be corrected by including terms for the energy resulting from Van der Waals forces (energy $= C/r^6$, where C is a constant) and for the zero point energy of the crystal. In addition, some theoreticians prefer to substitute for the Born repulsion term an expression derived from quantum mechanics which is known as the Born-Mayer repulsive potential energy. This term is $Ke^{-r/\rho}$, where K and ρ are constants, but whether it yields better results is a subject of lively debate.

Lattice energies calculated in the manner indicated here may be compared with those determined from the Born-Haber cycle, which is a thermodynamic cycle employed to account for the magnitude of the heat of formation of an ionic compound.

Using NaCl as an example, the Born-Haber cycle is

ΔH_f is the heat of formation, ΔH_S the heat of sublimation, ΔH_D the heat of dissociation, A the electron affinity, I the ionization energy, and U the lattice energy.[*]

These steps may be listed as:

(1)	$Na(s) \rightarrow Na(g)$	$\Delta H_{S(Na)}$
(2)	$Na(g) \rightarrow Na^+(g) + e^-$	I_{Na}
(3)	$\frac{1}{2} Cl_2(g) \rightarrow Cl(g)$	$\frac{1}{2} \Delta H_{D(Cl_2)}$
(4)	$Cl(g) + e^- \rightarrow Cl^-(g)$	A_{Cl}
(5)	$Na^+(g) + Cl^-(g) \rightarrow NaCl(s)$	U

(6)	$Na(s) + \frac{1}{2} Cl_2(g) \rightarrow NaCl(s)$	$\Delta H_{f(NaCl)}$

That is,

$$\Delta H_{f(NaCl)} = \Delta H_{S(Na)} + I_{Na} + \frac{1}{2} \Delta H_{D(Cl_2)} + A_{Cl} + U \quad .$$

The values for steps 1, 2, 3, and 6 are usually known. Although electron affinities (step 4) are difficult to measure, they have been measured accurately for halogens. With all but step 5 (the lattice energy) known, it is possible, in some cases, to compare theoretical and experimental values for the lattice energies. In general, these check each other quite well. For instance, the calculated lattice energy for NaCl is 7.94 eV, while the value determined from the Born-Haber cycle is 7.86 eV. Since this indicates that calculated values of U are dependable, one can use them to calculate unknown electron affinity values from the Born-Haber cycle.

---

[*]In terms of the convention adopted herein, numerical values of U will always be negative, as will A for the halogens and ΔH_f for the alkali halides; ΔH_D, ΔH_S, and I will carry positive signs. If values found in the literature are recorded in opposite fashion, the signs must be changed to be consistent with this discussion.

B. The Covalent Bond

The covalent bond, whose strength is of approximately the same order of magnitude as that of the ionic bond, is best treated by employing the valence bond method or the molecular orbital method of approximating a solution to the Schrödinger equation; of these, the latter appears to be currently favored. Modern discussions are normally in terms of wave functions, probability functions, overlap; bonding, antibonding, and nonbonding orbitals; σ bonds, π bonds, shapes of electron cloud orbitals, and hybridization. The interested reader is directed to a multitude of books on these topics, but for a short crystallography chapter, it appears sufficient simply to think in terms of the old Lewis 2-electron bond shared by two atoms, but remembering that hybridization, π bonding, shapes of electron cloud orbitals, etc., have important roles in determining the ways in which atoms are arranged within molecules. The covalent bond from the point of view of the energy band picture will be discussed in Chapters 3 and 4.

C. The Metallic Bond

Another strong bond is the metallic one, probably best discussed in terms of quantum mechanics and energy bands (see Chapter 3). For crystallographic purposes, it is probably sufficient to think of a metal as being an assemblage of positive ions suspended in a "sea" or "cloud" of valence electrons, with these electrons quite free to move about.

D. The Hydrogen Bond

The hydrogen bond is roughly 5-10% as strong as ionic and covalent bonds, and is formed between molecules which contain hydrogen and electronegative elements such as fluorine, oxygen, and nitrogen. Such bonds cause "polymerization" of molecules like HF, H_2O, and NH_3, with resulting abnormally high melting and boiling points for these compounds as compared with analogous ones where there is little or no hydrogen

bonding, because no highly electronegative elements are present. Hydrogen bonds are also often influential in determining crystal structures and
are apparently of vital importance in determining structures of biological
molecules, including DNA.

E. Van der Waals Bonds

Van der Waals bonds are much weaker than those described above
(i. e., about 1 % or less the strength of ionic and covalent bonds) and
are most prominently produced by forces of attraction between neutral
molecules with electric dipoles. In symmetrical molecules and noble
gas atoms, momentary random, fluctuating polarization called "dispersion"
results in a very weak Van der Waals force, which is nevertheless strong
enough to permit condensation at very low temperatures.

F.   Electronegativity and Percent Ionic Character

Although one can, and often does, speak of ionic and covalent
bonds as if they were absolute, the truth is that all bonds except those
between identical atoms are intermediate in character between the two
types. Even in a homonuclear molecule there exists some ionic character, though no net ionic character. In H_2, for instance, the electron
distribution is such that one can think of the covalent structure H : H
as having the greatest probability, but there are also small finite probabilities of electron distributions yielding the ionic structures H H :
(or H^+H^-) and : H H (or H^-H^+). There is, however, no net ionic character because the probabilities of these two species are identical.

In all but homonuclear molecules, there is a greater probability
that electrons will be in the vicinity of the more electronegative atom
of the two. This results in what is called "percent (net) ionic character," a concept to which we will return following a consideration of the
meaning of electronegativity.

Pauling defines electronegativity as "the power of an atom in a
molecule to attract electrons to itself." To calculate electronegativity

TABLE 2

Some Pauling Electronegativity Values

H 2.1						
Li 1.0	Be 1.5	B 2.0	C 2.5	N 3.0	O 3.5	F 4.0
Na 0.9	Mg 1.2	Al 1.5	Si 1.8	P 2.1	S 2.5	Cl 3.0
K 0.8	Ca 1.0	Sc 1.3	Ge 1.8	As 2.0	Se 2.4	Br 2.8
Rb 0.8	Sr 1.0	Y 1.2	Sn 1.8	Sb 1.9	Te 2.1	I 2.5

values, he made use of experimentally determined single bond energies expressed in electron volts. He first assumed that for a 100% covalent bond between atoms A and B, one should expect the arithmetic mean of the two bond energy values $D(A-A)$ and $D(B-B)$ to be equal to the energy $D(A-B)$ of bond A–B, but that if the bond has some ionic character (as a result of difference in electronegativities of A and B), then the bond energy will be greater than the arithmetic mean. The difference between $D(A-B)$ and the arithmetic mean is

$$\Delta = D(A-B) - \tfrac{1}{2}[D(A-A) + D(B-B)] \quad,$$

where Δ is simply the heat liberated in the reaction

$$\tfrac{1}{2}A_2(g) + \tfrac{1}{2}B_2(g) \rightarrow AB(g) \quad.$$

Pauling later found that his results were improved if he used the geometric mean instead of the arithmetic mean:

$$\Delta' = D(A-B) - [D(A-A)D(B-B)]^{\frac{1}{2}} \quad.$$

Δ' (or Δ) values should be additive if they are to be useful. For instance, $\Delta'(A-B) + \Delta'(B-C)$ should equal $\Delta'(A-C)$, but this does not work out very well. If, however, $\sqrt{\Delta'}$ is employed, fairly good additivity is found, and Pauling has chosen to express its relation to the difference in electronegativities of two atoms bonded together as

$$\Delta' = 30(X_A - X_B)^2 \quad \text{or} \quad X_A - X_B = 0.18\sqrt{\Delta'} \quad ,$$

where X_A and X_B represent individual electronegativity values of the two atoms.

Therefore, values of $0.18\sqrt{\Delta'}$ correspond to electronegativity differences, and their average values are obtained from all direct and indirect differences calculated from various bonds involving each element. An individual electronegativity value of 4.0 is arbitrarily assigned to the most electronegative element, fluorine, and values for the other elements then referred to this. Some electronegativities obtained by this method are listed in Table 2. Note that, as might be expected, there is an increase from left to right in periods, and a decrease from top to bottom in groups, of the periodic table.

From some experimental dipole moments and the values expected from complete separation of charge (100% ionic character), values of percent ionic character of several molecules were determined, and Pauling plotted percent ionic character vs the electronegativity difference $X_A - X_B$. He then fitted to these points a curve resulting from the expression

$$\text{Percent ionic character} = 1 - \exp\left[-\tfrac{1}{4}(X_A - X_B)^2\right] \quad .$$

This curve predicts percentages of ionic character for values of $X_A - X_B$ (see Table 3).

It is on the basis of the prediction of 50% ionic character when the electronegativity difference is 1.7 that many books state, incorrectly, that compounds in which the difference is greater than 1.7 are ionic and those with less than this value covalent. It should be clear, however,

TABLE 3

Predicted Percent Ionic Character from

Electronegativity Difference (Pauling)

$X_A - X_B$	% Ionic character
0.2	1
0.4	4
0.6	8
0.8	15
1.0	22
1.2	30
1.4	39
1.6	47
1.7	50
1.8	55
2.0	63
2.2	70
2.4	76

that this is far from the truth, for if a compound has 51% ionic character, it still has 49% covalent character. NaCl, normally thought of as
an ionic compound, is predicted to be about 66% ionic by this method
and is found experimentally (by measurement of the electric dipole moment of the gaseous molecule) to be 75% ionic. Although the curve just
described and the figures in Table 3 often predict accurate values, some
dipole moment measurements made more recently indicate considerable
deviation from the curve. For instance, 50% ionic character is predicted
for KI, but the measured value is 75%.

Mulliken reasoned that the electronegativity should be the average
of the first ionization energy (the energy of the reaction $A \rightarrow A^+ + e^-$;
that is, the average of the electron attraction of the atom and the positive

ion) and the electron affinity (the energy of the reaction $A + e^- \rightarrow A^-$; that is, the average of the electron attraction of the atom and the negative ion). For univalent atoms the treatment is straightforward and checks Pauling's values very well when a constant is employed to relate the two sets of figures to each other.

Many additional methods of obtaining electronegativity have been reported, and results are usually compared with those of Pauling. A currently popular approach is that of Allred and Rochow, who think of electronegativity as a force due to the partially screened nuclear charge acting on the atomic electrons at the covalent radius distance. This is expressed as

$$X = \frac{Z_{eff}\, e^2}{r_{cov}^2} \quad ,$$

or, to yield electronegativity values on a scale comparable with Pauling's,

$$X_{A-R} = 0.359 \frac{Z_{eff}}{r_{cov}^2} + 0.744.$$

II. CRYSTAL SYSTEMS, UNIT CELLS, AND AXES

All crystal structures belong to one of seven different crystal systems, each of which has its own basic symmetry. These seven systems are usually considered from the point of view of their unit cells and are so organized in Fig. 4. (The discussion following will, however, point out that this is a simplification not always valid.) The unit cell is the smallest parallelepiped unit which expresses the highest symmetry of the structure and which, when repeated in three dimensions, ultimately makes up a crystal. It is therefore the building block of the crystal.

It is often possible to choose a unit cell smaller or larger in size than the one which expresses the highest symmetry; an example is shown in Fig. 5, wherein it is seen that a face-centered cubic (fcc) structure

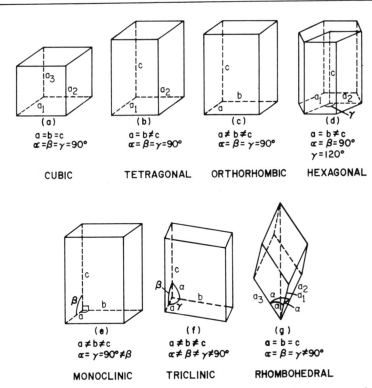

FIG. 4.   The seven crystal systems (a_1, a_2, and a_3 are used here in place of a, b, and c, respectively, when there is an equivalence of axes).

can be represented also by a smaller rhombohedral unit cell or a larger tetragonal unit cell.

The sides of the unit cell are called the axes and are designated a, b, and c. The angles between axes are α (between b and c), β (between a and c), and γ (between a and b). The faces of the unit cell are called A, B, and C (which intersect the a, b, and c axes, respectively).

The network formed by lines in three directions joining equivalent atoms (such as, for instance, the stacking of unit cells in three dimensions) is called a space lattice or simply a lattice. Only 14 types of arrangements of indistinguishable points in space are possible, and

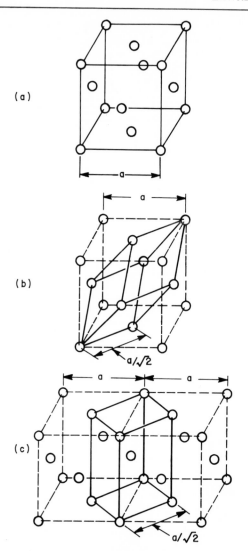

FIG. 5.   Three different possible unit cells of the fcc structure:
(a) fcc unit cell; (b) rhombohedral unit cell; (c) tetragonal unit cell.

they are known as the Bravais lattices.   These 14 lattices are illustrated in Fig. 6.

Referring again to Fig. 4 (in which the unit cells are not necessarily all drawn with the same perspective), it is helpful to consider
the derivations of the unit cells of the seven crystal systems in
terms of distortions of simpler ones to those of lower symmetry; such

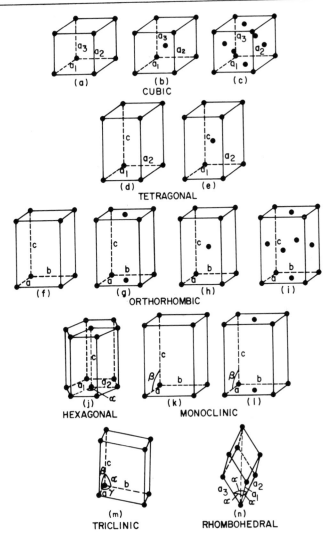

FIG. 6. The 14 Bravais space lattices (Hermann-Maugin symbols in parentheses). (a_1, a_2, and a_3 are used here in place of a, b, and c, respectively, when there is an equivalence of axes). (a) primitive (or simple) cubic (P); (b) body-centered cubic (bcc) (I); (c) fcc (F); (d) primitive (or simple) tetragonal (P); (e) body-centered tetragonal (I); (f) primitive (or simple) orthorhombic (P); (g) base-centered orthorhombic (C); (h) body-centered orthorhombic (I); (i) face-centered orthorhombic (F); (j) primitive (or simple) hexagonal [the symbol P is now used exclusively here; C was formerly employed because hexagonal crystals may also be regarded as base-centered orthorhombic (called "orthohexagonal")]; (k) primitive (or simple) monoclinic (P); (l) based-centered monoclinic (C); (m) primitive (or simple) triclinic (P); (n) primitive (or simple) rhombohedral (R).

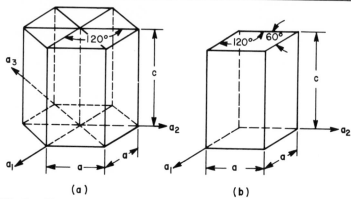

(a)   **(b)**

FIG. 7. The two types of hexagonal unit cell: (a) the large hexagonal type; (b) the small primitive type. The cell in (a) has exactly three
times the volume of that in (b).

a procedure enables easy retention of the information without deliberate,
detached memorization of the expressions under the drawings in Fig. 4.
Thus, if one initially visualizes a perfect cube [as in Fig. 4(a)], it is
obvious that cell edges are equal [a (or a_1) = b (or a_2) = c (or a_3)] and
all angles are right angles ($\alpha = \beta = \gamma = 90°$). If this cube is held at
opposite ends of the c (or a_3) axis and elongated or compressed, a
tetragonal unit cell [as in Fig. 4(b)] results, for one axis is now different
from the other two [a (or a_1) = b (or a_2) $\neq$ c], but all the angles are still
right angles ($\alpha = \beta = \gamma = 90°$). If the tetragonal unit cell [Fig. 4(b)] is
now held at opposite ends of the b (or a_2) axis and elongated or compressed, an orthorhombic unit cell [as in Fig. 4(c)] is obtained, for the
three axes are of different length (a $\neq$ b $\neq$ c), but the angles remain right
angles ($\alpha = \beta = \gamma = 90°$).

A hexagonal unit cell [as in Fig. 4(d)] may be produced by holding
a tetragonal unit cell [such as in Fig. 4(b)] at diagonally opposite c
axes and pulling or pushing until a parallelogram formed by the a and b
axes (the C face) has two 120° angles and two 60° angles. Thus, the
axes are unchanged [a (or a_1) = b (or a_2) $\neq$ c] as are two of the angles
($\alpha = \beta = 90°$; $\gamma = 120°$). Note that the 60° angle need not be specified
because it is not independent of the 120° angle in the parallelogram

which is the C face. This hexagonal unit cell is a true one and can be
used for all purposes. However, some crystallographers prefer to use
a cell exactly three times as large in which the C face takes the form of
a full regular hexagon. The relationship between these two unit cells
is shown in Fig. 7. In the larger unit cell, one refers to four axes
$(a_1 = a_2 = a_3 \neq c)$.

One can produce a monoclinic unit cell [as in Fig. 4(e)] by distortion
of an orthorhombic cell [such as in Fig. 4(c)] in a manner quite similar to
that used in producing the hexagonal cell. Thus, one holds the orthorhombic unit cell at diagonally opposite b axes and pushes or pulls so
that the angles in the B-face parallelogram are no longer right angles
but the axial length relationship remains the same $(a \neq b \neq c; \alpha = \gamma =$
$90° \neq \beta)$. The orthogonal axis has, by convention, been denoted b,
although in both the hexagonal and tetragonal cases the unique axis is
called c. However, many crystallographers are now using c and γ for
the orthogonal axis and the monoclinic angle, respectively. Both conventions are acceptable.

The triclinic unit cell [Fig. 4(f)] is not difficult to visualize for it
is simply the case where all axes are unequal as in the orthorhombic and
monoclinic cells but with distortions leaving behind no right angles
$(a \neq b \neq c; \alpha \neq \beta \neq \gamma \neq 90°)$.

Finally, the rhombohedral unit cell [as in Fig. 4(g)] is produced
from the perfect cube [Fig. 4(a)] simply by holding the cell at opposite
ends of any cube diagonal and elongating or compressing. This leaves
the cube axes unchanged and equal [a (or a_1) = b (or a_2) = c (or a_3)], but
changes the three angles so that they are no longer right angles but are
still equal to each other $(\alpha = \beta = \gamma \neq 90°)$. Actually, the rhombohedral
system is a special case of the hexagonal system and may therefore be
referred to hexagonal axes. This relationship is illustrated in Fig. 8.
Note that the long body diagonal of the rhombohedral unit cell is the
c axis of the hexagonal unit cell.

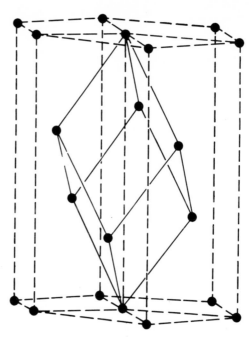

FIG. 8. The relationship between the rhombohedral unit cell and the hexagonal unit cell to which it may be referred.

Although the true symmetry of most crystals is the same as that expressed by one of the seven unit cells discussed here, it is <u>only</u> the true symmetry which determines the crystal system to which a structure belongs. A unit cell may be a perfect cube, but if the atoms in the cell are not arranged in such a way that the symmetry is cubic, then the structure is actually only pseudocubic. An example is β-Ag_2HgI_4, which is pseudocubic tetragonal. Similarly, a crystal can belong to the hexagonal system and have a c axis equal in length to the a axis, or it can belong to the monoclinic system and have equal a, b, and c axes. The $\neq$ sign, as used in Fig. 4 and in this section, does not therefore mean "does not equal in magnitude," but rather "usually but not necessarily unequal to in magnitude," or "does not equal in terms of symmetry but accidental equality of magnitude may occur" (except in the rhombohedral case where the angle really does not equal 90°).

III. MILLER INDICES OF PLANES

It is essential to consider planes in unit cells because the growth of crystals and their physical properties are directly related to planar density, stacking of atoms in planar fashion, etc. Also, as will be seen, x-ray crystallography is based on the fact that the planes form a diffraction grating for x-rays.

In order, then, to enable discussion of planes and to have a mathematical basis for calculations of interplanar spacings and of structures, "Miller indices" are used. To assign a Miller index, one must first choose an origin for the three-dimensional unit cell. Although this choice can actually be made arbitrarily, the usual convention is to choose the lower left rear corner of the unit cell and then to consider translation from this point to be positive if up, right, or toward the viewer, and negative if down, left, or away from the viewer. The Miller index of a parallel set of planes is then determined by counting the number of equal segments into which each of the three axes is divided by the set of planes in each unit cell. The numbers of segments along the a, b, and c axes are called h, k, and ℓ, respectively. When these three numbers, which must obviously be integers, though not necessarily only 0-9, are enclosed in parentheses, we have the Miller index notation. Thus, the general notation for a plane is $(hk\ell)$, and examples are (110), (312), and $(11\bar{2})$, which are read "one, one, zero," (or "one, one, oh"), "three, one, two," and "one, one, minus two," respectively. A bar over a numeral is therefore a minus sign. Whether a Miller index is positive or negative is determined by the orientation of the plane with respect to the origin, that is, whether the translation from the origin to the plane in the unit cell is positive or negative. It should be noted that, since all corners of a unit cell are crystallographically equivalent, each is the origin of its own unit cell, so that both positive and negative translations and orientations may be observed in just one unit cell by considering different corners as origins. The Miller index of a plane is also used to designate the crystal face parallel to the plane.

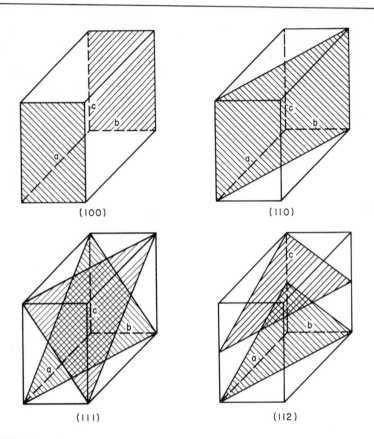

FIG. 9. The Miller indices (hkℓ) of some planes, here in orthorhombic unit cells.

Some examples of Miller indices of planes are given in Fig. 9. These
drawings are of orthorhombic unit cells, but exactly the same rules
apply to all crystal systems.

Just as parentheses are employed for Miller indices of a plane or
set of planes or a crystal face, braces signify a family of planes – that
is, sets of planes all of which have the same atomic configurations and
are therefore equivalent. For instance, in a cubic crystal (or unit cell),
all the cube faces (100), (010), (001), ($\bar{1}$00), (0$\bar{1}$0), and (00$\bar{1}$) are equivalent and belong to the same family of planes. These six sets of planes
as a group are therefore designated {100}.

It has been pointed out that, in the hexagonal system, one may choose either a unit cell with the C face in the form of a parallelogram or else a cell exactly three times as large whose C face is a regular hexagon, and that in the latter case four axes can be considered. The Miller index of a plane referred to these four axes is (hkiℓ), in which h, k, and ℓ are the same as the three indices required for the smaller unit cell. The additional index, i, is not, however, independent, for i = -(h + k). Transformation of indices from the small cell to the larger cell or vice versa is therefore extremely simple.

IV. LATTICE DIRECTIONS

Lattice directions are specified by a notation similar in appearance to the Miller index but different in its meaning and in the method of arriving at it. It consists of three numerals in brackets which represent consecutively the three crystal axes a, b, and c (e.g., [110]). Each numeral indicates the number of unit cells which a line must traverse in the a, b, or c direction from the origin to another unit cell corner. Several examples are given in Fig. 10. As with Miller indices, a bar over a numeral means "minus" and here represents translation in a negative direction. Note that if one wishes to know the direction from the origin to an atom in the unit cell, it is necessary to draw the line through the atom and continue until it reaches another unit cell corner. The number of unit cell distances in each direction is then counted.

In cubic crystals, numerically equal Miller indices and lattice directions are always perpendicular to each other. In all other systems, they intersect each other but are not necessarily perpendicular. For example, only some are perpendicular in the tetragonal case, and none are perpendicular in orthorhombic crystals.

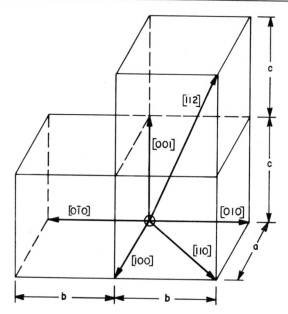

FIG. 10.   Some lattice directions.

V.   SYMMETRY ELEMENTS, POINT GROUPS,
AND SPACE GROUPS

A symmetry element describes an operation by which equivalent points are brought into coincidence.   It may also be defined as the geometrical locus about which operations of repetition take place. The symmetry elements are rotation axes and rotation-inversion axes (either of which may be one-, two-, three-, four-, or six-fold); screw axes (which combine rotation with translation parallel to the axis); symmetry planes [either reflection (mirror) planes or glide planes, which combine reflection planes with translation parallel to the plane]; and centers of symmetry (or of inversion).

Those symmetry elements that do not involve translation permit classification of crystals into only 32 symmetry classes or point groups, which are all the possible ways to arrange equivalent points symmetrically around a given point in space. Symmetry elements and point groups may be given Hermann-Mauguin notations in which symmetry axes parallel to and symmetry planes perpendicular to each of the principal directions are specified in order. When both an axis parallel to and a plane perpendicular to one direction are present, these are indicated by naming the axis first followed by a slash and the plane notation.

The symmetry elements and their notations are listed in Table 4, and the 32 point groups and their symbols in Table 5. The minimum symmetry elements in each of the seven crystal systems are given in Table 6. Although additional symmetry elements may occur along with these, they are not required for classification of a crystal into one of the seven systems.

TABLE 4

Symmetry Elements

Hermann-Mauguin notation	Element
1	1-fold rotation axis
2	2-fold rotation axis
3	3-fold rotation axis
4	4-fold rotation axis
6	6-fold rotation axis
$\bar{1}$	1-fold axis of rotation-inversion (equivalent to a center of symmetry)
$\bar{2}$	2-fold axis of rotation-inversion (equivalent to m)
$\bar{3}$	3-fold axis of rotation-inversion (equivalent to 3 + a center of symmetry)
$\bar{4}$	4-fold axis of rotation-inversion
$\bar{6}$	6-fold axis of rotation-inversion (equivalent to 3/m)

TABLE 4 (continued)

Hermann-Maugin notation	Element
2_1	2-fold screw axis (1/2 translation)
3_1	3-fold right-handed screw axis (1/3 translation)
3_2	3-fold left-handed screw axis (1/3 translation)
4_1	4-fold right-handed screw axis (1/4 translation)
4_2	4-fold screw axis with a 1/2 translation of a 2-fold rotation axis
4_3	4-fold left-handed screw axis (1/4 translation)
6_1	6-fold right-handed screw axis with a 1/6 translation
6_2	6-fold right-handed screw axis with a 1/3 translation of a 2-fold rotation axis
6_3	6-fold screw axis with a 1/2 translation of a 3-fold rotation axis
6_4	6-fold left-handed screw axis with a 1/3 translation of a 2-fold rotation axis
6_5	6-fold left-handed screw axis with a 1/6 translation
m	Reflection (mirror) plane
a	Glide plane with glide of a/2 along a axis
b	Glide plane with glide of b/2 along b axis
c	Glide plane with glide of c/2 along c axis
n	Glide plane with glide of 1/2 of a face diagonal along face diagonal
d	Glide plane with glide of 1/4 of a face diagonal along face diagonal

TABLE 5

Hermann-Mauguin and Schoenflies Symbols for the

32 Point Groups or Symmetry Classes

System	Hermann-Mauguin symbol	Schoenflies symbol[a]
Cubic	23	T
	432 (or 43)	O
	$2/m\ \bar{3}$ (or m3)	T_h
	$\bar{4}3m$	T_d
	$4/m\ \bar{3}\ 2/m$ (or m3m)	O_h
Tetragonal	4	C_4
	$\bar{4}$	S_4
	422 (or 42)	D_4
	$4/m$	C_{4h}
	4mm	C_{4v}
	$\bar{4}2m$	D_{2d} (or V_d)
	$4/m\ 2/m\ 2/m$ (or 4/mmm)	D_{4h}
Orthorhombic	222	D_2 (or V)
	mm2 (or mm)	C_{2v}
	$2/m\ 2/m\ 2/m$ (or mmm)	D_{2h} (or V_h)
Hexagonal	6	C_6
	$\bar{6}$	C_{3h}
	$\bar{6}m2$	D_{3h}
	622 (or 62)	D_6
	$6/m$	C_{6h}
	6mm	C_{6v}
	$6/m\ 2/m\ 2/m$ (or 6/mmm)	D_{6h}

TABLE 5 (continued)

System	Hermann-Mauguin symbol	Schoenflies symbol[a]
Monoclinic	m (or $\bar{2}$)	C_s (or C_{1h})
	2	C_2
	2/m	C_{2h}
Triclinic	1	C_1
	$\bar{1}$	C_i (or S_2)
Rhombohedral	3	C_3
	$\bar{3}$	C_{3i} (or S_6)
	32	D_3
	3m	C_{3v}
	$\bar{3}$ 2/m (or $\bar{3}$m)	D_{3d}

[a] Schoenflies symbols are often used by spectroscopists, but crystallographers now normally employ Hermann-Mauguin notations.

TABLE 6

Minimum Symmetry Elements in Crystal Systems

Crystal system	Symmetry elements
Cubic	Four 3's or $\bar{3}$'s (along the four cube diagonals)
Tetragonal	One 4 or $\bar{4}$
Orthorhombic	Three 2's or $\bar{2}$'s
Hexagonal	One 6 or $\bar{6}$
Monoclinic	One 2 or $\bar{2}$
Triclinic	1 or $\bar{1}$
Rhombohedral	One 3 or $\bar{3}$

Just as there can be only 32 point groups, so are there only 230 space groups – three-dimensional arrays of symmetry elements on space lattices. Although many of the space groups consist simply of point groups at the points of the 14 space lattices, not all space groups are so produced because there are certain symmetry elements possible in space groups that are not possible in point groups. The symmetry elements involving translation, not found in point groups, therefore occur in many space groups. Point groups and space groups are named either by the older Schoenflies system or the now preferred Hermann-Mauguin notations. In naming space groups with the latter, the type of lattice is followed by symmetry elements associated with special directions in the appropriate crystal system which make the space group unique. The lattice types are simple or primitive (P), face-centered (F), body-centered (I), base-centered (C), and rhombohedral (which is primitive) (R). One example is the monoclinic space group $P2_1/m$, which denotes a primitive unit cell with a 2-fold screw axis perpendicular to a mirror plane. In the monoclinic system the special direction is along the orthogonal axis. Another example is the cubic space group Fm3m, which denotes a face-centered unit cell with mirror planes perpendicular to the [100] and [110] directions and a 3-fold axis along the [111] axis. These axes are the special directions in the cubic system. A complete list of space group symbols and rules for interpretation of the symbols may be found in Volume 1 of International Tables for X-Ray Crystallography.

VI. X-RAY DIFFRACTION

It is well known that it is possible to diffract light by passing it through a grating consisting of lines drawn at distances from each other of approximately the same magnitude as the wavelength of the light. (Visible light has wavelengths from 4000 to 7000Å, or $4-7 \times 10^{-5}$ cm.) In 1912, von Laue reasoned that a crystal should act as a three-dimensional

grating for diffraction of x-rays because interatomic spacings are approximately equal to x-ray wavelengths. With Friedrich and Knipping, he
soon demonstrated that this was indeed true; x-ray diffraction has since
become an extremely important tool which makes possible measurement
of interplanar spacings and determination of crystal symmetry and actual
atomic positions. Von Laue was able to systematize the experimental
results quantitatively by the following three conditions which must be
met simultaneously for diffraction to occur:

$$a(\cos \alpha - \cos \alpha_0) = h\lambda$$

$$b(\cos \beta - \cos \beta_0) = k\lambda$$

$$c(\cos \gamma - \cos \gamma_0) = \ell\lambda \quad ,$$

in which a, b, and c are spacings in three directions; α_0, β_0, and γ_0
the angles between the incident beam and the rows of atoms in the three
directions; α, β, and γ the angles between the diffracted beam and the
three rows; h, k, and ℓ integers; and λ the wavelength of the x-rays.
These are now known as the Laue equations; they state that the path
difference in each dimension [e.g., $a(\cos \alpha - \cos \alpha_0)$] must be an integral number of wavelengths if the scattered waves are to be in phase.

Soon after this work was published, W. L. Bragg showed that
Laue's conditions were equivalent to reflection from a lattice plane and
that the effect could be described by the simple equation

$$n\lambda = 2d \sin \theta \quad \text{(Bragg's law)} \quad ,$$

in which n is an integer (the "order"), λ the wavelength of the x-rays,
d the interplanar spacing, and θ the angle of incidence of the x-ray
beam on the plane. It is easily derived from a drawing such as that in
Fig. 11. In that drawing, an x-ray wave strikes a crystal at point A at an
angle of incidence θ and is reflected at the same angle. The wave
which penetrates to the next plane just beneath the surface (at an interplanar distance AB or d) behaves similarly but strikes the second plane

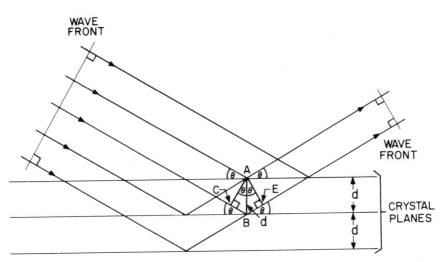

FIG. 11. The graphical derivation of Bragg's law, $n\lambda = 2d \sin \theta$.

at point B and must therefore travel farther than the wave that reflects off the surface by the distance CB + BE (where CB = BE). By simple geometry and trigonometry, it can be shown that CB and BE are each equal to $d \sin \theta$, so that CB + BE = $2d \sin \theta$. Since the condition for reinforcement is that waves be in phase, an x-ray reflection will be observed only at those angles of incidence where the additional path difference, CB + BE (or $2d \sin \theta$), is an integral number of wavelengths. (The top two planes shown in the drawing might appear to indicate that it would be possible to obtain diffracted waves of reduced intensity even when CB + BE is not an integral number of wavelengths as long as the reflection from the second layer is not completely opposite in phase to that from the top layer. However, consideration of reflection by additional sublayers (such as the third layer in the drawing) will show that destructive interference from some sublayer will occur if CB + BE is not an integral number of wavelengths. The cause is a systematic shift in phase of waves reflected from successive layers until destructive interference eventually occurs.) Using n to represent the integral order and λ the wavelength of the x-rays, we may therefore write $n\lambda = 2d \sin \theta$,

FIG. 12.   Philips 57.3-mm radius x-ray diffraction powder camera
with cover off.   (Photograph courtesy of Philips Electronic Instruments
Division of Philips Electronics and Pharmaceutical Industries Corp.)

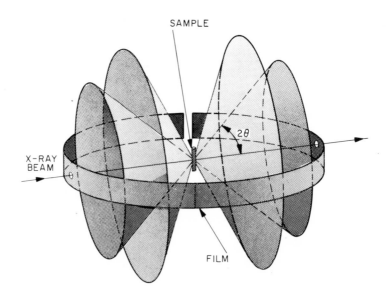

FIG. 13. Diffraction of x-rays as it occurs inside a powder camera.

which is Bragg's law. It is therefore possible to obtain diffracted waves
at any combination of θ, d, and n that fits this expression. In x-ray
diffraction studies, it is found more convenient to ignore the order of
the reflection and to work with d/n values rather than necessarily with
the true d. In fact, d/n is often referred to simply as d, in which case
some interplanar spacings reported to be observed experimentally are
really imaginary.

In a crystalline powder, the tiny crystals are oriented at random.
If such a powder is struck by an x-ray beam, many planes will be so
oriented that Bragg's law is satisfied and an x-ray diffraction pattern
obtained. To be certain that all possible planes are exposed to the
x-ray beam, a cylindrical specimen is usually employed and rotated on
its own axis during exposure. Most of the x-ray beam will pass directly
through the sample, and the diffracted beams will be of low intensity.
Therefore, when film is used to record the diffracted x-rays, exposure
times are normally of the order of several hours. With electronic
detectors, however, exposure times are short.

Figure 12 is a photograph of a cylindrical x-ray diffraction powder
camera which requires a cylindrical sample which may, for instance, be
a thin-walled glass capillary filled with powder. In this camera, a strip
of film is placed on the inside surface of the cylinder with the sample at
the center of the cylinder. A monochromatic x-ray beam is admitted
through a slit at one end and exits at the other end (into absorbing glass
coated with phosphor on the surface which the x-ray beam first strikes).
Because crystals in the sample are present at all orientations, diffraction occurs in cones. Figure 13 is a diagrammatic representation of what
occurs in the camera. The intersection of a cone with a flat surface
perpendicular to the cone axis is a circle, but similar intersection of a
cone with a cylinder yields an ellipse on flattening out the cylindrical
surface (with the major axis of the ellipse becoming infinitely long when
the cone angle is 180°). Since it is the diameter of the cone that carries
the desired information, it is necessary only to use a narrow (usually

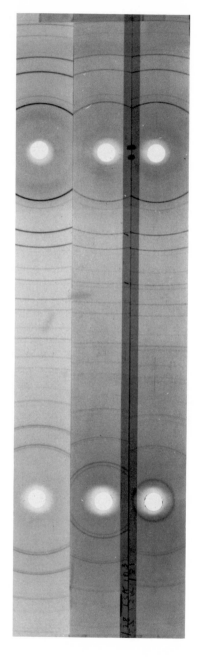

FIG. 14. X-ray diffraction powder patterns obtained with a Philips 57.3-mm radius powder camera. From top to bottom: ZnTe (zincblende structure), $Zn_3 \square PI_3$, and $Zn_3 \square AsI_3$ [the last two with disordered defect zincblende structure; see L. Suchow, M.B. Witzen, and N.R. Stemple, Inorg. Chem., $\underline{2}$, 441 (1963)].

35 mm) strip of film inside the cylinder. The cone's diameter is measured at the center of the strip after development and placement of the
film in a flat position. Several films in this position are shown in
Fig. 14. The holes punched in the film are for the collimators which
carry the x-ray beam in and out of the camera. Figure 15 gives the
geometry that makes possible determination of θ's (Bragg angles) and d-
spacings. If s is the measured distance on the film (that is, the cone
diameter at the film radius), then $s/r = 4\theta$ in radians, so that

$$\theta \text{ (in radians)} = \frac{s}{4r} \quad .$$

Since there are $360°/2\pi = 57.3$ degrees per radian, θ in degrees =
$s(57.3)/4r$. If, as is commonly done, the camera is machined precisely
so that its radius is 57.3 mm, then the Bragg angle θ is simply $s/4$ if
s is in millimeters. If one desires more significant figures it is not
difficult to make (by the Straumanis method) a very accurate determination of the actual radius of the film itself, which has perhaps shrunk
during the developing process. This is based simply on determining the
centers of the forward and back reflection cones; the distance between
these two centers is πr.

X-ray diffraction powder patterns thus yield interplanar spacings
very easily, and are extremely useful for identification purposes and for
the following of solid state reactions. With crystalline material of high
symmetry, it is possible also to determine lattice parameters and Miller
indices from the d- spacings. This is done by determining which of the
formulas in Table 7 fits all the data. It is especially easy to do with
cubic crystals, and graphical aids are available for some of the other
systems, but with the very complex powder patterns obtained from monoclinic or triclinic crystals, it is usually necessary to have additional
information, perhaps from isomorphous compounds whose lattice parameters are known. In addition, computers are now used effectively in
many cases. X-ray diffraction patterns from single crystals are easier
to interpret because they do not require determination of lattice

TABLE 7

Formulas Relating Interplanar Spacings (d) with

Lattice Parameters and Miller Indices (hkℓ)

Cubic	$d = \dfrac{a}{\sqrt{h^2 + k^2 + \ell^2}}$ or $\dfrac{1}{d^2} = \dfrac{h^2 + k^2 + \ell^2}{a^2}$
Tetragonal	$\dfrac{1}{d^2} = \dfrac{h^2 + k^2}{a^2} + \dfrac{\ell^2}{c^2}$
Orthorhombic	$\dfrac{1}{d^2} = \dfrac{h^2}{a^2} + \dfrac{k^2}{b^2} + \dfrac{\ell^2}{c^2}$
Hexagonal	$\dfrac{1}{d^2} = \dfrac{4}{3}\dfrac{h^2 + hk + k^2}{a^2} + \dfrac{\ell^2}{c^2}$
Monoclinic	$\dfrac{1}{d^2} = \dfrac{h^2}{a^2 \sin^2\beta} + \dfrac{k^2}{b^2} + \dfrac{\ell^2}{c^2 \sin^2\beta} - \dfrac{2h\ell \cos\beta}{ac \sin^2\beta}$
Triclinic	$\dfrac{1}{d^2} = [(abc)^2(1 - \cos^2\alpha - \cos^2\beta - \cos^2\gamma +$ $2\cos\alpha \cos\beta \cos\gamma)]^{-1}[(bc \sin\alpha h)^2 +$ $(ac \sin\beta k)^2 + (ab \sin\gamma\ell)^2 +$ $2abc^2(\cos\alpha \cos\beta - \cos\gamma)hk +$ $2a^2bc(\cos\beta \cos\gamma - \cos\alpha)k\ell +$ $2ab^2c(\cos\gamma \cos\alpha - \cos\beta)h\ell]$
Rhombohedral	$\dfrac{1}{d^2} = \dfrac{(h^2 + k^2 + \ell^2)\sin^2\alpha + 2(hk + k\ell + h\ell)(\cos^2\alpha - \cos\alpha)}{a^2(1 - 3\cos^2\alpha + 2\cos^3\alpha)}$

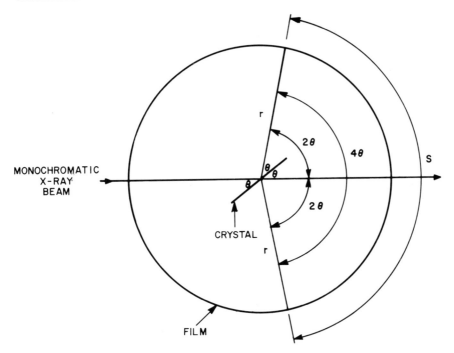

FIG. 15.   Geometry of x-ray diffraction powder camera.

parameters and Miller indices from the equations in Table 7, but rather
are usually interpreted with the use of graphical templates placed over
the films which here contain spots rather than lines. Single-crystal
patterns resolve reflections which are superimposed in powder patterns
(for example, 300 and 221, 110 and 011, or 111 and 1$\bar{1}$1 in the cubic
system). Therefore, single-crystal cameras are usually employed
wherever possible.

   Whereas the positions of x-ray reflections are a function of the
size, shape, and symmetry of the unit cell, the intensities of the
reflections are determined by the actual atomic positions and the numbers of electrons of the atoms in the cell. The intensities are proportional to the square of the structure factor (F), which is expressed by

$$|F|^2_{hk\ell} = \left\{ \sum_i f_i \cos[2\pi(hx_i + ky_i + \ell z_i)] \right\}^2 +$$

$$\left\{ \sum_i f_i \sin[2\pi(hx_i + ky_i + \ell z_i)] \right\}^2 ,$$

in which h, k, and ℓ are the Miller indices of the reflection; x_i, y_i, and z_i are the coordinates of all atoms in the unit cell; and f_i signifies scattering factors of all the atoms. $\sum_i$ indicates that the summation is carried out over all atoms in the unit cell. The scattering factor for an atom or ion at zero Bragg angle is simply the number of electrons it contains. At higher angles it falls off but remains a function of the number of electrons. More convenient expressions for the structure factors of each of the 230 space groups are given in Volume 1 of International Tables for X-Ray Crystallography.

$|F|^2_{hk\ell}$ must be calculated for all possible planes and compared with intensities of x-ray reflections observed after correction, wherever appropriate, for such items as temperature factor, absorption factor, multiplicity factor, Lorentz and polarization factor, and velocity factor. This is relatively easy to accomplish, though time-consuming, when one has other information which leads him to the correct values of the atomic coordinates. However, if nothing is known but experimental intensities, it is not possible to determine the x, y, and z values directly because a periodic function such as this contains both amplitudes and phase angles (or signs). Since it is the square of the structure factor which is proportional to intensity, one does not know whether the x, y, and z values are positive or negative. Fourier series methods are therefore employed to determine atomic coordinates. A number of tricks have been devised to simplify such calculations, and the computer is of course an extremely important tool since much trial and error is necessary.

One of the consequences of the structure factor equation given above is that certain x-ray reflections must have zero intensity. Face-centered (F) space lattices can have reflections only with h, k, and ℓ all even or all odd; body-centered (I) only those with the sum of h, k, and ℓ even; A-face-centered (A) with $k + \ell$ even; B-face-centered (B) with $h + \ell$ even; and C-face-centered (C) with $h + k$ even. Primitive (P) space lattices have no systematic absences. These are called extinction rules, and there are similar rules for most of the symmetry elements. The space group is determined experimentally by indexing the x-ray diffraction pattern and inspecting the calculated indices for systematic absences.

VII. ATOMS, PACKING, STACKING, AND HOLES

Although significant exceptions occur, it is found that a great many crystal structures (especially inorganic ones) may be considered to be a packing together of atoms or ions in the form of hard spheres, with radii of given species rather constant in different compounds of similar ionic-covalent character, but with small differences if the numbers of nearest neighbors (coordination numbers) are different. Radii commonly used are those of Goldschmidt, Pauling, Ahrens, and Shannon and Prewitt.

One may wish to visualize such packing of spheres of equal size in a plane by reference to 15 billiard balls in a rack (see Fig. 16). Any of the completely surrounded balls (5, 8, and 9 in the drawing) is seen to be enclosed by 6 other balls. Exactly the same effect can be

FIG. 16. Close-packed billiard balls in a rack.

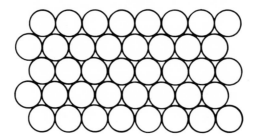

FIG. 17. A single close-packed plane of disks (such as coins) or of spheres.

demonstrated in two dimensions by using seven coins of the same denomination. Six of them will always fit exactly around the central coin so that each is in contact not only with the central one, but also with two others. This is called close packing, but the discussion to this point refers, of course, only to close packing in a plane, as illustrated further in Fig. 17. Crystals are actually three-dimensional, so that the billiard balls in Fig. 16 can also be surrounded by additional planes of balls above and below. If one attempts to fit additional balls above the billiard balls in a close-packed arrangement, it is quickly seen that putting them into the depressions in the base plane of balls (those in the rack) results in a triangle of three balls above each ball in the rack. Exactly the same situation would, of course, be found if we were to place balls beneath the base plane. (This can be demonstrated also by gluing table tennis balls together to form planes of spheres which may be easily manipulated.) The central sphere in close packing is therefore surrounded by twelve spheres if all are of the same size. It should easily be seen that the triangles of atoms above and below the central plane are really parts of full planes of their own, so that all atoms in the central plane have a coordination number of 12. If the reader has perhaps followed this argument with three-dimensional models, he has probably already realized that the triangles of balls may be placed in one of two nonequivalent orientations since there are actually six depressions around each atom. If the planes are stacked as shown in

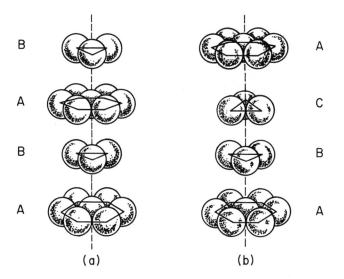

FIG. 18.   Exploded views of stacking of close-packed planes of spheres: (a) hexagonal close-packing; (b) cubic close-packing.

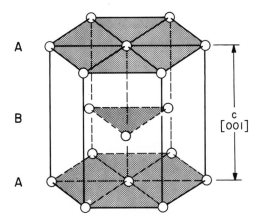

FIG. 19.   The close-packed hexagonal unit cell of a metal, showing the stacking layers parallel to the (001) plane.

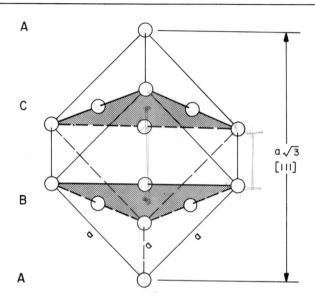

FIG. 20. Stacking layers [(111) planes] in the unit cell of a face-centered (close-packed) cubic metal.

exploded fashion in Fig. 18(a), that is, with the stacking sequence ···ABABAB···, then the result is hexagonal close packing, in which the stacking is parallel to the (001) plane (C face) of the close-packed hexagonal unit cell (Fig. 19). If, on the other hand, the planes are stacked as shown in exploded fashion in Fig. 18(b), that is, with the stacking sequence ···ABCABCABC···, then the result is cubic close packing, which is face-centered cubic (fcc). In this case, however, the stacking is parallel to the (111) planes of the fcc unit cell, as shown in Fig. 20.

It is obvious that close packing of hard spheres will still leave open spaces. One may easily calculate the volume of "empty space" in a crystal if the assumption is made that the atoms or ions are hard spheres. To do this, one need merely divide the volume of the atoms in the unit cell by the volume of the unit cell. For instance, for close-packed (face-centered) cubic metals

$$\text{Packing factor} = \frac{4(4/3\ \pi r^3)}{(4r/\sqrt{2}\ )^3} = 0.74\ .\quad FCC$$

Thus, only 74% of the space appears to be occupied by atoms in this case (and also in hexagonal close-packed metals). For comparison, body-centered cubic (bcc) metals have a packing factor of 0.68.

In crystallography the open spaces between atoms are called interstitial holes, and the two most important types are octahedral holes and tetrahedral holes, but there are, in addition, the smaller spaces surrounded by planar triangles of atoms.

A regular octahedron has eight sides but only six corners; an octahedral hole is the open space surrounded by six lattice atoms. In both hexagonal and cubic close packing, there is one octahedral hole for each lattice atom. It is useful to realize that a regular octahedron is obtained by connecting points at the centers of the six faces of a cube (see Fig. 21); if atoms are at the points connected, the octahedral hole is at the center of the octahedron and of the cube.

A regular tetrahedron has four sides and four corners; a tetrahedral hole is the open space surrounded by four lattice atoms. There are two such holes for each lattice atom in hexagonal and cubic close packing. A regular tetrahedron may be constructed by connecting alternate corners of a cube (see Fig. 22); if atoms are at the four cube corners connected, the tetrahedral hole is at the center of the tetrahedron and of the cube.

It is easily shown that the radius (see Fig. 23) of the largest spherical atom which will fit into an octahedral hole in a close-packed lattice without forcing the atoms apart is 0.414 times the radius of the lattice atom. One may make similar calculations for holes with other coordination numbers; the results are tabulated in Table 8. These numbers are also called "minimum radius ratios."

It will be demonstrated that different coordination numbers may be found in different compounds. The prime reasons for a given coordination number being chosen by a compound are probably bond type, availability of orbitals, and relative ionic sizes (though the crystal field also has an effect in compounds of many transition elements). Certain

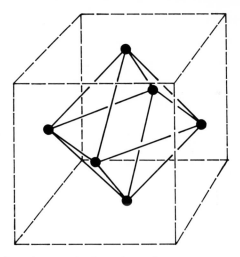

FIG. 21. Regular octahedron formed by connecting the centers of the six faces of a cube. If atoms are present at the face centers, there is an octahedral hole at the center of the octahedron and of the cube.

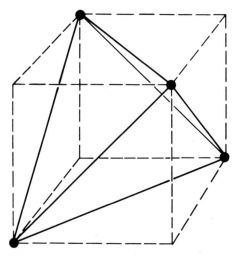

FIG. 22. Regular tetrahedron formed by connecting alternate corners of a cube. If atoms are present at the connected corners, there is a tetrahedral hole at the center of the tetrahedron and of the cube.

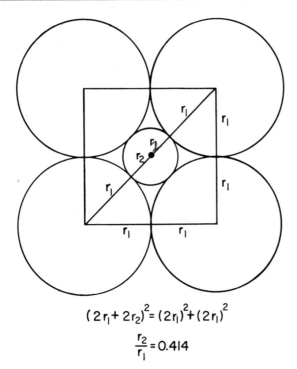

$$(2r_1 + 2r_2)^2 = (2r_1)^2 + (2r_1)^2$$

$$\frac{r_2}{r_1} = 0.414$$

FIG. 23. Determination of the minimum radius ratio for octahedral coordination (showing only four planar atoms of the octahedron).

TABLE 8

Hole Sizes (Minimum Radius Ratios)

Coordination number of hole or central atom	Ratio of radius of largest sphere which can fill hole to radius of coordinating atoms (minimum radius ratio)
3 (planar)	0.155
4 (tetrahedral)	0.225
6 (octahedral)	0.414
8 (cubic)	0.732
12[a]	1

[a]This is the case of a vacancy in a close-packed lattice.

structures are more common among ionic compounds than covalent ones, and vice versa. For instance, the NaCl, CsCl, TiO_2, CaF_2, and ReO_3 structures are normally assumed by ionic compounds, while CdI_2 and $CdCl_2$ layer structure compounds are more covalent.

In addition, the minimum radius ratio has, in many cases, a direct bearing on the coordination number assumed. In an analogous series of ionic compounds, there is a general rule that the structure assumed will be such as not to permit the enclosed ion to "rattle around" in the hole formed by the surrounding ions. In order to accomplish this, the structure will usually be one where the ratio of the radius of the surrounded atom to that of the surrounding ones will be equal to or greater than the minimum radius ratio value.

For instance, NaCl and NaBr, whose radius ratios of 0.52 and 0.49, respectively, fall between 0.414 and 0.732 (see Table 8), have octahedral structures. CsCl and CsBr, whose radius ratios of 0.93 and 0.87, respectively, fall between 0.732 and 1, have cubes of eight anions surrounding the cesium (and vice versa).

There are, however, some significant exceptions to this rule. LiCl, LiBr, and LiI have respective radius ratios of 0.33, 0.31, and 0.28 and might therefore be expected to assume tetrahedral coordination. In actual fact, they have the octahedral NaCl structure, although the unit cell edge appears to be determined more by contact between anions across face diagonals than by contact between anion and cation along the cell edge (see discussion on pages 115-117). Other exceptions are RbF, RbCl, RbBr, and CsF, all of which are found to have the NaCl structure at normal temperature and pressure.

There have been many studies and considerations of the reasons for the existence of such exceptions, and they may perhaps very simply be summarized by stating that, much as the crystallographer might like ions to be as hard, undistortable, chargeless, and constant in radius as billiard balls, they really are not and will sometimes express their independence of our oversimplifications.

VIII.   REPRESENTATIVE STRUCTURES AND
THEIR INTERRELATIONSHIPS

Certain structures are especially interesting because (1) they can be used to show interrelationships among different structures, (2) they occur very commonly, and (3) a great number of the compounds of theoretical and practical interest in solid state physics and chemistry have these structures.   Many of them are cubic.

To show the interrelationships, let us first refer to the four types of structures found in metals.   The hexagonal close-packed metal structure (as in Mg, Y, and Ru, for example) is depicted in Fig. 19.   There are 6 atoms in the large hexagonal cell, all atoms have a coordination number of 12, $a = 2r$ (where r = atomic radius), and the c/a axial ratio is 1.63.   The fcc metal structure (as in Cu, Ag, and one form of Ni, for example) has already been depicted in Figs. 6(c) and 20.   In this case, there are 4 atoms per unit cell, the coordination number is 12, and contact between atoms is along the face diagonal, whose length is therefore $4r$ so that the lattice constant a is $4r/\sqrt{2}$.   In the bcc metal [Fig. 6(b)], there are 2 atoms per unit cell, the coordination number is 8, and contact is along the body diagonal ($4r$) so that $a = 4r/\sqrt{3}$.   K, V, and Cr are examples of this structure.   In the simple cubic metal [Fig. 6(a)], which has been observed only in one form of polonium, there is but one atom in each unit cell, the coordination number is 6, and contact is along the cell edge, so that $a = 2r$.

In the fcc metal structure, octahedral holes occur at the centers of the unit cell edges and at the cell's body center.   The one at the body center has already been illustrated in Fig. 21; the others are exactly like this and each could be moved to the center of the cube by proper translation of the origin of the unit cell.   The positions of the octahedral holes are indicated by X's in Fig. 24; there are four per unit cell.   There are also eight tetrahedral holes per unit cell, and these occur along the four cube diagonals at distances of one-fourth the cube

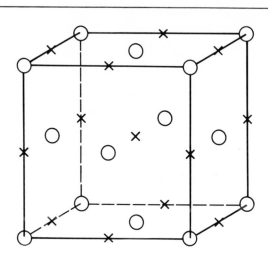

FIG. 24.   Location of octahedral holes (X's) in the unit cell of a
face-centered (close-packed) cubic metal, the atoms of which are represented by open circles.

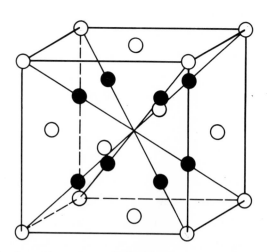

FIG. 25.   Location of tetrahedral holes (filled circles) along the
body diagonals of the unit cell of a face-centered (close-packed) cubic
metal, the atoms of which are represented by open circles.

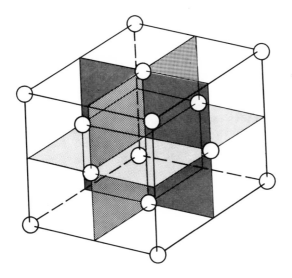

FIG. 26. The unit cell of a face-centered (close-packed) cubic
metal subdivided into eight cubelets by bisecting the unit cell edges
with mutually perpendicular planes. There is a tetrahedral hole at the
center of each of the eight cubelets because the cubelets have atoms
at alternate corners.

diagonal from each cube corner. This is illustrated in Fig. 25. Note

that these eight tetrahedral holes are at the corners of a simple cube,

of which use will be made later. It is also useful to see that the eight

tetrahedral holes are at the centers of the eight cubelets obtained on

bisecting the unit cell edges with mutually perpendicular planes, as

illustrated in Fig. 26. The eight cubelets here are seen to have lattice

atoms on alternate corners which, when connected as in Fig. 22, yield

a tetrahedron within each cubelet.

Now, referring to Fig. 24, if the lattice sites are filled with

chloride ions and the octahedral holes (the X positions) with sodium

ions, the NaCl (rocksalt) structure results (see Fig. 3). Each Na^+ and

Cl^- here is surrounded octahedrally by six of the other. Because the

Na^+ ions are larger than the octahedral holes left by an fcc chloride

lattice, the chlorides are forced apart and the unit cell size is deter-

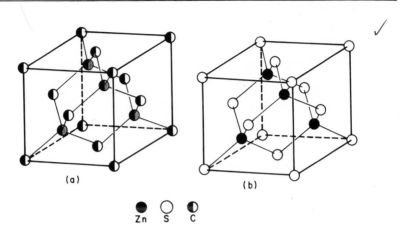

FIG. 27. The relationship of the diamond and zincblende structures: (a) the fcc unit cell of diamond; (b) the fcc unit cell of zinc
sulfide (the zincblende or sphalerite structure).

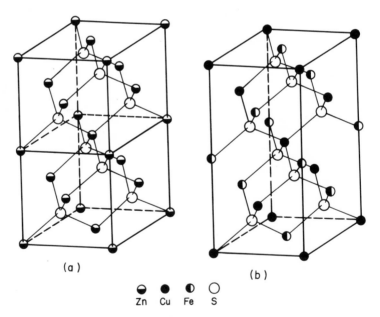

FIG. 28. The relationship of the zincblende and chalcopyrite
structures: (a) two unit cells of cubic ZnS (zincblende); (b) the tetragonal unit cell of $CuFeS_2$ (chalcopyrite). [Note that these have been drawn
with cell origins at cations rather than at anions as in Fig. 27(b).]

mined by contact along the cell edge; that is, $a = 2(r_{Na^+} + r_{Cl^-})$. There are four formula units of NaCl per unit cell. The infrared-sensitive photoconductors, PbS, PbSe, and PbTe, all crystallize in the rocksalt structure. The thermoelectric compound, $AgSbTe_2$, also has this structure but with the Ag and Sb disordered on the cation sites, giving what is called the $LiFeO_2$ structure.

If carbon atoms are on all the lattice sites and in half the tetrahedral positions (alternately), the result is the diamond structure [see Fig. 27(a)] with eight carbon atoms per unit cell, in which all carbons are surrounded tetrahedrally by four others. Because the atoms placed in the tetrahedral positions are of course too large to fill these positions without forcing the lattice atoms apart, the lattice constant here is determined by contact along the body diagonal; that is, $2r = \frac{1}{4}$(body diagonal) $= \frac{1}{4}(a\sqrt{3})$, so that

$$a = \frac{8r}{\sqrt{3}} .$$

This structure is also found in silicon and germanium, which are commonly used as semiconductors.

If, instead, sulfur atoms are on the lattice sites and zinc atoms in half the tetrahedral positions (alternately), the zincblende or sphalerite structure with four formula units of ZnS results [see Fig. 27(b)]. All Zn and S atoms are surrounded tetrahedrally by four of the other type, and contact is again along the body diagonal. Therefore,

$$(r_{Zn} + r_S) = \frac{1}{4}(\text{body diagonal}) = \frac{1}{4}(a\sqrt{3}) ,$$

so that

$$a = \frac{4(r_{Zn} + r_S)}{\sqrt{3}} ,$$

in which r_{Zn} and r_S are covalent radii. This structure is also adopted by the Group III - Group V compound semiconductors, such as GaAs and InP. Cubic ZnS itself, when properly doped with impurities, is an

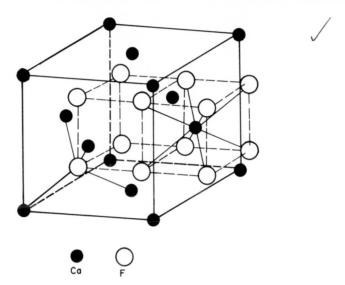

FIG. 29. The unit cell of fcc CaF_2 (fluorite) with several extra atoms to show the 8-fold cubic coordination about a representative calcium ion. Also shown are both filled and empty cubelets of volume one-eighth that of the unit cell, and the tetrahedral coordination about a representative fluoride ion.

excellent photoluminescent, electroluminescent, and cathodolumines-cent phosphor.

If the Zn atoms in the zincblende structure are replaced alternately by Cu and Fe, the result is the tetragonal chalcopyrite structure, with a doubled unit cell edge in one direction (see Fig. 28). The semiconduct-ing compound $CdSnAs_2$ also has this structure.

If the fcc lattice sites of Fig. 25 or 26 contain calcium ions, and fluoride ions are in all eight tetrahedral positions, the CaF_2 (fluorite) structure, with 4 formula units of CaF_2 per unit cell, is obtained (see Fig. 29). In this case, each fluoride is surrounded tetrahedrally by calciums and each calcium by a cube of eight fluorides. The unit cell size is again determined by contact along the body diagonal; that is,

$$r_{Ca^{2+}} + r_{F^-} = \tfrac{1}{4}(\text{body diagonal}) = \tfrac{1}{4}(a\sqrt{3}) \ ,$$

so that

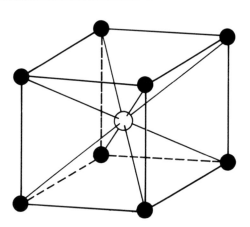

FIG. 30.   The primitive (simple) cubic unit cell of CsCl.   Either
the open or filled circles may be taken to represent Cl⁻; the remaining
circles then represent Cs⁺.

$$a = \frac{4(r_{Ca^{2+}} + r_{F^-})}{\sqrt{3}}.$$

Let us at this point digress a bit to the CsCl structure (Fig. 30).
This is like the bcc structure illustrated in Fig. 6(b), but with chloride
ions at the cube corners and a cesium ion at the body center (or vice
versa).   This is, however, simple cubic (or primitive) rather than bcc
because body-centering requires that the atom at the cell center be
identical with those at the corners.   Contact is along the body diagonal,
whose length is $2(r_{Cs^+} + r_{Cl^-})$.   The lattice constant a is therefore
$2(r_{Cs^+} + r_{Cl^-})/\sqrt{3}$.   There is only one formula unit of CsCl per unit
cell, and all ions are surrounded by eight of the other type.

Returning once more to the fluorite structure, subcells of volume
one-eighth that of the true unit cell are pointed out in Fig. 29.   It will
be seen that all such subcells are cubelets with fluoride ions at the
eight corners, but that the cubelet centers are alternately empty and
filled with calcium ions.   Those which are filled are exactly like CsCl
unit cells, but the larger cube must of course be chosen as the unit cell
because the subcell is not the repeat unit.

One might expect from the foregoing that it should be possible to prepare a compound of type MX_3 (or M_4X_{12} per unit cell), in which atoms or ions, M, are on the fcc lattice sites and all the tetrahedral and octahedral positions are filled with atoms or ions, X (of opposite charge, of course, if ions). Such a structure, while not common, is indeed found in intermetallic phases such as $BiLi_3$ and ordered $AlFe_3$, as well as in some rare earth hydrides, such as CeH_3, which tend to be nonstoichiometric. In addition, single phase compositions with solid CaF_2 as solvent can be prepared containing from 0 to 55 mole % YF_3. The YF_3 is known to go into the CaF_2 structure by substitution of Y^{3+} for Ca^{2+} and with two of the fluoride ions on two normal tetrahedral fluoride sites and the third on what is called an interstitial site and is in fact the hole which is surrounded both by an octahedron of calcium (and yttrium) ions (where the cation-anion distance is $a/2$), and by a cube of fluoride ions [where the F^--F^- distance is $(\sqrt{3}/4)a$]. Similar effects are found with rare earth trifluorides in place of YF_3 and with CdF_2 in place of CaF_2, with which it is isomorphous. CaF_2 crystals doped with trivalent rare earths for use as lasers usually achieve the required charge compensation by incorporation of the extra fluoride in this manner. However, it is also possible to bring about the compensation by means of a divalent oxide ion substituting for a univalent fluoride next to the rare earth ion. In these two cases, however, there are differences in the emission process which result in laser wavelengths slightly offset from each other.

The spinel structure is an interesting one and has been widely studied, especially because of the interesting physical properties exhibited by a number of compounds with this structure. Although there are many exceptions, spinels commonly contain divalent and trivalent cations in a 1:2 ratio in a lattice structure built of divalent anions. An example is the mineral spinel itself, $MgAl_2O_4$. Let us, however, consider a more general type, $A^{II}B_2^{III}C_4^{VI}$. The C anions pack in the fcc positions of the unit cell and the A and B ions go into octahedral and tetrahedral holes but in such a way that it is necessary to double the unit cell edge in all three directions, thereby increasing the volume of the unit cell by a factor of eight and yielding a cubic unit cell with 8 formula units of AB_2C_4. In this cell, then, the 32 C anions are packed as already described and the 8 A and the 16 B cations fill a total of 24 of the 32 octahedral and 64 tetrahedral holes. The filling is systematic, however, and there are two primary ways in which it occurs. If all the A cations fill 8 (of 64) tetrahedral holes and all the B cations fill 16 (of 32) octahedral holes, the structure is called a normal spinel. On the other hand, if half the B cations fill 8 (of 64) tetrahedral holes and the other half fill 8 (of 32) octahedral holes, and all the A cations fill the remaining 8 spinel octahedral sites, the structure is called an inverse spinel. Many compounds are a hybrid of the normal and inverse structures, and in fact, it is sometimes possible to shift the equilibrium between them by changing preparative conditions. The following summarizes the description of the two structures:

	Tetr.	Oct.	
Normal spinel	(A)	$[B_2]$	C_4
Inverse spinel	(B)	$[AB]$	C_4

Structural or solid state inorganic chemistry has perhaps achieved its greatest success to date by permitting a very simple explanation of the magnetic behavior of spinel ferrites such as $MnFe_2O_4$ and Fe_3O_4. It has been found that there is a superexchange across oxygen involving the transition metal ions with unpaired 3d electrons in such a way that the ions on the tetrahedral sublattice and those on the octahedral sublattice are aligned in each domain as small magnets all pointing in the same direction on each of the sublattices, but with the directions within the two sublattices opposite to each other. The effect is called ferrimagnetism, and the observed magnetic moment is simply the net difference between the expected moments of the two sublattices. In compounds where the magnetic moments of two groups of ions are exactly equal, the effect is termed antiferromagnetism. This picture of ferrimagnetism and antiferromagnetism is due to Néel. For comparison, ferromagnetism is the case where all magnetic dipoles in a single domain point in the same direction. Examples are metallic iron and also the normal spinels $CdCr_2S_4$ and $CdCr_2Se_4$.

Another interesting cubic structure related to those above is that called the C-M_2O_3 structure, which is one of the structures of the rare earth sesquioxides and is also found in α-Mn_2O_3. In this case, we may consider it to be built up of 64 distorted cubelets ($4 \times 4 \times 4$) with oxide ions on six of the eight corners of each cubelet. Cations are found in the centers of alternate cubelets; three-quarters of the cations are in cubelets with the two missing oxide ions at the ends of a face diagonal, and one-quarter in cubelets with the missing anions at the ends of a body diagonal. There are therefore 16 formula units of M_2O_3 per unit cell. Note that the cubelets here are related to those in the CaF_2 and $CsCl$ structures except that the cubelet corners are not all filled; they are also related to cubelets of the zincblende structure because they result from addition of two anions at vacant corners of cubelets in which tetrahedra are inscribed (as in Fig. 22). This structure is therefore intermediate between those of fluorite and zincblende, and all three of these structures have cations in alternate cubelets, but the overall symmetry in C-M_2O_3 requires a larger unit cell. The result is actually body-centered cubic. The cathodoluminescent material, Y_2O_3: Eu^{3+}, which is sometimes employed as the red phosphor in color television picture tubes, has this structure.

The perovskite structure is one which has been widely studied. (See Fig. 31, which illustrates the simple cubic perovskite unit cell with two different origins in order to indicate coordination numbers of both cations present.) In this structure, large cations and anions of comparable size in an atomic ratio of 1:3 form roughly close-packed layers like those found in fcc metals (see Fig. 20), though in the metals all atoms are of course identical. One of the octahedral holes is then

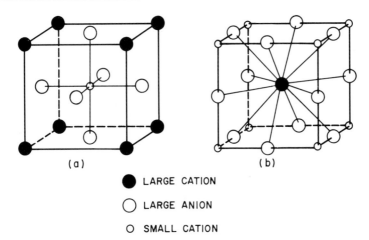

(a)   (b)

● LARGE CATION

○ LARGE ANION

o SMALL CATION

FIG. 31. The primitive (simple) cubic unit cell of a perovskite compound such as $SrTiO_3$ drawn with two different origins to show coordination about the large and the small cations: (a) the small cation is surrounded by an octahedron of six anions; (b) the large cation is surrounded by twelve anions.

filled with a small cation. In terms of the unit cell shown in Fig. 31(a), the anions are at the centers of the six faces (C sites), the small cation at the body center (octahedral B site), and the large cations at the eight cube corners (A sites, with coordination number of 12, which is very unusual for ionic compounds). It therefore has one formula unit, ABC_3, per unit cell. If the A and C ions are in contact with each other, and the B ion has exactly the same radius as the resulting octahedral hole at the cube center, then (if r is the radius) the face diagonal = $2r_A$ + $2r_C$ and the cell edge $a = 2r_B + 2r_C$. Since the face diagonal of a cube is $a\sqrt{2}$,

$$2r_A + 2r_C = (2r_B + 2r_C)\sqrt{2}$$

or

$$\frac{r_A + r_C}{(r_B + r_C)\sqrt{2}} = 1 \quad .$$

Actually, it is found that perovskites will still form if the numeral 1 in this equation is replaced by the Goldschmidt tolerance factor t, which for cases where A and B are divalent and tetravalent, respectively, can have values between about 0.77 and 1. Usually the ideal cubic perovskite will form only if t is greater than 0.89 and a distorted perovskite if t is between 0.77 and 0.89, but this is probably an oversimplification of the reasons for the formation of an ideal or a distorted perovskite. Other factors of importance are the polarizability of the cations and (for appropriate cations) the Jahn-Teller effect. If t is less than the minimum value for perovskite, the hexagonal ilmenite structure, which will be described later, usually forms.

The name of the perovskite structure type is taken from that of the mineral of formula $CaTiO_3$, which was originally thought to have the ideal cubic structure. It has, however, been found more recently to be distorted to orthorhombic. $SrTiO_3$ is cubic at room temperature, but $BaTiO_3$ is tetragonal at room temperature and cubic only above 120°C. In the tetragonal form of $BaTiO_3$ at room temperature, the Ti^{4+} ion is displaced in the direction of one oxygen of the octahedron surrounding it; this leads to a reversible spontaneous polarization known as ferroelectricity.

The ReO_3 structure is like the perovskite structure in that the oxide ions are in the same positions in both and the Re^{6+} ion fills the same octahedral site as, for instance, the Ti^{4+} ion in $SrTiO_3$. The site corresponding to that of Sr^{2+} in $SrTiO_3$ is, however, vacant. The structure is therefore like that in Fig. 31(a) but with the cube corners unoccupied so that it is simply a regular array of ReO_6 octahedra sharing all six corners. It is also like that in Fig. 31(b) but with the cube center empty. In tungsten bronzes, a fraction of these cube-center sites contain large ions such as those of the alkali metals, resulting in defect compounds like Na_xWO_3, where x < 1, so that the structure is intermediate between those of ReO_3 and perovskite.

The structure of nickel arsenide, NiAs, is determined essentially by a hexagonal close-packed[*] (hcp) arrangement of arsenic atoms with nickel atoms in all the resulting octahedral holes. The Ni atoms are therefore surrounded by octahedra of As atoms, but there are also two additional Ni atoms at an only slightly greater distance, so that metal-metal bonds are present and probably necessary for the stability of the structure. (In other, more metallic compounds with this structure, e.g., CrSb, the two distances are nearly identical.) The As atoms are coordinated by six Ni atoms at the corners of a trigonal prism. The structure of corundum, α-Al_2O_3, is related to the NiAs structure in that it is based on a hexagonal close-packed[*] arrangement of oxide ions, but the Al^{3+} ions are regularly placed in only two-thirds of the octahedral holes. Other compounds with this structure are α-Fe_2O_3 and Cr_2O_3. If alternate Al^{3+} ions are replaced by Fe^{2+} and Ti^{4+} ions, the result, already mentioned, is called the ilmenite structure. In addition to $FeTiO_3$ (the mineral ilmenite), $MgTiO_3$, $CoTiO_3$, and $NiTiO_3$ are examples of compounds assuming this structure.

One more hexagonal structure appropriate of mention here is that of wurtzite, the hexagonal form of ZnS. As in the case of cubic ZnS (zincblende), half the tetrahedral holes in the close-packed sulfide lattice are filled with zinc atoms, and coordination numbers of Zn and S are both 4 (of the other atoms). The two structures differ only in that there is cubic stacking of layers in one case and hexagonal stacking in the other. Because of this close relationship, hexagonal ZnS, which is stable at equilibrium only above 1020°C, is rather easily obtained

---

[*] The term "close-packed" is not always used literally, but is often also employed loosely to describe structures of compounds in which lattice-forming ions are actually not in contact with each other because the ions of opposite charge are too large to fit the interstitial holes, and so force the lattice-forming ions slightly apart while permitting them to remain in the same crystallographic positions relative to each other.

metastably at room temperature by quenching from temperatures above
the transition point; it is, in fact, difficult to remove all traces of hexagonal ZnS from preparations which have been above the transition temperature, and there has been much discussion of whether the presence
of some hexagonal ZnS has a role in the Destriau electroluminescence
of activated ZnS.

IX.   LINE AND PLANE DEFECTS

Zero-dimensional or point defects such as vacancies will be
considered in Chapter 8 (Volume 2).   One also finds one-dimensional
or line defects and two-dimensional or plane defects in solids, and
these will be discussed here.

A.   Line Defects

Line defects are of two general types: edge dislocations and
screw dislocations. An edge dislocation occurs when a plane of atoms
abruptly comes to an end while parallel planes in the crystal continue
in normal fashion. Figure 32 is a perspective drawing of such a case,
and Fig. 33 a cross section of the same situation; both show that, where
such a plane ends, a larger than usual space is left because the atoms
at the edge of this plane are not as well surrounded as the other atoms.
Therefore, the electrical forces around these atoms are abnormal and,
much as in the case of a grain boundary, the atoms are at higher energy
than others elsewhere in the crystal. It is the edge of this extra plane,
perpendicular to the page, which is the line defect or edge dislocation.
The drawings also attempt to show that, only several atomic distances
away from the edge dislocation, things are essentially back to normal.
If a dislocation such as this comes to the surface of a crystal, a suitable

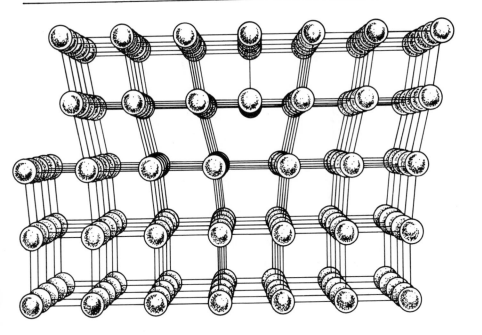

FIG. 32. Perspective drawing of an edge dislocation.

etchant will cause an etch pit to form at that point because the atoms there are at higher energy and therefore more reactive.

Also in Fig. 33, there is shown what is called a Burgers circuit around the dislocation. One chooses an origin and moves around the dislocation as far up as later down and as far to the right as later to the left. With one plane missing, the circuit ends, of course, one repeat distance away from the starting point. This distance, taken as a vector from the end point to the origin, is known as the Burgers vector, which, in the edge dislocation case, is obviously perpendicular to the dislocation line. The Burgers vector length must be a crystallographic repeat distance. Slip in crystalline materials is possible with much less force where edge dislocations are present than in their absence. Where atoms in two planes are ideally close-packed, slip can occur only if all atoms in a plane simultaneously move over to the next close-packed position, and this requires considerable force. If, how-

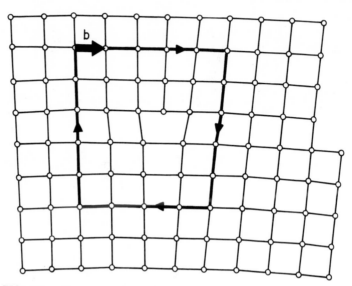

FIG. 33.   Cross section of an edge dislocation showing the
Burgers circuit and the Burgers vector b.

ever, a dislocation is present, it is as if the atoms at the edge of the
extra plane were already halfway over, so that slip can occur by a
wormlike movement where the line dislocation moves in the slip
direction.

The screw dislocation is illustrated in Fig. 34.   Here we have a
twist effect where, as in the edge dislocation, the defect is localized
around a line and its effects vanish only several atomic distances away.
As seen in Fig. 34, the Burgers vector is here parallel to the screw dislocation.   As with edge dislocations, screw dislocations permit slip to
occur with much less applied force.   They are also important to crystal
growth because additional atoms and unit cells can be added to the step
of the screw more easily than to an ideal flat surface because they are
bound on more than one side while growth is in progress.   In the case
of a flat surface, the bond to only one side is so weak that there is a
tendency for atoms to escape before becoming permanently fixed.

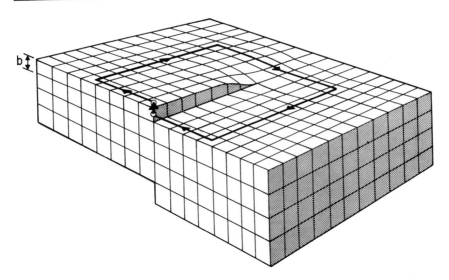

FIG. 34. Schematic drawing of a screw dislocation showing the Burgers circuit and the Burgers vector b.

B. Plane Defects

As has already been stated above, fcc crystals consist of close-packed planes in the stacking sequence $\cdots$ABCABCABC$\cdots$, while hexagonal close-packed crystals have the stacking sequence $\cdots$ABABAB$\cdots$. If during the growth of (for example) a cubic crystal, one layer is placed incorrectly, a growth fault is produced. That is, the stacking $\cdots$ABCABCABCAB$\cdots$ is disrupted by the formation of another A layer instead of C. To return then to cubic stacking, the next two layers must be different so that the sequence becomes $\cdots$ABCABCABCABACBACBAC$\cdots$, and continues as shown. The BAC sequence is, of course, equivalent to the ABC sequence but a plane defect remains between these two sequences.

Another type of plane defect is the deformation fault. In this case (again employing cubic stacking as an example), a perfect crystal is deformed in shear by sliding half of it past the other half along the close-packed stacking planes. The resulting stacking is then

···ABCABCABCAB ABCABCABC···. That is, both sections of the crystal remain as before, but there is now a plane defect between them.

C. Polytypism

In addition to the existence of such random stacking faults, many compounds contain such faults repeated at constant intervals, and a given compound may exist in forms with varying values of these constant intervals. What we have here, then, is a type of polymorphism in which all forms have two dimensions of the unit cell which are the same, while the third is a variable integral multiple of a common unit. Such one-dimensional polymorphism is known as polytypism, and polytypes may belong to different crystal systems and space groups. All compounds exhibiting polytypism are, of course, of the layer-structure type, and include SiC (the most commonly studied example), ZnS, CdI_2, $CdBr_2$, PbI_2, $K_3Co(CN)_6$, and some clay minerals and other layer silicates.

In discussing stacking layers (···ABC···) of such compounds, it should be pointed out that A, B, and C can represent more than single plane layers of atoms. For example, in SiC, which has tetrahedral structures like those of ZnS, the unit layer is composed of a close-packed plane of silicon atoms with a carbon atom directly above each silicon at a constant distance; the unit is therefore actually a double layer. Polytypes may of course be represented simply by writing the ABC sequence found in each true unit cell, but this becomes rather unwieldy as the cell increases in height. Several more convenient notation systems have therefore been developed, of which the most commonly used is the Ramsdell notation, which consists of a numeral representing the number of layers in a unit cell followed by a capital letter to specify the crystal system (e.g., 15R). In some cases, two types can have the same crystal system and repeat distance but differ in the stacking sequence within the unit cells. These are distinguished from each other by attaching subscripts a, b, c, etc. (e.g., $51R_a$ and

TABLE 9

Some Polytypes of α-SiC

Ramsdell notation	Layer sequence	Early notation
2H	AB (wurtzite structure)	-
4H	ABCB	SiC III
6H	ABCACB	SiC II
15R	ABCBACABACBCACB	SiC I
21R	ABCACBACABCBACBCABACB	SiC IV
33R	ABCACBABCACBCABACB - CABACABCBACABCB	SiC VI
51R	ABCACBABCACBABCACB - CABACBCABACBCABACA - BCBACABCBACABCB	SiC V

$51R_b$). In Table 9, which lists a number of examples, the column "Early notation" gives the names used when SiC polytypes were first discovered. The Roman numerals simply indicate the order in which the polytypes were discovered. As additional polytypes were found, this system became impractical; moreover, it did not offer structural information. More than 45 different polytypes of SiC have been discovered (of which over 25 have known structures) ranging in cell height from about 5Å in Type 2H to about 1500Å in Type 594R, and some of the polytypes with unidentified structures are even larger. The most common polytype of SiC is 6H with the $\cdots$ABCACB$\cdots$ sequence. All SiC polytypes have hexagonal unit cell dimensions very close to $a = b = 3.078$Å , $c = n \times 2.518$Å , where n is the number of layers in the unit cell. In addition to the many hexagonal (and rhombohedral) polytypes, there is also a cubic modification, β-SiC, with the zincblende structure $\cdots$ABCABC$\cdots$, which is known as polytype 3C (Ramsdell notation).

X. GLASS

Glasses are noncrystalline solids which have the same sort of atomic arrangement found in liquids, and may, in fact, be considered to be frozen liquids. The x-ray diffraction patterns of most glasses consist of a few diffuse haloes which indicate that there is not complete absence of order. Phosphates, arsenates, germanates, borates, and even some organic compounds are glass-formers, but silicates are of course the most common and useful. SiO_2 is therefore a good compound to choose for the purpose of discussing glass structure. All silicates are built of SiO_4 tetrahedra; in SiO_2 all four corners are shared by other tetrahedra so that each tetrahedron can be thought of as being $SiO_{4/2}$ (or SiO_2). If the tetrahedra are arranged so that there is long-range order, a form of crystalline SiO_2 results. Actually there are a number of such crystalline forms, including quartz, cristobalite, and coesite; which structure results depends only on the arrangement of the tetrahedra with respect to each other. If there is no long-range order, vitreous (or fused) silica (also known as vitreous or fused quartz) results, although the tetrahedra are still present and shared. The softening point of vitreous silica is quite high (> 1300 °C) and it is therefore useful for many purposes. Other silicate glasses to which oxides such as Na_2O, CaO, and PbO have been added have lower softening points because there is less sharing of tetrahedra corners when positive ions of charge lower than that of Si^{4+} are present to neutralize negative charge at these corners.

XI. CRYSTALLINITY OF POLYMERS

X-ray diffraction patterns of polymers often consist of both very diffuse haloes and sharper but still rather diffuse rings. The haloes are typical of those produced by a glass or liquid, while the less diffuse rings indicate the presence of crystallites about 100-200Å in size. Some polymers, such as random copolymers, for example, appear to be entirely amorphous because they produce only haloes. Optical examination of partially crystalline polymers which have been solidified from the melt indicates the presence of mosaic-type aggregates of crystallites called spherulites, which appear to be embedded in an essentially amorphous matrix. However, even the spherulites are not entirely crystalline. They form when crystals grow from various nuclei by spreading out more or less spherically from each nucleus until the growth fronts come into contact with each other. When the spherulite appears to be filled with crystallites it still contains amorphous and poorly crystallized material between the crystals.

The term percentage (or degree) of crystallinity, which is often used, is of some value but is not an absolute number for any given material; it varies somewhat with the measurement method employed. It may be determined in a number of ways including suitable integration of the intensities of x-ray reflections due to the two types of material, or more simply by comparing the integrated intensity of the amorphous band with that of a completely amorphous sample.

Before the discovery that polymers were long-chain molecules, the indicated crystallite size of about 100-200Å was assumed to correspond to the length of the molecules. When, however, most polymer molecules were found to be much longer than this, it was postulated that, although there was random, spaghetti-like interaction of polymer chains, there would be occasional alignment of segments of a number of molecules such that the crystallites resulting from the alignment were approximately 100-200Å in size. This proposed crystallite with its fringelike molecule ends extending from it randomly was known as the "fringed micelle."

More recent work, especially with single crystals, indicates, however, that the fringed micelle model is incorrect for crystalline polymers although it may have some validity for polymers quenched to a glassy state. When special care is taken to prepare polymer crystals by recrystallization from solution or from the melt, electron microscopy shows most of them to be platelets, lamellae, or terraces with thicknesses or step heights of about 100Å and diameters of 10-100 μ. The molecules in these layers are usually folded back and forth on themselves (that is, folded through 180°) with the chain perpendicular to the crystal faces at intervals which increase with the temperature of crystallization. Each fold length in polyethylene usually contains about 80 carbon atoms along the chain, but the intervals between folds can be much longer in polytetrafluoroethylene. It is even possible to increase the fold length to the point where the chains are fully extended by crystallizing at high pressure and temperature; the effect of the pressure may be simply to increase the maximum possible melt temperature. The crystals formed contain a rather high concentration of lattice imperfections to which the x-ray reflections owe some of their diffuseness. Since the estimate of 100-200Å crystallite size in spherulites is based on such crystallites being perfect or nearly so, this estimated size is probably somewhat low.

Normally, one would expect a unit cell of any molecular compound to be large enough to contain at least one molecule. Some crystallographers have therefore considered one true unit cell dimension here to be the thickness of the entire lamella, which contains a monolayer of folded molecules. Since the intensity of x-ray scattering from such a large spacing would expectedly be so weak as to be nonobservable, x-ray measurement would indicate a much shorter unit cell dimension which would really be due only to a very important subcell. Somewhat similar considerations can be applied to the other two dimensions.

However, molecules in polymers are not of uniform length. If one considers the polymer chain to be effectively infinitely long, but with "defects" at molecule ends, then it is possible to think in terms of a unit cell containing crystallographic repeat units somewhat comparable with those found in metallic selenium and tellurium, the structures of

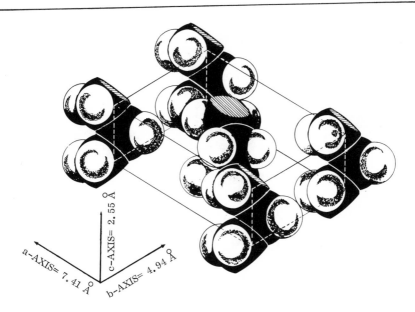

FIG. 35. The unit cell of polyethylene as determined by x-ray diffraction. (Drawing by E.S. Clark made available by P.H. Geil and slightly modified.)

which are based on "infinite" helical chains of atoms. Also, there is evidence that most polymer crystals have high concentrations of many kinds of defects, one of which is the folds not being in crystallographic register; that is, adjacent fold lengths may vary somewhat so that the platelet thickness cannot be a true unit cell dimension. (This is especially so in the crystallites formed from the melt in spherulites; crystals with the highest degree of perfection are grown from solution.) Most polymer crystallographers therefore now appear to share the conviction that the classical idea of the unit cell must be discarded in the case of polymers and that the useful unit cell and the only realistic one is that actually observed by x-ray diffraction.

The apparent x-ray unit cell of polyethylene, which contains two "repeat units" (not molecules), is illustrated in Fig. 35. Repeat units (whose lengths are called "repeat distances") are also illustrated in different fashion in Fig. 36 for polyethylene and two forms of polytetra-

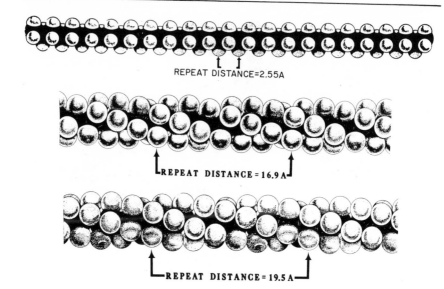

REPEAT DISTANCE=2.55A

REPEAT DISTANCE = 16.9 A

REPEAT DISTANCE = 19.5 A

FIG. 36. Conformation of molecules (showing repeat units and repeat distances) of polyethylene (at top) and two forms of polytetra-fluoroethylene. (Illustrations by E. S. Clark as displayed in Polymer Single Crystals, by P. H. Geil, Wiley-Interscience, New York, 1963.)

fluoroethylene. Polyethylene is seen to contain two $-\underset{\underset{\text{H}}{|}}{\overset{\overset{\text{H}}{|}}{\text{C}}}-$ groups in a planar zigzag repeat unit, while polytetrafluoroethylene has either 13 or 15 $-\underset{\underset{\text{F}}{|}}{\overset{\overset{\text{F}}{|}}{\text{C}}}-$ groups in its helical repeat unit. Nylons are all planar zigzag, while isotactics like polypropylene and polystyrene are helical. Such geometrical arrangement within a molecule is known as the conformation of the molecule.

Unit cell dimensions are determined from layer lines of x-ray diffraction patterns, and placement of the atoms is then made by one of several methods. The classical Fourier series method of crystal structure determination is of course always the best, where it is possible, but has actually been used to determine only a very few polymer crystal structures (including polyethylene and isotactic polypropylene) because the data obtained from polymer crystals are, as a rule, insufficient to permit this. Atomic placement is usually carried out, instead, by filling the cell with molecules whose conformation is deduced from knowledge of the chemical structure or configuration of the molecule, known bond angles and distances, and model-building; or, for helices, by the "helical structure method," in which Bessel functions are employed.

It should be clear that most, if not all, of what has been stated herein also applies to biological polymers such as proteins and DNA, in which helical structures have also been found.

One other type of crystallization or ordering in polymers is worthy of mention — crystallization induced by stress. If a polymer sample is stretched, the random molecules tend to straighten out and line up parallel to the direction of the applied force. Such alignment is equivalent to crystallization, as confirmed by x-ray diffraction studies. Actually, nearly every absent-minded schoolchild has discovered this for himself while playing with rubber bands, though he probably never realizes the significance of his discovery. If a rubber band is elongated while in contact with one's lip, heating is noted which is due to stress-induced crystallization or increase in order. On removing the stress and thus permitting the rubber band to return to its original position while it is still touching the lip, a coolness is felt which is now due to relaxation of the molecules to random orientations. These heat effects correspond qualitatively to what is observed on freezing and melting, respectively. The process itself is known as entropy elasticity.

BIBLIOGRAPHY

1. L. Pauling, The Nature of the Chemical Bond, 3rd ed., Cornell
   Univ. Press, Ithaca, New York, 1960.

2. W. Kauzmann, Quantum Chemistry, Academic Press, New York,
   1957.

3. E. Cartmell and G. W. A. Fowles, Valency and Molecular Structure,
   3rd ed., Butterworth, London, 1966.

4. C. S. Barrett and T. B. Massalski, The Structure of Metals, 3rd ed.,
   McGraw-Hill, New York, 1966.

5. L. V. Azaroff, Elements of X-Ray Crystallography, McGraw-Hill,
   New York, 1968.

6. International Tables of X-Ray Crystallography, 3 volumes, Kynoch,
   Birmingham, England, 1952.

7. A. F. Wells, Structural Inorganic Chemistry, 3rd ed., Oxford Univ.
   Press, London, 1962.

8. J. Weertman and J. R. Weertman, Elementary Dislocation Theory,
   Macmillan, New York, 1964.

9. A. R. Verma and P. Krishna, Polymorphism and Polytypism in Crystals, Wiley, New York, 1966.

10. P. H. Geil, Polymer Single Crystals, Wiley-Interscience, New York,
    1963.

11. B. K. Vainshtein, Diffraction of X-Rays by Chain Molecules,
    Elsevier, Amsterdam, 1966.

12. A. Keller, "Polymer Crystals," Rept. Progr. Phys., 31, 623-704
    (1968).

Chapter 3

BONDING MODELS OF SOLIDS

Billy L. Crowder

IBM Thomas J. Watson Research Center
Yorktown Heights, New York

I. INTRODUCTION

This chapter considers the bonding models appropriate for crystalline solids. This class of materials is characterized by the existence of long-range order in the location of the constituent atom cores. Two approaches to describe the binding in such solids are available. One approach focuses attention upon the chemical bonds formed between the atoms in the solid and will be designated the <u>bond</u> approach. The other approach considers the electronic states that arise from introducing the valence electrons of the atoms of the solid into the periodic potential produced by the ion cores (nuclei and filled inner shell electrons) located at their fixed positions in the solid lattice. The latter approach is designated the <u>band</u> model. The band model, including free electron theory, will be discussed in Sec. II; the bond model, in Sec. III.

A description of d electrons in partially filled d orbitals merits special attention. At one extreme, the d electrons may be assigned to a particular transition metal atom (or ion) in the solid. The interaction between the set of d electrons belonging to one metal atom and those belonging to its neighbors is small. An adequate description of such localized d electrons is provided by crystal field theory (alternatively known as ligand field theory). Such a theory is also applicable to solids containing partially filled f orbitals. At the other extreme, the d electrons may participate in the binding of the solid and become delocalized. The concepts appropriate to understanding systems with d electrons are presented in Sec. IV.

II. BAND THEORY OF SOLIDS

A. General Considerations

The behavior of electrons in crystals is obviously a many-electron problem. A convenient approximation to reality is to consider the one-electron solutions of the energy operator H (the Hamiltonian)

$$H\psi(\underline{r}) = [(-\hbar^2/2m)\nabla^2 + V(\underline{r})]\psi(\underline{r}) = E\psi(\underline{r}) \ . \tag{1}$$

In this equation

$$\nabla^2 \equiv \frac{\delta^2}{\delta x^2} + \frac{\delta^2}{\delta y^2} + \frac{\delta^2}{\delta z^2} \ ,$$

$\hbar = h/2\pi$, and $V(\underline{r})$ is a suitably average potential in which the electron in question moves. The eigenvalue E represents the total energy and is constant. Depending upon the form of the potential $V(\underline{r})$ and the nature of the boundary conditions imposed upon solutions of Eq. (1) the values of E for which solutions to Eq. (1) exist may form a discrete set or a continuous set [1]. The form of the crystal potential $V(\underline{r})$ is of crucial importance in band calculations. For illustrative purposes, we shall consider two limiting cases: (1) free electrons (including nearly free electrons) and (2) tightly bound electrons.

B. Free Electron Model

The crystal potential due to the ion cores, $V(\underline{r})$, in Eq. (1) is completely neglected in the free electron (or empty lattice) approximation. The resulting equation

$$(-\hbar^2/2m)\nabla^2\psi(\underline{r}) = E\psi(\underline{r}) \tag{2}$$

has solutions that depend upon the nature of the boundary conditions imposed. For a free particle in a volume V, the solutions are of the form

$$\psi(\underline{r}) = (1/V)^{\frac{1}{2}} e^{i\underline{k}\cdot\underline{r}} \tag{3}$$

where the wave vector $\underline{k}$ is related to the de Broglie wavelength λ associated with the electron by $k = |\underline{k}| = 2\pi/\lambda$, i.e., the wave vector $\underline{k}$ is related to the electron momentum $\underline{p}$ by $h/2\pi$. The direct substitution of (3) into (2) yields

$$(\hbar^2 k^2/2m)\psi = E\psi \ .$$

The allowed energy eigenvalues appropriate for the free electron model are thus

$$E = \hbar^2 k^2 / 2m \; . \tag{4}$$

Equation (4) is a central result of the free electron model.

In the case of a completely free particle, the energy eigenvalues form a continuous set. In the limit of classical theory, Eq. (4) is just the familiar energy of a particle of mass m moving with velocity v, $\frac{1}{2} mv^2$ (remember that $\underline{p} = \hbar \underline{k}$).

In solids of finite extent, the appropriate solutions to the wave equation are subject to certain restrictions, the boundary conditions. Boundary conditions can be introduced in various ways, but the simplest to understand physically is to consider the electron as confined to the region of space occupied by the solid, e. g., to the cube determined by $0 \le x \le \ell; \; 0 \le y \le \ell; \; 0 \le z \le \ell$.

The general solution of Eq. (2) is of the form

$$\psi(\underline{r}) = A e^{i\underline{k} \cdot \underline{r}} + B e^{-i\underline{k} \cdot \underline{r}} \tag{5a}$$

or alternatively,

$$\psi(\underline{r}) = (A_1 e^{ik_1 x} + B_1 e^{-ik_1 x})(A_2 e^{ik_2 y} + B_2 e^{-ik_2 y})(A_3 e^{ik_3 z} + B_3 e^{-ik_3 z}) \tag{5b}$$

where

$$k^2 = k_1^2 + k_2^2 + k_3^2 \; . \tag{6}$$

The wave equation $U(\underline{r})$ must vanish at the boundaries of the solid, which means

$$A_j + B_j = 0$$
$$\qquad\qquad\qquad\qquad j = 1, 2, 3 \; . \tag{7}$$
$$A_j e^{ik_j \ell} + B_j e^{-ik_j \ell} = 0$$

Therefore, $A_j = -B_j$ and $\sin k_j \ell = 0$, or

$$k_j = n_j \pi / \ell \quad , \quad j = 1, 2, 3 \ . \tag{8}$$

The boundary conditions imposed have restricted the allowed values of k to

$$k^2 = (\pi/\ell)^2 (n_1^2 + n_2^2 + n_3^2) \tag{9}$$

where n_1, n_2, n_3 are positive integers. The energy eigenvalues given by Eq. (4) become

$$E = (h^2/2m)(\pi/\ell)^2 (n_1^2 + n_2^2 + n_3^2) \ . \tag{10}$$

For every triplet of positive integers (n_1, n_2, n_3) there corresponds a wave function

$$\psi_{k_x k_y k_z} = (8/V)^{\frac{1}{2}} \sin(k_1 x/\ell) \sin(k_2 y/\ell) \sin(k_3 z/\ell) \tag{11}$$

where the normalization has been accomplished by integrating over all space and requiring the result to equal unity. The volume of the solid is $\ell^3 = V$. The energy values as given by Eq. (10) form an infinite but discrete set. To every triplet $(n_1 n_2 n_3)$ there corresponds an energy level of the system. These triplets $(n_1 n_2 n_3)$ may be used in conjunction with the spin quantum number m_s ($= \pm \frac{1}{2}$) as the quantum number appropriate to the free electron model, i.e., the state of the electron is specified when the values of n_1, n_2, n_3, and m_s are given.

The wave functions (11) are standing waves and can be used to describe the free electron model. An alternative choice of boundary conditions, utilizing the translational symmetry of the lattice, yields plane wave solutions which are more convenient than (11). If $\underline{a}_1$, $\underline{a}_2$, and $\underline{a}_3$ are three primitive noncoplanar vectors connecting lattice points, any translation operator will be of the form

$$T = n_1 \underline{a}_1 + n_2 \underline{a}_2 + n_3 \underline{a}_3 \tag{12}$$

For Fig. 1, we have drawn the primitive vectors for the simple cubic lattice and the face centered cubic (fcc) lattice. Such operations

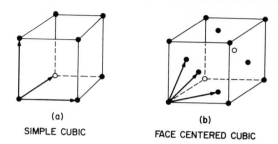

(a)   (b)
SIMPLE CUBIC   FACE CENTERED CUBIC

FIG. 1.   Primitive lattice vectors.

will take the crystal lattice into itself except for displacing the boundaries. We then consider our sample to be a crystal of size $N_1 a_1$ by $N_2 a_2$ by $N_3 a_3$ where N_1, N_2, N_3 are large numbers. If we imagine an infinite number of such samples stacked side by side, we can require that acceptable solutions to our wave equation conform to

$$\psi(r) = \psi(r + N_1 a_1) = \psi(r + N_2 a_2) + \psi(r + N_3 a_3) \tag{13}$$

Let us consider the special case of a simple cubic lattice of lattice spacing a. The solutions to Eq. (2) are of the form

$$\psi_k(r) = \left(\frac{1}{V}\right)^{\frac{1}{2}} e^{ik \cdot r} \tag{14}$$

provided

$$k_x = \frac{2\pi n_1}{N_1 a}, \quad k_y = \frac{2\pi n_2}{N_2 a}, \quad k_z = \frac{2\pi n_3}{N_3 a}, \tag{15}$$

$$(n_1, n_2, n_3 = 0, \pm 1, \pm 2, \text{ etc.})$$

The volume of the sample, V, is $(N_1 N_2 N_3 a^3)$.

The vector k has the dimensions of a reciprocal length and can be considered to be a vector in reciprocal space.[*] The values of k that

---

[*]Associated with our direct lattice (which can be generated from the primitive translation vectors a_1, a_2, and a_3) is a reciprocal lattice. The primitive translation vectors b_1, b_2, b_3 in this reciprocal lattice are given by the condition

$$a_i \cdot b_j = 2\pi \delta_{ij} = 0, \quad i \neq j \quad (i, j = 1, 2, 3)$$
$$= 2\pi, \quad i = j.$$

correspond to acceptable wave functions are restricted to certain values which form a discrete lattice in reciprocal space or $\underline{k}$ space. In our free electron approximation, the solutions that correspond to an energy E have $\underline{k}$ vectors which lie on a sphere in $\underline{k}$ space centered at $\underline{k} = 0$.

The spacing between the allowed $\underline{k}$ values is such that $\underline{k}$ space would be filled if, associated with each discrete $\underline{k}$ value in $\underline{k}$ space, there were a volume $(8\pi^3/N_1 N_2 N_3 a^3)$, i.e., the number of states per unit volume in $\underline{k}$ space is $1/8\pi^3$, not including spin degeneracy $(N_1 N_2 N_3 a^3$ is the volume of the sample). The volume around each allowed $\underline{k}$ value is very small for microscopic crystals and vanishes for a sample of infinite dimensions. The origin of discrete k values is thus a consequence of Heisenberg's uncertainty principle [1]. The coordinates of an electron in our crystal are known within $N_1 a$, $N_2 a$, and $N_3 a$ for x, y, and z, respectively. Heisenberg's uncertainty principle then requires that

$$\Delta x \cdot \Delta p_x = (N_1 a) \cdot \hbar \, \Delta k_x \geq \hbar$$
$$\Delta k_x \geq \frac{1}{N_1 a}$$

which is consistent with the result obtained in (15).

Since the volume around each allowed $\underline{k}$ value in $\underline{k}$ space is small, the number of states within a volume of radius $\underline{k}$ is, per unit volume of the crystal,

$$\frac{4}{3}\pi k^3 \div 8\pi^3 = \frac{k^3}{6\pi^2} \quad .$$

The number of allowed $\underline{k}$ values per unit volume within a range $d\underline{k}$ of wave vector is then simply $(k^2/2\pi^2) d\underline{k}$. The number of allowed electron states including spin degeneracy is twice this value.

The density of states per unit volume per unit energy or per unit wave vector can then be defined as

$$g(E) dE = g(k) dk = (k/\pi)^2 dk \tag{16}$$

Since the energy and wave vector are connected by Eq. (4),

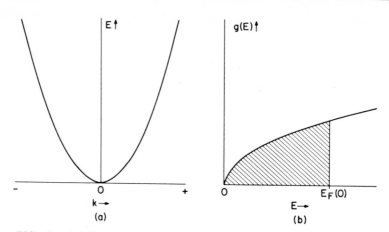

FIG. 2.   (a) Energy vs wave vector for free electrons.
          (b) Density of states vs energy for free electrons.

$$g(E) = \frac{1}{2\pi^2} \left(\frac{2m}{h^2}\right)^{\frac{3}{2}} E^{\frac{1}{2}} .$$

(17)

Figure 2 shows the relationship in one dimension of E vs k and g(E) as a function of E for the free electron model.

For each allowed value of $\underline{k}$, there exists a solution ψ_k of the form (14) with a corresponding energy eigenvalue. The components of $\underline{k}$, (k_x, k_y, k_z) together with the spin quantum number $m_s = \pm\frac{1}{2}$ can be used to specify the state of the electron. The wave function for the system as a whole Ψ must be constructed from the one-electron wave functions $\psi_a^{(1)}$, $\psi_b^{(2)}$, etc. (including spin), in such a way that Ψ is antisymmetric with respect to the interchange of the coordinates of any pair of electrons, as required by the Pauli exclusion principle. This restriction requires that no two electrons have the same values of k_x, k_y, k_z, and m_s.

For indistinguishable particles, when filling quantum states in accord with the Pauli exclusion principle, the distribution of particles over a set of energy levels at thermal equilibrium is given by Fermi-Dirac statistics. In a system of particles which obey the Pauli exclusion

principle, the thermal equilibrium probability that a state of energy E
is occupied is given by[*]

$$f(E) = \frac{1}{1 + \exp[(E - E_F)/kT]} \; .$$
(18)

The quantity E_F is known as the Fermi level or electrochemical potential. For an energy equal to E_F, f(E) is 0.5; for energies higher or
lower by 2kT, $f(E) = (1 + e^2)^{-1} = 0.12$ or $f(E) = (1 + e^{-2})^{-1} = 0.88$. The
probability that a level is occupied is appreciably different from 1 or 0
only for energies within a few kT or E_F.

At $T = 0$, the energy levels below the Fermi level will be filled
with electrons; above the Fermi level the energy levels will be empty.
The relationship between the value of the Fermi level at $T = 0$ and the
electron concentration n can be easily derived. The number of electron
states per unit volume allowed by the periodic boundary conditions which
have k less than some value k_c was shown to be (including spin)
$2(k_c^3/6\pi^2) = k_c^3/3\pi^2$. If the number of electrons per unit volume is n,
the Fermi level is determined by the value of k_c for which

$$\frac{k_c^3}{3\pi^2} = n$$
(19a)

or

$$k_c = (3\pi^2 n)^{\frac{1}{3}} \; .$$
(19b)

The Fermi level at $T = 0$, using Eq. (4), is

$$E_F(0) = \frac{\hbar^2 k_c^2}{2m} = \frac{\hbar^2}{2m} (3\pi^2 n)^{\frac{2}{3}} \; .$$
(19c)

For electron densities characteristic of metals, the value of $E_F(0)$
is approximately 5 eV. The fraction of the electron population which
lies within kT of unoccupied states is small and of the order of
$kT/E_F(0)$; the system is described as degenerate. The number of

---

[*]For details of the derivation of this result see Ref. [2].

electrons that can contribute to the electronic specific heat and to the magnetic susceptibility is of the order of $nkT/E_F(0)$. The thermal energy acquired per electron which is within kT of the Fermi level is of the order of kT. The electronic thermal energy is therefore

$$U \approx nk^2 T^2 / E_F(0) \qquad (20a)$$

or the heat capacity

$$C_v = \frac{\delta U}{\delta T} \approx \frac{nk^2 T}{E_F(0)} \qquad (20b)$$

which is proportional to T and smaller than the classical value of $(3/2)nk$ by two orders of magnitude, as is observed experimentally. Similarly, the classical value of the electronic magnetic susceptibility χ is given by

$$\chi = n\mu_B^2 / 3kT \qquad (21a)$$

where μ_B is the Bohr magneton. This equation predicts much larger values than observed for simple metals at room temperature and a T^{-1} temperature dependence instead of the experimentally observed temperature independent susceptibility. The number of electrons sufficiently near unoccupied levels so that they can align in the applied magnetic field is not n but of the order of $nkT/E_F(0)$. The susceptibility then becomes of the order

$$\chi \approx \frac{nkT}{E_F(0)} \frac{\mu_B^2}{3kT} \approx \frac{n\mu_B^2}{3E_F(0)} \qquad (21b)$$

i.e., independent of temperature and appreciably smaller than the predicted classical value.

The Fermi level for metals is a weak function of temperature; the change is very small at physically attainable temperatures [2]:

$$E_F(T) - E_F(0) = -\frac{\pi^2 k^2 T^2}{12 E_F(0)} \quad , \quad kT \ll E_F(0) \quad .$$

C. Wave Functions in a Periodic Lattice:

Nearly Free Electrons

The theory as developed to this point does provide a reasonable model for the simple monovalent metals, but it cannot explain the existence of materials that are either insulators or semiconductors. We have completely neglected the potential due to the ion cores and have not fully exploited the consequences of the periodicity of the lattice.

If the crystal potential $V(\underline{r})$ in Eq. (1) has the periodicity of the lattice, the acceptable wave functions are of the form

$$\psi_{\underline{k}}(r) = U_{\underline{k}}(r)\, e^{i\underline{k}\cdot\underline{r}} \tag{22}$$

where $U_{\underline{k}}(r)$ is a function that depends upon the wave vector $\underline{k}$ and that reflects the periodicity of the lattice. This result was proved by Bloch [3, 4], and such solutions are known as Bloch functions.

The value of $\underline{k}$, for a given wave function, is not uniquely defined since $\underline{k}$ may be replaced by $\underline{k} + \underline{G}$, where $\underline{G}$ is a reciprocal lattice vector, without changing the wave function $\psi_{\underline{k}}$. For example, in a one-dimensional monatomic lattice of spacing a, k may be replaced by $k + 2\pi n/a$ without destroying the periodicity of $U(x)$. It is therefore convenient to restrict the values of $\underline{k}$ to a region which is a unit cell of the reciprocal lattice. The particular unit cell chosen by convention is the first Brillouin zone defined by bisecting the lines joining $\underline{G} = 0$ to the nearest reciprocal lattice points. The various ψ's and U's belonging to the same value of k' are labeled with different indices, $n = 1, 2, 3$, etc.:

$$\psi_{n\underline{k}'}(r) = U_{n\underline{k}'}\, e^{i\underline{k}'\cdot\underline{r}} \quad . \tag{23}$$

The vector $\underline{k}'$ defined in this manner is called the reduced wave vector. The indices n are considered to be band numbers. Our boundary conditions [Eq. (13)] restrict the total number of $\underline{k}$ values to be $N_1 N_2 N_3$, the

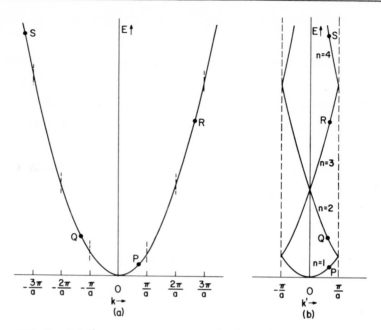

FIG. 3.   (a) Energy vs wave vector for free electrons.
          (b) Energy vs reduced wave vector for free electrons.

number of unit cells in the sample.[*]

Figure 3 shows the relationship between the description in terms of the wave vector $\underline{k}$ as previously used (extended zone) and the reduced wave vector introduced for free electrons in a monatomic one-dimensional lattice of spacing a. The relationship between points P, Q, R, and S is clear in the reduced wave vector labeling; they all correspond to the same value of k'. For periodic wave functions of the Bloch form $(\psi_{n\underline{k}} = U_{n\underline{k}} e^{i\underline{k}\cdot x})$, the quantity

$$\lim_{L \to \infty} \frac{1}{L} \int_0^L \psi_{n'\underline{k}'} U(x) \psi_{n\underline{k}} \, dx \qquad (24)$$

for any function U which has the periodicity of the lattice will vanish

---

[*]If the origin of the crystal is at its center, the values of n_1, n_2, n_3 in Eq. (14) are restricted to $\pm N_1/2$, $\pm N_2/2$, $\pm N_3/2$, respectively.

unless $\underline{k} = \underline{k}'$. The points P, Q, R, and S can combine in the sense of (24).

The utility of the reduced wave vector representation is more apparent when we go to the case of nearly free electrons. As we shall see, the periodicity of the lattice may introduce discontinuities in the allowed energy eigenvalues at wave vectors which lie on the surface of the first Brillouin zone.

For simplicity, we consider the case of nearly free electrons in a one-dimensional monatomic lattice of spacing a and size N_1a. The crystal potential $V(x)$ is considered (1) to be small in magnitude relative to the average kinetic energy of the electrons, (2) to have the periodicity of the lattice, and (3) to have a mean value of 0. Under these conditions, Schrödinger's equation can be written as

$$(H_0 + H')\psi = \left(-\frac{h^2}{2m}\frac{\delta^2}{\delta x^2} + V(x)\right)\psi = E\psi \tag{25}$$

where the crystal potential $V(x)$ is treated as a small perturbation H'.

The solutions to $H_0\phi = E\phi$ with periodic boundary conditions were previously derived and in one dimension are

$$\phi_{0k}(x) = (N_1a)^{-\frac{1}{2}}e^{ik_xx} \quad . \tag{26}$$

The solutions to Eq. (25) are then linear combinations of the ϕ_{0k}:

$$\psi_k(x) = \sum_n C_{nk}\phi_{0n}(x) \quad . \tag{27}$$

The crystal potential $V(x)$ has the periodicity of the lattice and therefore can be expanded in a Fourier series:

$$V(x) = \sum_{n=-\infty}^{\infty} V_n e^{-2\pi nx/a} \quad . \tag{28}$$

The wave functions $\psi_k(x)$, of course, must be of the form

$$\psi_k(x) = U_k(x)e^{ikx} \quad , \tag{29}$$

as required by Bloch's theorem.

Using standard perturbation theory, the matrix elements required to determine the coefficients C_{jk} in Eq. (27) are

$$H'_{k'k} = (\psi_{k'}^* \, | V(x) | \, \psi_k) = \frac{1}{N_1 a} \int_0^{N_1 a} e^{-ik'x} V(x) e^{ikx} dx \qquad (30)$$

or, using the expansion (28),

$$H'_{k'k} = \frac{1}{N_1 a} \sum_{n=-\infty}^{\infty} V_n \int_0^{N_1 a} e^{-ik'x} e^{-2\pi inx/a} e^{ikx} dx \qquad (31)$$

which vanishes unless

$$k' = k - \frac{2\pi n}{a} . \qquad (32)$$

The wave function for nearly free electrons is, in first order perturbation theory,

$$\psi_k(x) = \frac{1}{(N_1 a)^{\frac{1}{2}}} \left[e^{ikx} + \sum_{\substack{k' \\ (k' \neq k)}} \frac{H'_{k'k}}{E_{0k} - E_{0k'}} e^{ik'x} \right] \qquad (33a)$$

where

$$E_{0k} = \frac{\hbar^2 k^2}{2m} \qquad \text{and} \qquad E_{0k'} = \frac{\hbar^2 k'^2}{2m}$$

as required by the free electron solutions ϕ_{0k}. Using the values of $H'_{k'k}$ from Eq. (30) and the restriction that $H'_{k'k}$ vanishes unless $k' = k - 2\pi n/a$ yields

$$\psi_k(x) = \frac{1}{(N_1 a)^{\frac{1}{2}}} e^{ikx} \left[1 + \sum_n \frac{V_n}{E_{0k} - E_{0k'}} e^{-2\pi inx/a} \right] \qquad (33b)$$

which is of the Bloch form, as required. To first order, the energy of the wave function $\psi_k(x)$ is unchanged. To second order, the energy is

$$E_k = E_{0k} + \sum_{\substack{k' \\ (k \neq k')}} \frac{|H'_{k'k}|^2}{E_{0k} - E_{0k'}}$$

$$= \frac{\hbar^2 k^2}{2m} + \sum_n \frac{|V_n|^2}{E_{0k} - E_{0k'}} \qquad (34)$$

where again $k' = k - (2\pi n/a)$.

Since V_n is assumed to be small compared with the energy eigenvalues, the nearly free electrons behave much like free electrons except when the value of k approaches $\pi n/a$. At these points, the term $(E_{0k} - E_{0k'})$ for the corresponding values of n vanishes, and the assumptions used in deriving (33) and (34) are not valid. When the value of k is equal to or nearly equal to $\pi n/a$, the wave function is of the form

$$\psi_k = (N_1 a)^{-\frac{1}{2}} (e^{ikx} + c e^{-ik'x}) \tag{35}$$

where k' is equal to $k - (2\pi n/a)$ and the constant c is not small relative to unity. Under these conditions, it can be shown that the energy associated with this wave function (35) is

$$E_k = \frac{1}{2} [E_{0k} + E_{0k'} \pm \sqrt{(E_{0k} - E_{0k'})^2 + 4|V_n|^2}] \quad . \tag{36}$$

The negative sign is appropriate for $k < n\pi/a$; the positive sign, for $k > n\pi/a$ [5].

The value of E as a function of k for this case is plotted in Fig. 4 in both the extended wave vector representation and the reduced wave vector respresentation. A discontinuity occurs at $k = \pi n/a$ of magnitude $2|V_n|$. Energies lying between $(h^2/2m(n\pi/a)^2 \pm (|V_n|)$ do not correspond to acceptable wave functions, i.e., there exists a forbidden gap between the bands of allowed energy eigenvalues. In the reduced wave vector representation, the bands are labeled to distinguish the different ψ's for a given value of k. Each band, as shown previously, can accommodate $2N_1$ electrons. We shall return to this point after considering the physical origin of this energy discontinuity.

When $k = \pi n/a$, the constant c in Eq. (35) is ± 1, and ψ_k is of the form

$$\psi_k = (N_1 a)^{-\frac{1}{2}} (e^{ikx} \pm e^{-ikx}) \quad , \tag{37}$$

i.e., for this value of k, two solutions exist.

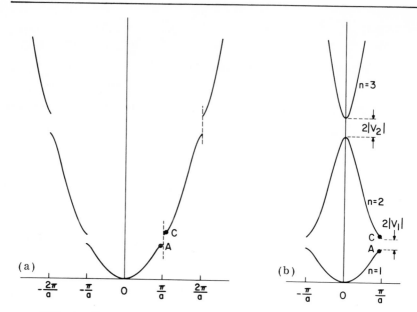

FIG. 4.   Energy vs wave vector (a) and reduced wave vector (b) for nearly free electrons.

If we place the origin of our coordinate system at a lattice point, i. e., at the location of a positive ion core (the value of $|\psi_k|^2$ is the same at all lattice points), the average values of the potential energy over the three charge distributions would be expected to be of the order P. E.(ψ_k^+) < P. E.(ϕ_{0k}) < P. E.(ψ_k^-) and, at $k = \pi n/a$, a gap in the allowed energy eigenvalues would exist.

The condition $k = \pi n/a$ is also just that for Bragg reflection of a plane wave e^{ikx} in one dimension from the lattice of spacing a; i. e., whenever the Bragg condition is satisfied, the acceptable solutions cannot be running values of the form $e^{i\mathbf{k}\cdot\mathbf{r}}$ since these waves would suffer Bragg reflections but must be standing waves of the form $(e^{i\mathbf{k}\cdot\mathbf{r}} \pm e^{-i\mathbf{k}\cdot\mathbf{r}})$.

A comparison of Figs. 3 and 4 shows clearly that the presence of the periodic crystal potential gives rise to discontinuities or energy gaps in the allowed energy eigenvalues for a one-dimensional model. The total number of states in any band is $2N_1$ (including spin). If our monatomic one-dimensional solid has one valence electron per atom, then clearly the N_1 electrons will fill only half the states available in the band, and the nearly free electrons behave much as if they were free electrons. If there are <u>two</u> electrons per atom, all the available states in a band would be filled.

Due to the existence of the forbidden energy gap, the solid will be an insulator if the gap is very large compared to kT or a semiconductor if thermal energy is sufficient to excite electrons into the next higher band. Our simple model thus provides a qualitative understanding of the existence of materials which are nonmetals.

We now return to our model and consider the behavior of an electron just above the first discontinuity, at a point C in Fig. 4, just above $k = \pi/a$. We expand Eq. (36) to first order in $(E_{0k} - E_{0k'})^2$ and let $k'' = k - \pi/a$, obtaining

$$E \cong \frac{h^2}{2m} \left[\left(\frac{\pi}{a} \right)^2 + k''^2 \left(1 + \frac{2E_a}{V_1} \right) \right] + V_1 \quad ,$$

where

$$E_a \equiv \frac{h^2}{2m} \left(\frac{\pi}{a} \right)^2 \quad .$$

In band 2, the energy at $k'' = 0$ is just $E_a + V_1$, so that near $k'' = 0$

$$E'' \equiv E - (E_a + V_1) = \frac{h^2}{2m} k''^2 \left(1 + \frac{2E_a}{V_1} \right) \quad , \tag{38}$$

i.e., the electrons in band 2 behave as if the <u>effective mass</u> m^* were such that

$$\frac{m^*}{m} = \left(1 + \frac{2E_a}{V_1} \right)^{-1} \quad . \tag{39}$$

The concept of effective mass is an important factor in band theory. We shall return to this point later.

D. Tight Binding Approximation

In the nearly free electron approximation, the potential due to the ion cores was treated as a perturbation on the free electron wave functions. In the tight binding approximation, the influence of incorporating the atom into a solid lattice will be treated as a perturbation upon the wave functions of the electron in the isolated atom. For an atom of the solid at a position $\underline{R}_j$ and an electron at $\underline{r}$, the atomic orbitals will be specified by $\phi(\underline{r} - \underline{R}_j)$.

Schrödinger's equation (1) can then be written as

$$(-\frac{h^2}{2m} \nabla^2 + V_a + V')\psi = (H_0 + V')\psi = E\psi \tag{40}$$

where V_a is the potential the electron would have in a single isolated atom and V' is the additional potential energy acquired when the atom is incorporated into a crystal. Of course,

$$H_0\phi = E_a\phi \tag{41}$$

where E_a is the energy eigenvalue associated with the atomic orbital ϕ. If perturbation theory can be applied, the wave functions which are solutions of Eq. (40) can be written as linear combinations of the atomic orbitals ϕ:

$$\psi_{\underline{k}}(\underline{r}) = \sum_j e^{i\underline{k}\cdot\underline{R}_j} \phi(\underline{r} - \underline{R}_j) \tag{42}$$

where the sum runs over all the atoms in the solid. The coefficients in Eq. (42) are determined by the requirement that the acceptable wave function must be of the Bloch form; i.e., a translation by a lattice vector $\underline{R}_a$ yields

$$\psi_{\underline{k}}(\underline{r} + \underline{R}_a) = \sum_j e^{i\underline{k}\cdot\underline{R}_j} \phi(\underline{r} + \underline{R}_a - \underline{R}_j)$$

$$= e^{i\underline{k}\cdot\underline{R}_a} \sum_j e^{i\underline{k}\cdot(\underline{R}_j - \underline{R}_a)} \phi(\underline{r} - (\underline{R}_j - \underline{R}_a)) \tag{43}$$

or

$$\psi_{\underline{k}}(\underline{r} + \underline{R}_a) = e^{i\underline{k}\cdot\underline{R}_a} \psi_{\underline{k}}(\underline{r}) , \tag{44}$$

as required. If we multiply Eq. (40) by $\psi_{\underline{k}}^*$ and integrate over all space, we obtain

$$\int \psi_{\underline{k}}^{*} H_0 \psi_{\underline{k}}\, d\tau \;+\; \int \psi_{\underline{k}}^{*} V' \psi_{\underline{k}}\, d\tau \;=\; E \int \psi_{\underline{k}}^{*} \psi_{\underline{k}}\, d\tau \quad . \tag{45}$$

If we substitute Eq. (42) into this result, we obtain for the first term an integral of the sum of products of atomic orbitals, since $H_0 \phi$ yields $E_a \phi$. Assuming that the atoms are widely enough spaced, integrals involving functions centered around two different atoms will vanish. The integrals $\int \phi^{*}(\underline{r} - \underline{R}_j) d(\underline{r} - \underline{R}_j) d\tau$ will equal unity for normalized functions. Since there are N such integrals (one for each atom of the solid), the first term becomes simply NE_a, where N is the number of atoms in the solid. By similar reasoning, the term to the right of the equality in Eq. (45) is simply NE.

In the second term, we neglect all terms in the summation except those involving nearest neighbors. We define the parameters

$$\alpha = -N \int \phi^{*}(\underline{r} - \underline{R}_j) V' \phi(\underline{r} - \underline{R}_j)\, d\tau$$

$$\beta = -N \int \phi^{*}(\underline{r} - \underline{R}_m) V' \phi(\underline{r} - \underline{R}_j)\, d\tau \tag{46}$$

where j and m are neighboring lattice sites. The parameters α and β are positive since the potential V' represents an increased binding energy relative to the free atom. Using (46) together with the values of NE_a for the first term and NE for the last term in Eq. (45), we have

$$E = E_a - \alpha - \beta \sum_{m} e^{i\underline{k}\cdot(\underline{R}_j - \underline{R}_m)} \tag{47}$$

provided we assume that the ϕ's are spherically symmetric (atomic $\underline{s}$ functions). In this case, the β integrals are the same for all nearest neighbors.

For a simple cubic lattice, the nearest neighbor positions are given by $(\pm a, 0, 0)$; $(0, \pm a, 0)$; $(0, 0, \pm a)$. Eq. (47) becomes simply

$$E = E_a - \alpha - 2\beta(\cos k_x a + \cos k_y a + \cos k_z a) \quad . \tag{48}$$

The energies are thus confined to a band with limits $\pm 6\beta$. Each state

of an electron in the free atom corresponds to a band of energies in the
crystal. For a nondegenerate state of the free atom, each band contains
2N states (including spin), where N is the number of atoms in the solid.

For small values of $\underline{k}$ in a simple cubic lattice,

$$E \cong E_a - \alpha - 6\beta + \beta k^2 a^2 \tag{49}$$

which is similar to the result for free and nearly free electrons; the
energy is independent of the direction of motion and depends upon k^2.
The effective mass is

$$m^* = h^2/2\beta a^2 \quad . \tag{50}$$

The width of the bands depends directly upon the overlap between orbitals on adjacent atoms in the solid, and the effective mass near the
bottom of a band depends inversely upon this overlap, i. e., narrow
bands are characterized by large effective masses.

The tight binding approximation can also be extended to describe
degenerate atomic orbitals (p, d, f) and the situation in which the band
width 12β is comparable to the separation between atomic s and p
states [5].

From the above considerations, we expect to have bands in the
solid that correlate with corresponding atomic orbitals, i. e., one band
from atomic 1s states, one band from atomic 2s states, three bands
from atomic 2p states, etc. The electrons in inner shell states in the
isolated atom are still tightly bound in the solid, and the bands associated with these electrons are exceedingly narrow. Such electrons are
localized; they are closely associated with a given nucleus in the
solid. The bands corresponding with the outermost electrons <u>may</u> cover
a broad range of energy; these electrons are delocalized and belong to
the entire crystal. In the latter case, overlap between bands is likely.
In a given direction in $\underline{k}$ space, there is an energy gap as shown in
Fig. 4; but in a <u>different</u> direction in $\underline{k}$ space, it is likely that states
in the upper band may lie lower than the upper states in the lower band

shown in Fig. 4. When all directions in $\underline{k}$ space are considered for this case, there is no forbidden gap and the bands are said to overlap. This situation exists, for example, in the divalent metal Mg. If the band arising from atomic 3s states of Mg were lower in energy than any states of the 3p bands, Mg would not be a metal. The actual situation in Mg is such that bands arising from 3s and 3p atomic states do overlap slightly and metallic behavior is possible. In many materials, the bands corresponding to the outermost electrons exhibit behavior between these two extremes.

E. The Effective Mass

In the previous two sections, we investigated simple models for the limiting cases of nearly free electrons and tightly bound electrons. We turn our attention to some of the consequences of the E-$\underline{k}$ relationships derived from these models.

The relationship between E and $\underline{k}$ for free electrons was shown to be

$$E = \frac{\hbar^2 k^2}{2m} \ . \tag{51}$$

The energy E is independent of the direction of $\underline{k}$ and, in $\underline{k}$ space, the curves of constant energy are spheres centered at $\underline{k} = 0$. For $\underline{k}$ values near a band minimum E_0 (associated with a wave vector $\underline{k}_0$), we have observed that a functional form similar to Eq. (51) is valid for both nearly free electrons and tightly bound electrons,

$$E - E_0 = \frac{h^2}{2m^*} (\underline{k} - \underline{k}_0)^2 \ , \tag{52}$$

provided we define an effective mass m^* in place of the electron rest mass of Eq. (51) [Eqs. (38) and (50)]. For states sufficiently near $\underline{k}_0$, the energy relative to the band minimum is again independent of the direction of $\underline{k}$, and, in $\underline{k}$ space, the curves of constant energy are spheres centered at $\underline{k}_0$.

A general expression for the effective mass may be derived by considering the response of an electron in the lattice to an applied electric field. The acceleration of an electron wave packet in an applied electric field $\underline{F}$ is given by [2]

$$\underline{a} = -\frac{1}{h^2} \underline{\nabla}_k \underline{\nabla}_k E \cdot \underline{F} \tag{53a}$$

or, in component form, by

$$a_i = -\frac{1}{h^2} \sum_j \frac{\partial^2 E}{\partial k_i \partial k_j} F_j \qquad (i, j = 1, 2, 3) \quad . \tag{53b}$$

By analogy with the Newtonian equations of motion ($\underline{F} = m\underline{a}$),

$$\left(\frac{1}{m^*}\right)_{ij} = \frac{1}{h^2} \frac{\partial^2 E}{\partial k_i \partial k_j} \qquad (i, j = 1, 2, 3) \tag{54}$$

defines an effective mass tensor m^*. The reciprocal of the effective mass is directly related to the E-$\underline{k}$ relationship of the solid by (54). For those cases in which the surfaces of constant energy in $\underline{k}$ space are spheres, the effective mass is a scalar quantity.

We digress for a moment and return to Fig. 4. At the point A in the lower band, the E-$\underline{k}$ relationship for a point near π/a can be derived in the same fashion used to derive Eq. (38). The result is

$$(E - E_a - V_1) = -\frac{h^2}{2m}\left(1 + \frac{2E_a}{V_1}\right)\left(k - \frac{\pi}{a}\right)^2 \quad . \tag{55}$$

The effective mass required to convert (55) into the form of (52) is

$$m^* = -m \bigg/ \left(1 + \frac{2E_a}{V_1}\right) \quad . \tag{56}$$

The effective mass is a negative quantity. Under the influence of an applied field $\underline{F}$, the behavior of this state is that of a positive particle with a positive mass, $-m^*$. The acceleration is in the direction of the electric field rather than opposite the field, as for negatively charged electrons.

(a) (b) (c)

FIG. 5. Tight binding approximation for the simple cubic lattice. Constant energy surfaces at the Fermi level in the first Brillouin zone for (a) band 1/6 full, (b) band 1/3 full, and (c) band 5/6 full.

A similar result is also obtained for the other limiting case, the tightly bound electron model, when the band is nearly filled with electrons. In both limiting models, the functional form of $E - \underline{k}$ near the top of a band E_t (wave vector $\underline{k}_t$) is

$$(E - E_t) = - \frac{h^2}{2m_n^*} (\underline{k} - \underline{k}_t)^2 , \qquad (57)$$

where m_n^* is the hole effective mass (a positive quantity). The energy of the hole is independent of the direction of the wave vector $\underline{k}$. Constant energy surfaces in $\underline{k}$ space are spheres centered at $\underline{k} = \underline{k}_t$. These spheres represent states occupied by holes, i.e., not occupied by electrons.

The description of a large assembly of electrons in a nearly filled band in terms of a comparatively small number of positive entities (holes) is an interesting feature of band theory. Experimental evidence for the existence of such entities comes most directly from the Hall effect, which will be discussed fully in Chapter 4.

We now return to our consideration of $E - \underline{k}$ relationships. The effective mass tensor m^*, as mentioned, reduces to a scalar quantity only for the case in which surfaces of constant energy are spheres in $\underline{k}$ space. This situation does occur in both of our simple models for $\underline{k}$ values sufficiently near a band minimum. The shapes of $E - \underline{k}$ for our tight binding model of Sec. II.D for various degrees of band filling are shown in Fig. 5. The total energy spread in this band is 12β. For the degree of band filling in Fig. 5(a), ($E = 2\beta$), the $E - \underline{k}$ curves are almost spherical, and the effective mass tensor m^* can be approximated by a scalar quantity. When the band is only $1/3$ filled ($E = 4\beta$), the constant energy surfaces are definitely not spherical. The constant energy curve is an octahedron which touches the zone boundaries along the coordinate axes. The volume enclosed by this constant surface is $1/6$ of the volume of the Brillouin zone. The effective mass is definitely a tensor quantity. When the band is almost filled ($E = 10\beta$), the situation is similar to that in Fig. 5(a), except the surfaces of constant energy which are almost spherical enclose the empty states in the band. (The corners of the Brillouin zone for the simple cubic structure are equivalent points.) The effective mass can be approximated by a scalar quantity which is negative, and the description is in terms of holes, rather than electrons.

Another quantity derived from the E-k relationship which is required in the calculation of many properties of solids (specific heat, paramagnetism, optical absorption, etc.) is the density of states per unit volume per unit energy range, $g(E)$. For free electrons, we derived an expression for $g(E)$ [Eq. (17)]. When the functional form of E-k is similar to that of free electrons, e.g., Eq. (52), $g(E)$ is proportional to $(E - E_0)^{\frac{1}{2}}$ and to $(m^*)^{\frac{3}{2}}$,

$$g(E) = \frac{1}{2\pi^2} \left(\frac{2m^*}{h^2}\right)^{\frac{3}{2}} (E - E_0)^{\frac{1}{2}} \ . \tag{58}$$

When the effective mass is a negative quantity and E-k has the functional form of Eq. (57), $g(E)$ refers to the hole states and is given by

$$g(E) = \frac{1}{2\pi^2} \left(\frac{2m_n^*}{h^2}\right)^{\frac{3}{2}} (E_t - E)^{\frac{1}{2}} \ . \tag{59}$$

The density of states, $g(E)$, for our tight binding model of Sec. II.D is shown in Fig. 6. For comparison, the curves for nearly free electrons (with our effective mass m^* equal to that for our tightly bound electrons at $k = 0$) and nearly free holes ($m_n^* = m^*$) are also shown.

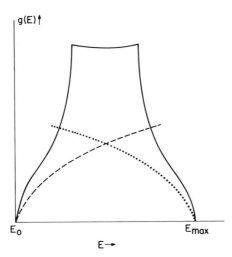

FIG. 6. Density of states for the tight binding model (solid line), for free electrons (dashed line), and for free holes (dotted line).

The number of carriers in a band is obtained by integrating over the band

$$n = \int g(E) f(E) \, dE \tag{60}$$

where $f(E)$ is the Fermi-Dirac distribution function introduced earlier. [It is clear that $f(E)$ is replaced by $[1 - f(E)]$ when holes are the carriers.]

The Fermi level at absolute zero depends upon $(m^*)^{-1}$. As a consequence, the electronic specific heat and the electronic magnetic susceptibility are proportional to m^*. Since the acceleration in an applied field is inversely proportional to the effective mass, the conductivity mobility is also proportional to m^{*-1}. Narrow bands are characterized by a high density of states, a high effective mass, and low conductivity mobility.

F. Summary

In the previous sections, we have considered two simple models representing limiting cases for one-electrons solutions of Schrödinger's

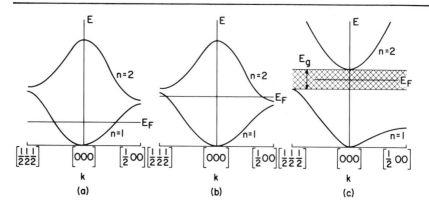

FIG. 7.   Energy band diagrams for simple cubic crystals with
(a) one valence electron and nearly free electrons, (b) two valence
electrons and nearly free electrons, and (c) two valence electrons
and the tight binding model.

equation (1) with a periodic crystal potential $V(\underline{r})$. Both models (nearly

free electrons and tightly bound electrons) are characterized by the

existence of forbidden gaps separating bands of quasicontinuous energy

eigenvalues. The existence of such gaps in allowed eigenvalues is a

direct consequence of the translational symmetry of the lattice and the

assumption that the crystal potential $V(\underline{r})$ reflects this transitional

symmetry. This general result is applicable to all wave propagation in

periodic structures (elastic, electromagnetic, etc.) [6].

The energy discontinuities occur at certain planes in reciprocal

lattice space for which the condition of Bragg reflection of the propa-

gating wave with wave vector $\underline{k}$ is satisfied ($2\underline{k}\cdot\underline{R} + k^2 = 0$, where $\underline{R}$

is a vector connecting reciprocal lattice points [2, 6]). The location of

the energy discontinuities in $\underline{k}$ space is determined by the translational

symmetry of the lattice; the magnitude, by the strength of the crystal

potential $V(\underline{r})$.

The number of states in each allowed band was shown to be twice

the number of atoms in a crystal. If each atom has only one valence

electron, the band will be only half-filled, and the solid will be a

conductor. This situation is depicted in Fig. 7(a). We have assumed a monatomic simple cubic lattice with a small perturbation due to the crystal potential (nearly free electron approximation).[*]

If each atom has two valence electrons, overlap between adjacent bands must be considered. Such overlap is likely in the nearly free electron model [Fig. 7(b)] and not likely in the tight binding model of Sec. II. D [Fig. 7(c)]. In Fig. 7(b), the Fermi level E_F is such that band 1 is partially empty and band 2 partially filled. The solid in this case is a metal. In Fig. 7(c), the Fermi level lies within the forbidden zone, shown as the shaded area. Band 1, completely filled at $T = 0$, is designated as the valence band. Band 2, completely empty at $T = 0$, is designated as the conduction band. The forbidden zone separating these two bands is designated the energy gap (E_g). In the absence of any imperfections which disrupt the perfect periodicity of the lattice, a solid with the band structure shown in Fig. 7(c) would be an insulator if $E_g \gg kT$, an intrinsic semiconductor if $E_g \gtrsim kT$, and a semimetal if $E_g < kT$. In the latter two cases, thermal energy is sufficient to excite electrons from the valence band into the conduction band, producing mobile carriers in both bands (holes in the valence band, electrons in the conduction band).

Imperfections, which disrupt the periodicity of the lattice, can lead to the presence of levels in the forbidden zone. The electrons associated with these levels are generally localized near the site of the imperfection. A relatively simple imperfection would be produced in our monatomic solid by replacing one of its atoms with a different atom. In general, if the foreign atom has one additional valence electron (but less than a filled shell), the imperfection produces a donor

---

[*]In Fig. 7, we are plotting the energy eigenvalues E as a function of the wave vector $\underline{k}$ along two directions in reciprocal space. For the simple cubic lattice, the reciprocal lattice spacing is $2\pi/a$. The boundary of the first Brillouin zone is encountered at $[\pi/a, 0, 0]$ along the positive x axis of reciprocal space and at $[\pi/a, \pi/a, \pi/a]$ along the $\langle 111 \rangle$ direction. The factor $2\pi/a$ is often omitted for convenience. The point $[\pi/a, 0, 0]$ is expressed simply as $[\frac{1}{2}, 0, 0]$, and $[\pi/a, \pi/a, \pi/a]$ becomes $[\frac{1}{2}, \frac{1}{2}, \frac{1}{2}]$.

level in the forbidden gap. The separation between this level and the conduction band minimum may be appreciably less than E_g, and thermal energy may be sufficient to excite electrons from this level into the conduction band. Solids, which otherwise would be insulators or intrinsic semiconductors, can become n-type extrinsic semiconductors if they contain these donor levels. In a similar fashion, the substitution of a foreign atom with one fewer valence electrons (but at least one outside a filled shell) may produce an acceptor level near the valence band maximum. At temperatures sufficient to thermally excite electrons from the top of the valence band to this localized acceptor level, the solid will have holes in the valence band and will exhibit extrinsic p-type semiconductivity. The topic of imperfections in solids is the basis of Chapter 8 (Volume 2).

Another feature of our simple model [Fig. 7(c)] is that the conduction band minimum and the valence band maximum do not necessarily occur at the same value of k, i.e., these states do not combine in the sense of Eq. (24). In optical transitions, conservation of momentum requires that, in the absence of cooperation with lattice phonons (quantized lattice vibrations), the k values of the initial and final states be equal (photons have very small momenta). The band gap depicted in Fig. 7(c) is designated indirect because transitions from the valence band maximum to the conduction band minimum require the cooperation of phonons to conserve momentum. In many actual cases, the minimum of the conduction band and the maximum of the valence band occur at the same value of k (most frequently $k = 0$). The band gap E_g in these cases is designated direct since "direct" optical transitions are allowed.

Figure 8 depicts a more realistic E-k relationship. This figure is similar to the band structure calculations of Herman for Ge [7]. The valence band is degenerate at its maximum at $k = 0$ and a split-off valence band is also present. The conduction band has several minima, but the lowest shown is at $[\frac{1}{2}, \frac{1}{2}, \frac{1}{2}]$.

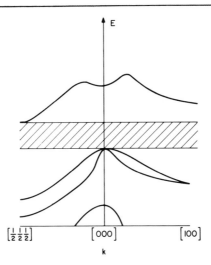

FIG. 8.   Energy band diagram typical of a semiconductor such as Ge.

The existence of minima in the conduction band at higher energies (such as at [0, 0, 0] and [1, 0, 0] in Fig. 8) can influence the properties of the solid, if the energy separation is not too large. An important example of a phenomenon which involves two conduction band minima is the Gunn effect [8].

In semiconductors, the number of carriers is usually quite small relative to the total number of states in a band, which is of the order of 10^{22} to 10^{23} cm^{-3}. The single models presented in Secs. II. C and II. D predict a parabolic behavior of E - $\underline{k}$ with a scalar effective mass for $\underline{k}$ sufficiently small. In real semiconductors, the assumption of a scalar effective mass may be quite poor. For example, in both Si and Ge the constant energy surfaces near the conduction band minima are elongated ellipsoids. In Ge, the four ellipsoids are located at the zone boundary along the $\langle 111 \rangle$ direction; half an ellipsoid occurs at each of the eight locations. In Si, the six ellipsoids are about 75% of the distance from the origin to the zone boundary along the $\langle 100 \rangle$ directions [7]. Fortunately, in discussing carrier statistics, a "density of states" effective mass, which is a scalar quantity, can be derived [5].

The discussion of band theory presented in this chapter is of necessity brief. Several important points have been neglected, but two further concepts require mention.

In ionic crystals, the electron is capable of polarizing the lattice, and a description, even in terms of the one-electron approximation, must involve this interaction. The electron plus the accompanying lattice polarization is designed as a "polaron." In many cases, it appears that this interaction can be treated as a perturbation of the static (i. e., unpolarized) lattice and is called the "large" polaron [9]. In other cases, the polaron may have a very large effective mass (relative to that calculated from a static lattice). This latter case has been designated a "small" polaron [10, 11] and has properties markedly different from the corresponding static lattice model.

N. F. Mott has pointed out that the Bloch model and the Heitler-London [12] model are not different approximations to the same exact wave function [13]. In a solid with large separation between atoms, the Heitler-London model is valid; below a certain critical spacing, the Bloch model is appropriate. The proposed Mott transition is expected to be abrupt, characterized by a sharp increase in the number of carriers available for conduction when the Bloch model is applicable.

III. THE BOND MODEL

A. General Considerations

In the previous section the existence of a periodic crystal potential, related to the translational symmetry of the lattice, was shown to be sufficient to produce forbidden zones in the energy eigenvalue spectrum of one-electron solutions to Schrödinger's equation. The presence of long-range order is not a necessary condition for the existence of energy bands. Many metals and semiconductors exhibit little or no change in electrical properties upon melting. Others, such as Ge, do exhibit changes upon melting, but these changes are associated with a modification of the short-range order in the material. In Ge, the coordination changes from four nearest neighbors to six nearest neighbors, and the electrical behavior changes from that of a semiconductor to that of a metal. Gray tin (which is a semiconductor and exhibits fourfold coordination) and white tin (which is a metal and has sixfold coordination) represent another example. The most important factor in determining whether or not a given solid is a metal or a semiconductor (or an insulator for sufficiently large energy gaps) is the manner in which it is bonded to its neighbors.

Solids can be approximately classified according to the nature of their chemical bonding. The usual limiting cases considered are (1) ionic, (2) covalent, (3) metallic, (4) molecular, and (5) hydrogen bonded.

Molecular crystals are held together by van der Waals bonds. These bonds are the result of forces between inert gases or saturated molecules. The main contribution to the attractive potential responsible for such bonds is from the "dispersion" forces [14]. These forces are not directional in nature, and the crystal structure is determined by geometrical considerations in which molecules (or atoms) tend to pack together as closely as possible.

The hydrogen bond is believed to be largely ionic in character and exists with only the most electronegative atoms (O, F, etc.). This bond is important in such solids as ice, inorganic and organic acids, proteins, salt hydrates, and in a few ferroelectric crystals (e.g., KH_2PO_4) [15].

The ionic bond is formed between a very electropositive atom and a very electronegative atom. Enough electrons are transferred to the electronegative species to give each ion a closed shell electron configuration characteristic of the inert gases. The resulting ions are almost spherically symmetric. The ionic bond results from Coulombic interactions, with the ions arranged to maximize the attraction between

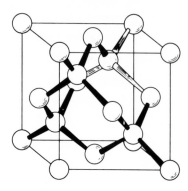

FIG. 9.   The diamond lattice.

unlike ions and to minimize the repulsion between like ions. Since
Coulombic forces are not directional, the arrangement of ions consistent with the above criterion may be dictated by geometrical considerations (size of ions, etc.). In the simple ionic compounds the most
common structures are the rock-salt structure (NaCl), in which each
ion is surrounded by six ions of opposite charge, and the CsCl structure, in which each ion is surrounded by eight ions of opposite charge.

The covalent bond is formed when two atoms share a pair of
electrons of opposite spin. The electron density between the bonded
cores is high, and the covalent bond exhibits marked directional properties. The most familiar example is the tetrahedral bond configuration
characteristic of carbon. Figure 9 illustrates the crystal structure of
diamond. The diamond lattice may be considered to be two interpenetrating fcc lattices; an atom of one sublattice is tetrahedrally bonded
to four atoms of the other sublattice, resulting in a very open structure.
Many important crystalline solids exhibit tetrahedral bonding similar to
that of diamond. The zincblende structure results when cations occupy
one sublattice and anions the other sublattice.

Metals have certain characteristics in common: high electrical
and thermal conductivity, metallic luster, ductility, etc. Most of the
metallic elements crystallize in highly coordinated structures: face
centered cubic (fcc) [12], hexagonal close packed (hcp) [12], or body

centered cubic (bcc) [8]. With such a large number of nearest neighbors, there are more stable orbitals available for bond formation than valence electrons to fill the orbitals completely. The concept of fractional bonds was introduced to account for this situation, and a full discussion may by found in Pauling's book [15].

B.   Major Bond Models

Most of the nonmetallic inorganic compounds exhibit bonding intermediate between the extremes of ionic bonding and covalent bonding. It is this class of materials to which we devote most of this section and the final section.

In discussing chemical bonding, experience tells us that we usually need consider only the s and p orbitals of the valence shells of the constituent atoms, and, in transition metal compounds, the d orbitals of the next lower shell.[*] Three theories have been instrumental in formulating a description of the bonding in solids:

(1) The valence bond theory, first employed by Heitler and London [12] and fully explored by Pauling [15] in his classic book.

(2) The molecular orbital theory, based upon the ideas of Mulliken [16].

(3) The crystal field theory, the foundations of which were laid by Bethe [17].

In the valence bond method, the wave function describing the electrons involved in a shared pair bond between two atoms is such that the two electrons tend to remain on the two different atoms. In the molecular orbital method, the two electrons are introduced into an orbital which encompasses two (or more) atomic cores. Crystal field theory, as originally developed, treats metal ions in a crystalline

---

[*]In some molecules, the d orbitals of the valence shell must also be considered; e.g., PF_5 [15].

environment as a type of inorganic molecule. The nearest neighbor anions (ligands) are considered to give rise to a constant electric potential, which reflects the symmetry of the arrangement of the ligands around the metal ion. The "crystalline field" thus produced destroys the spherical symmetry of the isolated metal ion. The consequences of crystal field theory are most important for d and f electrons, and a discussion of this model will the postponed until the next section.

The valence bond method offers a satisfactory understanding of the bonding in elements that crystallize in the diamond structure and compounds which crystallize in the zincblende structure. The strength of covalent bonds formed between two atoms depends upon the overlap between orbitals on adjacent atoms. If one constructs hybrid orbitals from the s and the three p orbitals, the combinations which give rise to the strongest bond (i. e., allowing the maximum amount of overlap) are [15]:

$$t_{111} = \tfrac{1}{2}(s + p_x + p_y + p_z)$$
$$t_{11\bar{1}} = \tfrac{1}{2}(s + p_x - p_y - p_z)$$
$$t_{\bar{1}1\bar{1}} = \tfrac{1}{2}(s - p_x + p_y - p_z)$$
$$t_{\bar{1}\bar{1}\bar{1}} = \tfrac{1}{2}(s - p_x - p_y - p_z)$$

$$(61)$$

These sp^3 hybrid orbitals are directed toward the corners of a regular tetrahedron. The wave function which describes the binding between two atoms A and B can be obtained from a linear combination of these hybrid sp^3 orbitals of atoms A and B

$$\psi = \phi_A + \phi_B \qquad (62)$$

or, in the case of heteropolar bonds,

$$\psi = \lambda\phi_A + \phi_B \qquad (63)$$

where the parameter λ is a measure of the degree of ionic character in the bond ($\lambda = 0$ corresponds to completely ionic character; $\lambda = 1$, to completely covalent character). At this point, we must note that, for a compound $A^N B^{8-N}$, a bond which is completely covalent would have marked polar properties. This may be seen as follows: The fraction of time spent by each bonding electron described by (63) around atom A is simply $(\lambda^2/1 + \lambda^2)$. The number of electrons involved in bonding each atom is eight (the number of stable orbitals in Eq. (61) is four), thus the charge due to the valence electrons around atom A is simply $(-8e\lambda^2/1 + \lambda^2)$. Since the charge on the atom core of atom A is Ne, the effective charge associated with atom A is thus

$$e_A^*/e = N - (8\lambda^2/1 + \lambda^2)$$
$$= -e_B^*/e \quad . \tag{64}$$

For the elements which crystallize in the diamond structure (C, Si, Ge, and α-Sn), the effective charge is 0, i. e., the bonding can be described as both covalent and "neutral." For compounds $A^N B^{8-N}$ crystallizing in the zincblende structure, the effective charge would vary from $(N-4)e$ for completely covalent bonding ($\lambda = 1$) to Ne for completely ionic bonding; e. g., for the III-V compounds, the range is $-e$ to $+3e$; for II-VI compounds, $-2e$ to $+2e$; and for I-VII compounds, $-3e$ to $+e$.

Coulson et al. [18] have applied the technique of LCAO to obtain values for λ in Eq. (63), and the effective charge for all compounds which crystallize in the zincblende structure. Their results indicate that most of the compounds tend to have effective charges rather closer to the "neutral" point than to either extreme, namely, completely ionic or completely covalent bonding.

The hybrid sp^3 orbitals [Eq. (61)] have the same energy eigenvalue in the free atom. In the solids which have bonding derived from these sp^3 hybrids (elements which crystallize in the diamond lattice and compounds $A^N B^{8-N}$ which crystallize in the zincblende structure), the number of electrons per atom involved in the bonding is eight for the four stable orbitals. The band of energies derived from these sp^3 hybrid orbitals will then be a filled band. In the absence of overlap with

TABLE 1

Band Gap and Effective Ionic Charge for Isoelectronic Series
(Zincblende Structure)[a]

| | IV | | III-V | | | II-VI | | | I-VII | |
Element	E_g (eV)	Compound	E_g (eV)	e_A^*/e	Compound	E_g (eV)	e_A^*/e	Compound	E_g (eV)	e_A^*/e
Ge	0.67	GaAs	1.35	-0.51	ZnSe	2.58	$|0.10|$	CuBr	2.9	$|1.0|$
α-Sn	0.07	InSb	0.18	-0.45	CdTe	0.76	$|0.08|$	AgI	2.8	—

[a]The energy gaps refer to values observed at 300 °K and were obtained from P. Aigrain and M. Balkanski, Selected Constants Relative to Semiconductors, Pergamon, New York, 1961.

III-V compounds: C. Hilsum, in Semiconductors and Semimetals (R. Willardson and A. Beer, eds.), Vol. 1, Academic, New York, 1966, p. 13.

II-VI compounds: D. Berlincourt, H. Jaffe, and L.R. Shiozawa, Phys. Rev., 129, 1009 (1963).

I-VII compounds: B. Szigeti, Trans. Faraday Soc., 45, 155 (1949).

TABLE 2

Band Gap and Effective Ionic Charge for Selected III-V and
II-VI Compounds (Zincblende Structure)[a]

III-V			II-VI		
Compound	E_g (eV)	e_A^*/e	Compound	E_g (eV)	e_A^*/e
GaP	2.24	|0.58|	ZnS	3.54	+0.27
GaAs	1.35	-0.51	ZnSe	2.58	|0.10|
GaSb	0.8	|0.33|	ZnTe	2.26	+0.07

[a] The energy gaps refer to values observed at 300 °K and were obtained from P. Aigrain and M. Balkanski, Selected Constants Relative to Semiconductors, Pergamon, New York, 1961.

The references from which effective charges were obtained are:

III-V compounds: C. Hilsum, in Semiconductors and Semimetals (R. Willardson and A. Beer, eds.), Vol. 1, Academic, New York, 1966, p. 13.

II-VI compounds: D. Berlincourt, H. Jaffe, and L.R. Shiozawa, Phys. Rev., 129, 1009 (1963).

bands derived from higher energy configurations of the free atoms, these solids will be semiconductors or insulators in which an energy gap exists.

Diamond, silicon, germanium, and gray tin have energy gaps of 5.4 eV, 1.1 eV, 0.66 eV, and 0.08 eV, respectively. Tables 1 and 2 are compilations of energy gaps at 300°K and effective charges experimentally determined[*] for selected elements (diamond structure) and compounds (zincblende structure). Table 1 lists the isoelectronic series of both Ge and α-Sn. Table 2 is a comparison of the III-V Ga compounds with the isoelectronic II-VI Zn compounds. The trend toward increasing band gap with increasing ionic character is apparent from these two tables.

Another trend also exists. The higher the degree of ionic character associated with the bonding, the less likely is the occurrence of both n- and p-type semiconductivity in the compound. For example, all the Ga III-V compounds listed in Table 2 exhibit appreciable n- and p-type semiconductivity when appropriately doped with impurities. In the case of the Zn II-VI compounds listed in the same table, ZnTe exhibits significant semiconductivity of only p-type while ZnSe and ZnS exhibit significant semiconductivity of only n-type. This trend is undoubtedly related to the fact that deviations from stoichiometry are also more likely for more ionic materials and stoichiometric defects are electrically active [21].

---

[*]The effective charge may be determined experimentally from (1) reflectivity measurements of the restrahlen bands [19], which yield only the magnitude of the effective charge, (2) piezoelectric measurements, which yield both the magnitude and sign but which require high resistivity material for accurate results [20], and (3) electron density plots obtained from x-ray diffraction experiments, which also yield the magnitude and sign of the effective charge.

C. Elemental Semiconductors and Polycompounds

The existence of a completely filled band in a solid is the result
of at least one of the components of the solid attaining a completed
valence shell. For elemental semiconductors, structures in which the
valence shell can be completed are characterized by the fact that the
coordination number (number of nearest neighbors), C, and the group in
which the element belongs, N, are related by

$$C = 8 - N \quad . \tag{65}$$

For group IV elements, the structure which exhibits a filled valence
band is the three-dimensional tetrahedral network of the diamond lattice.
For group V elements, this structure is a puckered double layer (or, for
P, P_4 tetrahedra); for group VI, rings or chains; and for group VII, pairs
(molecules which condense into molecular crystals). For the elements
occurring lower in the periodic table, account must be taken of the
possibility of overlap from higher lying bands, leading to semimetallic
behavior rather than semiconductivity.

A generalized 8 - N rule can be obtained for polycompounds, as
demonstrated by Hulliger and Mooser [22]. Their approach is a marriage
of the tight binding approximation (Sec. II. D) with valence concepts
relating to wave functions derived from linearly combined atomic orbitals. The wave functions ψ_n obtained in such a manner have certain
properties (see Coulson [23]):

(1) The number of ψ_n is equal to the number of atomic orbitals
involved in their construction, i. e., each ψ_n may be said to correlate
with an atomic state.

(2) Three types of ψ_n can be distinguished: bonding, nonbonding,
and antibonding, in order of increasing energy.

(3) The energy differences between orbitals of the same bonding
character are normally small. For nonbonding orbitals, as a rule,

energy differences are small only between nonbonding orbitals of the same atom.

(4) If bonding interactions occur between like atoms, the corresponding bonding and antibonding orbitals correlate with the same atomic orbitals. Formally, we may state that each atomic orbital participating in such an interaction splits into half a bonding and half an antibonding orbital.

From our results in Sec. II. D, each ψ_n corresponds to a band in the solid with a finite bandwidth. From point 3 above, all ψ_n of the same bonding character therefore constitute one composite band. The condition that the band be filled is thus [22]:

$$n_a + n_c - 2N_c O_c = n_a + n'_c = 2(4N_a + N_a B_a + N_c B_c) \qquad (66)$$

where there are $n_a + n_c$ anion and cation valence electrons in the unit cell. There are N_a anions per unit cell, which utilize 4 bonding and nonbonding orbitals and, on the average, B_a orbitals in anion-anion bonds. Likewise, there are N_c cations per unit cell with B_c orbitals involved in cation-cation bonding and O_c nonbonding or cation orbitals. The term n'_c represents the number of cation valence electrons per unit cell, not counting any nonbonding ones among them. Rearranging terms yields

$$\frac{n_a + n'_c - N_c B_c}{N_a} + B_a = 8 \qquad (67)$$

Assuming (1) that no cation-cation interactions are involved, (2) that anion-anion interactions can be described in terms of a shared electron pair, and (3) that, in the anion sublattice, the coordination of each anion is such that each anion has C_a nearest anion neighbors, Eq. (67) simplifies, with rearrangement, to

$$C_a = 8 - \frac{n_a + n'_c}{N_a} . \qquad (68)$$

The similarity of Eq. (68) to Eq. (65) indicates that semiconductors (or insulators) may be found in which the anion sublattice exhibits structures similar to those of the elemental semiconductors, e.g., $C_a = 4$ for structures in which $(n_a + n'_c)/N_a = 4$ with a three-dimensional anion sublattice; $C_a = 3$ for structures in which $(n_a + n'_c)/4a = 5$ with puckered double layers or tetrahedra of anions, etc. This building principle is discussed in full detail in the review article by Hulliger and Mooser [22].

In this discussion, we have tacitly assumed that the nonbonding cation orbitals are not involved in electrical transport properties of these compounds. This assumption will be reviewed in the next section which deals with d electrons in solids.

IV. MATERIALS WITH INCOMPLETE INNER ELECTRON SHELLS

When considering materials with d (or f) electrons, one must confront the question of "localized" vs "delocalized" or "itinerant" electron states. In the previous sections, we have concentrated primarily upon the latter concept whereby the valence electrons in particular are assigned to the whole crystal in various bands derived from atomic orbitals. Localized levels associated with crystal imperfections and impurities were discussed. In compounds of the transition elements, one may well have "localized" and "itinerant" outer electrons present simultaneously as, e.g., in magnetic semiconductors. The existence of such materials does pose theoretical problems. Simple band theory can explain insulating (or semiconducting) behavior in materials with an even number of electrons per unit cell on the basis of an energy gap which may exist between filled bands and higher lying empty conduction bands. Simple band theory can also accommodate the existence of metallic materials (such as Mg) which have an even number of electrons per unit cell because of the possibility of overlap between filled and empty

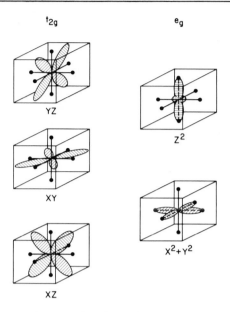

FIG. 10.   The angular dependence of the five d orbitals.

bands (cf. Sec. II. D). However, the existence of insulators (or semi-conductors) such as NiO cannot be explained by simple band theory, which would predict a partially filled band derived from Ni 3d orbitals giving rise to metallic behavior.

An explanation of the failure of simple band theory for NiO was proposed by Mott [23]. For a particularly narrow band, the reduction in energy brought about by band formation might not overcome the increase in electrostatic repulsion which results from conduction electrons moving in Bloch states. The electron-electron repulsion is an electronic correlation which is neglected in all one-electron treatments used in band calculations and which cannot be analyzed from the results of such band calculations. The Mott result is that whenever one expects a narrow, partially filled band from band structure calculations, one must be alert to the possibility that such calculations may not be valid and that such electrons cannot be described by Bloch functions. A more accurate starting point for materials such as NiO is to assume that the

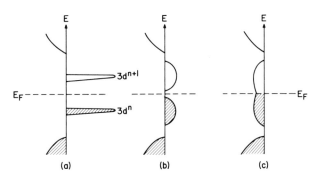

FIG. 11. Schematic density of states for a system with n 3d electrons where (a) 3d levels separated by Coulomb repulsion, (b) 3d levels broadened until they nearly overlap, and (c) 3d levels broadened until they overlap. The dotted line represents the Fermi level. Itinerant band states are drawn to the left and localized states to the right.

d (or f) electrons are localized on the cation cores. A natural theory to use for such an ionic model is the crystal field theory mentioned in an earlier section [17]. In this theory, the energy degeneracy of the d (or f) electrons is removed by the electrostatic interactions with the surrounding anions. For example, Fig. 10 illustrates the orbital angular distributions of the five 3d wave functions for a central cation surrounded by a regular octahedron of anions. From electrostatic interactions with the anions, the orbitals designated as e_g are raised in energy relative to those designated as t_{2g}. These splittings are very important in understanding the optical and magnetic properties of such "ionic" compounds involving cations with incomplete d (or f) shells. Another important factor is the energy of the d (or f) electrons relative to the energy of the wide bands derived from overlap of the outer shell s and p wave functions (this overlap leads to the formation of wide bands even in the ionic alkali halides). For example, in a free Ni atom the 3d and 4s levels have approximately the same energy while for Ni^{2+} the 4s levels lie at considerably higher energy than the 3d. When deriving band structure from a linear combination of atomic orbitals, it obviously makes a great deal of difference as to which starting point is used.

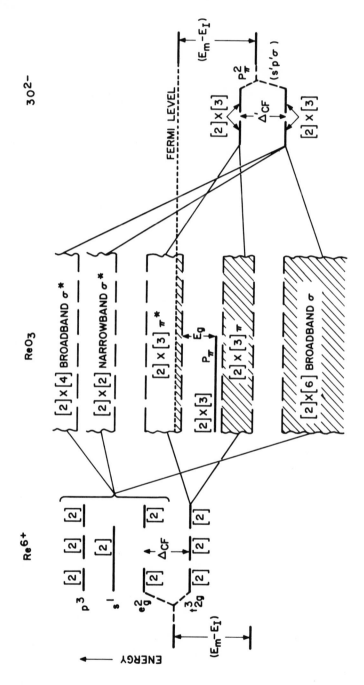

FIG. 12. Elementary band structure for ReO₃ (after Ref. 24).

Qualitative energy band schemes may be derived by combining one-electron band states (for wide bands) and localized many-electron orbitals as shown in Fig. 11. In this figure, itinerant band states are drawn to the right and localized states to the left of 0. The system shown is assumed to have n 3d electrons, which completely fill the $3d^n$ levels. In Fig. 11(a), the $3d^{n+1}$ levels are separated from the $3d^n$ levels by Coulomb repulsion. If the 3d levels are broadened by some mechanism, the separation between $3d^n$ and $3d^{n+1}$ may eventually disappear [Figs. 11(b) and 11(c)]; i.e., the energy gained by band formation overcomes the electron-electron repulsion mentioned by Mott leading to itinerant band states [Fig. 11(c)]. Mechanisms which can broaden d (or f) levels are direct cation-cation overlap and increased overlap between cation and anion wave functions.

The best example of a system in which cation-anion overlap involving d orbitals leads to band formation is ReO_3, a metallic transition metal oxide [24]. Goodenough proposed the model shown in Fig. 12, which represents a one-electron band structure scheme based on the overlap of various Re 5d, 6s, and 6p orbitals with the various O 2s and 2p orbitals as dictated by crystal symmetry. On the left and right of Fig. 12 are the atomic levels of Re^{6+} and of O^{2-}, including the crystal field splitting appropriate for the cation in an octahedron of O anions. If the compound were considered to be completely ionic, the O levels would be completely filled, and there would be a single electron in the 5d (t_{2g}) Re levels, which do not overlap enough to provide band formation as required by the existence of metallic conductivity. Goodenough postulates that the overlap of Re and O orbitals produces levels characteristic of ReO_6 clusters, which are further broadened into bands by the large orbital overlap extending throughout the entire crystal, as shown in the center of Fig. 12. When the 25 valence electrons per ReO_3 unit are accommodated, all bands and levels are filled up to the π^* antibonding band, which is one-sixth full and therefore can account for the observed metallic characteristics of ReO_3.

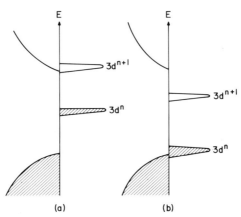

FIG. 13. Schematic density of states for magnetic semiconductors: (a) n-type and (b) p-type.

Overlap between orbitals can be modified by experimental conditions, as, e.g., applying pressure, heat, etc. For systems in which the Coulomb repulsion is nearly counterbalanced by level broadening [e.g., Fig. 11(b)], it is conceivable that by changing experimental conditions one could increase the overlap sufficiently so that band formation would overcome electron-electron repulsion [Fig. 11(c)]. These electronic Mott transitions can occur without any changes in lattice symmetry. Mott's controversial position is that such transitions from localized to itinerant states in periodic solids must be relatively sharp [13, 23].

In the previous paragraphs, we have given qualitative energy band schemes for magnetic insulators [Fig. 11(a)] and for metallic compounds [Figs. 11(c) and 12]. We now turn our attention to semiconductors. Figure 13(a) represents an n-type magnetic semiconductor, because the empty $3d^{n+1}$ states merge with the band states at the bottom of the conduction band. If the Fermi level can be brought to the bottom of the conduction band by doping or by thermal excitation, n-type semiconductivity results. A p-type doping with acceptors would bring holes into the occupied $3d^n$ states. In such a system, holes could move from

cation to cation by activated hopping. If the band structure were qualitatively as in Fig. 13(b), doping with acceptors would produce a p-type
magnetic semiconductor because the $3d^n$ levels merge with the levels
at the top of the valence band. Doping this system with donors would
lead to electrons in the $3d^{n+1}$ levels, which could move from cation
to cation by activated hopping. Activated hopping processes are characterized by low mobilities (10^{-4} to 10^{-8} cm^2/V-sec range).

The simple approach of combining one-electron band schemes
with many-electron localized levels does offer qualitative energy band
structure diagrams useful in interpreting the properties of materials
with incomplete d (or f) shells. Unfortunately, the electrical transport properties of such compounds do not present a clear picture. A
difficult problem is that small concentrations of impurities or deviations from stoichiometry can markedly affect the experimental results
and subsequent theoretical interpretations. A second consideration is
that using heavy doping of such materials in order to minimize the
effects of nonstoichiometry and uncontrolled impurities can actually
strongly modify the energy band scheme. For example, Adler [25] has
shown that the 2p band in NiO moves up over 1 ev relative to the
localized $3d^8$ Ni levels for heavily Li-doped NiO as compared to pure
NiO. This happens because large concentrations of Ni^{3+} are produced
to compensate the Li^+ and the excitation of a 2p electron onto Ni^{3+}
is much easier than onto a Ni^{2+}. In heavily Li-doped NiO, this relative variation of band energies results in the fact that wide band 2p
hole conduction is the dominant conduction mechanism. Had one
simply taken the band structure based on pure NiO and introduced the
Li levels at their expected location relative to the $3d^8$ Ni levels, one
would have dismissed 2p band conduction as negligible.

In summary, theoretical problems associated with band structure
calculations of compounds with incomplete inner shells (d or f) are
complicated by the fact that many-electron considerations must be
employed. A judicious blend of one-electron bands and many-electron

localized levels can provide a qualitative understanding of the properties of these compounds. Progress is being made in this area. Equally
important is the need for careful characterization of the materials under
study before a full understanding of this extensive research field can
be realized.

REFERENCES

1. L. I. Schiff, Quantum Mechanics, McGraw-Hill, New York, 1955.

2. C. Kittel, Introduction to Solid State Physics, 2nd ed., Wiley,
   New York, 1957.

3. F. Bloch, Z. Physik, 52, 555 (1928).

4. E. T. Whittaker and G. N. Watson, Modern Analysis, 4th ed.,
   Cambridge Univ. Press, Cambridge, England, 1927, p. 412.

5. N. F. Mott and H. Jones, The Theory of the Properties of Metals
   and Alloys, Dover, New York, 1958.

6. L. Brillouin, Wave Propagation in Periodic Structures, 2nd ed.,
   Dover, New York, 1953.

7. F. Herman, Proc. IRE, 43, 1703 (1955).

8. J. B. Gunn, Solid State Comm., 1, 88 (1963).

9. H. Frohlich, H. Pelzer, and S. Zienau, Phil. Mag., 41, 221 (1950).

10. S. Pekar, J. Exptl. Theoret. Phys. USSR, 16, 341 (1946).

11. G. L. Sewell, Phys. Rev., 129, 597 (1963).

12. W. Heitler and F. London, Z. Physik, 44, 455 (1927).

13. N. F. Mott, Can. J. Phys., 34, 1356 (1956); Proc. Phys. Soc.,
    A62, 416 (1949); Phil. Mag., 6, 287 (1961).

14. M. Born, Atomic Physics, 5th ed., Hafner, New York, 1951.

15. L. Pauling, Nature of the Chemical Bond, 3rd ed., Cornell Univ.
    Press, Ithaca, New York, 1960.

16. R. S. Mulliken, Phys. Rev., 40, 55 (1932).

17. H. A. Bethe, Ann. Physik, 3, 133 (1929).

18. C.A. Coulson, L.B. Redei, and D. Stocker, Proc. Roy. Soc. (London), 270, 357 (1962).

19. B. Szigeti, Trans. Faraday Soc., 45, 155 (1949).

20. D. Berlincourt, H. Jaffe, and L.R. Shiozawa, Phys. Rev., 129, 1009 (1963).

21. G. Mandel, Phys. Rev., 134, A1073 (1964).

22. F. Hulliger and E. Mooser, in Progress in Solid State Chemistry (H. Reiss, ed.), Vol. 2, Pergamon, New York, 1965, p. 330.

23. C.A. Coulson, Valence, Oxford Univ. Press, Oxford, England, 1960.

24. J.B. Goodenough, Bull. soc. chim. France, 1965, 1200.

25. D. Adler, IBM J. Res. Develop., 14, 261 (1970).

PART II

PHYSICAL PROPERTIES AND IMPERFECTIONS

Chapter 4

ELECTRICAL PROPERTIES OF SOLIDS

Jerome H. Perlstein[*]

Department of Chemistry

The Johns Hopkins University

Baltimore, Maryland

[*]Present address: Research Laboratories, Eastman Kodak Company, Rochester, New York.

I.   INTRODUCTION

In the earlier chapters of this book the concepts of metals, semi-conductors, and insulators were introduced, and the properties of these materials in light of their practical significance were discussed. The presence of electrons and holes in these materials was shown to be of vital importance in understanding the chemical interactions that take place in solids as well as being important in understanding the physical properties of most material solids.

In this chapter, we would like to discuss in more detail the origin of the electrical properties of solids. We will see that the idea of an electric current as being simply the flow of electrons in the presence of an electric field needs to be modified to account for the existence of holes. We will also want to look at the details of the mechanism for current flow, viz., from where and how do the electrons and holes originate and how do they "move" from one point to another through the sample? Once we understand something about the mechanism of elec-trical conduction in solids, we will be able to make some predictions as to how to chemically synthesize new materials which will possess novel solid state properties of practical importance. Some of these

will be mentioned at the end of this chapter and are considered further
in later sections of the book.

We can begin by mentioning some simple facts about the electrical
properties of materials. We know that copper is a very good conductor
of electricity; we also know that crystals of alkali halides like NaCl are
very poor conductors of electricity. What do we mean by a good conductor or poor conductor? A good conductor is one that passes an electric current easily; it has a low resistance to the flow of electric charge.
A poor conductor has a high resistance to the flow of electric charge.
The difference then between good conductors and poor conductors of the
same geometric shape as well as all the gradations between good and
poor is the different resistance that materials have. Obviously, then,
a material that has no resistance would be the best conductor of all and,
in fact, such materials do exist. We will discuss these freaks of
nature, called superconductors, later on.

The idea of resistance can be quantified by doing a very simple
experiment. If we take a battery of known voltage V and place it in

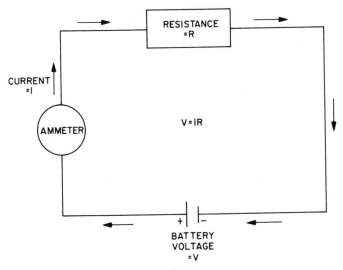

FIG. 1. Block diagram for measuring the value of resistance.
Ammeter measures current I; voltage V, across resistance R, is provided by the battery.

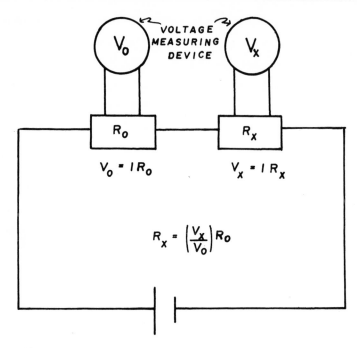

FIG. 2. Measurement of an unknown resistance R_x in terms of a known resistance R_0.

series with a resistance R and an ammeter which measures current I, as shown in Fig. 1, then it is found that the voltage is proportional to the current with the proportionality constant equal to R:

$$V = IR \qquad (1)$$

This is a statement of Ohm's law, named after its discoverer. Although it is a very simple statement, we shall see that any atomic theory explaining the nature of electrical conduction in solids must be able to account for Ohm's law.

Ohm's law allows us to determine the value of an unknown resistance in terms of a known resistance as shown in Fig. 2. In this figure R_x is the unknown resistance to be determined, R_0 is the known resistance. Neither the battery voltage nor the current through the circuit need be known in order to determine R_x. We need only apply Ohm's

law to each resistance and realize that the current, which is continuous, must be the same going through both resistances. To apply Ohm's
law to each resistance we simply place a voltage measuring device
(e. g., voltmeter, potentiometer, electrometer) across each resistance,
measure the voltage, and apply Ohm's law to the results. If V_x is the
voltage across R_x and V_0 is the voltage across R_0, then Ohm's law
says

$$V_x = IR_x \quad , \tag{2}$$

$$V_0 = IR_0 \quad . \tag{3}$$

Taking the ratio of Eq. (2) to Eq. (3) gives R_x in terms of known
quantities:

$$\frac{V_x}{V_0} = \frac{R_x}{R_0}$$

or

$$R_x = \left(\frac{V_x}{V_0}\right) R_0 \quad . \tag{4}$$

In Eq. (4), V_x and V_0 are measured quantities and R_0 is a known
resistance. Thus, R_x can be calculated.

Notice what happens when we add Eqs. (2) and (3). We get

$$V_x + V_0 = I(R_x + R_0) \tag{5}$$

or

$$V_{total} = IR_{total} \quad .$$

If R_x and R_0 are the only resistances in the circuit, then the sum
of the potentials across them must equal the battery voltage. This fact
is of great experimental importance, for if the voltages across the known
and unknown resistances do not add up to the battery voltage, one can
conclude that there are other resistances in the circuit which have not
been accounted for. We shall see later on when we describe the experimental techniques in more detail that the presence of this additional

resistance occurs more often than not and has led to a great deal of erroneous electrical resistance measurements.

As a practical example consider a 1.35-V battery in series with $R_0 = 100$ ohms (abbreviated 100Ω) and R_x to be determined. The potential measured across R_0 is 0.63 V and the potential measured across R_x is 0.72 V. What is R_x? Applying Eq. (4) we find

$$R_x = \left(\frac{0.72}{0.63}\right) \times 100 = 114\Omega .$$

In comparing materials of different resistance it is necessary to compare samples that have the same geometry. Thus, for example, two pieces of copper wire, one twice as long as the other, will have different resistances. In fact the resistance is proportional to the length of the sample measured, so that the longer wire will have twice the resistance of the shorter. However, if we compare two copper wires of the same length but one with a cross-sectional area twice as great as the other, then the wire with twice the cross-sectional area will, in fact, have one-half the resistance of the other. Resistance is then inversely proportional to cross-sectional area. Thus

$$R \propto L \quad , \tag{6}$$

where L is the length of the specimen, and

$$R \propto \frac{1}{A} \quad , \tag{7}$$

where A is the cross-sectional area of the specimen. Combining Eqs. (6) and (7):

$$R \propto \frac{L}{A}$$

or

$$R = \rho \frac{L}{A} \quad . \tag{8}$$

In Eq. (8), ρ is the proportionality constant which relates the resistance R to the length: area ratio and has been given the name resistivity. The resistivity is a quantity which is independent of the dimensions of a material. Thus, in comparing two materials one should compare their

resistivities. As an example, consider a sample of Cu which is 10 cm

long by 0.1 cm wide by 0.01 cm thick and whose measured resistance R

is 0.0167Ω. The L/A ratio for this sample is then

$$\frac{L}{A} = \frac{10}{(0.1)(0.01)} = 10^4 \text{ cm}^{-1} \quad .$$

Hence, from Eq. (8), the resistivity of copper (regardless of

dimensions) is

$$\rho = \left(\frac{A}{L}\right) R = R \times 10^{-4} = 0.0167 \times 10^{-4} = 1.67 \times 10^{-6} \text{ }\Omega\text{-cm} \quad .$$

As a comparison, Table 1 lists the resistivities of a number of

materials, some of chemical interest, showing the range over which ρ

can vary. Also in Table 1 is listed the conductivity σ for each mate-

rial. The conductivity is defined simply as the reciprocal of the

resistivity:

$$\sigma = 1/\rho \tag{9}$$

Cond. dec. as T inc.
which means Resist.
is increasing.

or

$$\sigma = \left(\frac{L}{A}\right)\frac{1}{R} \quad .$$

Thus, those materials with very high conductivity are good con-

ductors, those with very low conductivity are good insulators. The

interesting point about the conductivities in Table 1 is that the range

of values observed cover more than 20 orders of magnitude! It has

been tempting in the past to call those materials at the top of the table,

having the highest conductivities, metals, those in the center, semi-

conductors, and the ones near the bottom with the lowest conductivity,

insulators. This division is rather arbitrary and, in fact, mechanistic-

ally incorrect. We shall see that the mechanism for conduction can be

quite different even for materials which have the same order of magni-

tude conductivity.

As an example of the difference in behavior between a metal and

a semiconductor, Figs. 3(a) and 3(b) show the temperature dependence

TABLE 1

Resistivity ρ and Conductivity $\sigma = 1/\rho$ for
Various Materials at Room Temperature

Material	ρ (Ω cm)	σ (Ω^{-1} cm^{-1})
Copper	1.7×10^{-6}	6×10^5
$Na_{0.9}WO_3$	5.6×10^{-6}	1.8×10^5
Ce_3S_4	5.0×10^{-4}	2×10^3
$Na_{0.33}V_2O_5$	7.1×10^{-3}	1.4×10^2
$(SN)_x$ polymer	7.7×10^{-3}	1.3×10^2
$KTaO_{3-\delta}$	1×10^{-2}	1×10^2
n-type Si[a]	2.5×10^{-2}	4×10^1
$K_2Pt(Cn)_4Br_{0.3}$	10^{-1}	1×10^1
$Li_{0.02}NiO$	1.6×10^{-1}	6.3
CdF_2-0.1 mole % TbF_3	2.7	3.7×10^{-1}
$[(CH_3CH_2)_3NH](TCNQ)_2$	10^3	1×10^{-3}
UO_2	2×10^4	5×10^{-5}
Cu_2O	10^6	10^{-6}
Ce_2S_3	10^{10}	10^{-10}
Cu-phthalocyanine	$> 10^{13}$	$< 10^{-13}$
Teflon	$> 10^{14}$	$< 10^{-14}$

[a] 10^{18} Phosphorus donors/cm^3.

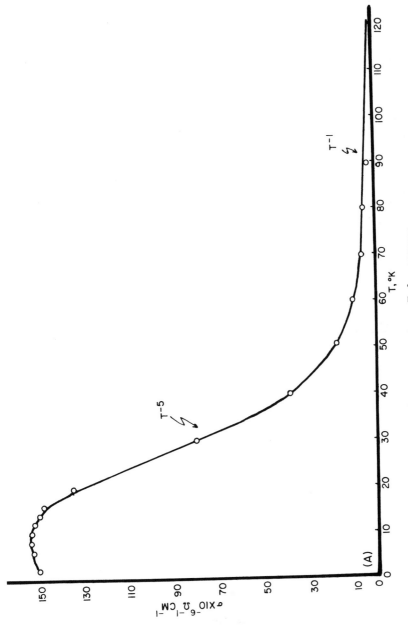

FIG. 3a. Conductivity σ vs absolute temperature T for copper.

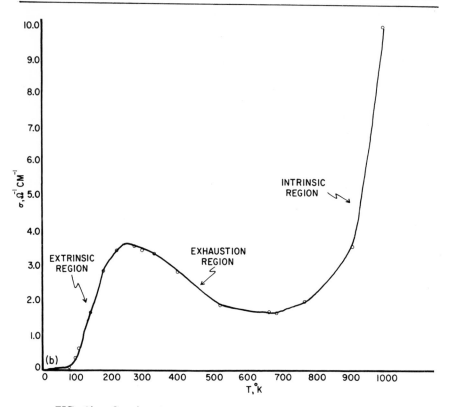

FIG. 3b.   Conductivity vs temperature for n-type silicon.   Compare with copper in the same temperature region.

of the conductivity for a typical metal like copper [Fig. 3(a)] and a typical semiconductor like silicon [Fig. 3(b)].   It is at once apparent that the behavior of the conductivity is vastly different for the two materials.   Near absolute zero, the conductivity of copper is quite high with a value of $\sigma = 10^8 \ \Omega^{-1} \ cm^{-1}$.   The conductivity of silicon, on the other hand, is very low, approaching zero as the temperature approaches absolute zero.   Thus, the first distinction we can make between a metal and a semiconductor is the value of the conductivity at $0 \, °K$.   At absolute zero, the conductivity of a semiconductor is zero, while that for a metal is greater than zero.

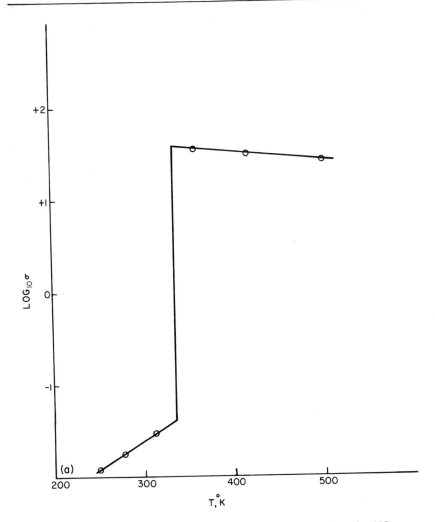

FIG. 4a. Logarithm of conductivity vs temperature for VO_2.

Another distinguishing feature between a metal and a semiconductor is what happens to the conductivity as the temperature increases. For the metal, it can be seen in Fig. 3(a) that the conductivity decreases as the temperature increases, whereas for the semiconductor in Fig. 3(b) the conductivity increases with increasing temperature. Thus, the

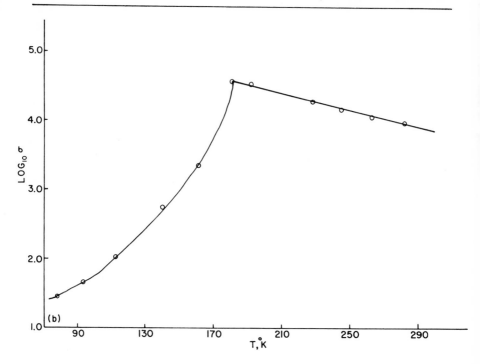

FIG. 4b.   Logarithm of conductivity vs temperature for $K_{0.30}MoO_3$.

second distinguishing feature between a metal and a semiconductor is
that the temperature coefficient for conduction, $d\sigma/dT$, is negative for
a metal whereas it is positive for a semiconductor.   It should be mentioned at this point that there are materials which show both semiconducting and metallic behavior in different temperature regions.
Figures 4(a), 4(b), and 4(c) show three examples.   In Fig. 4(a), the
conductivity of VO_2 shows semiconductor behavior ($d\sigma/dT > 0$) below
339°K but metallic behavior ($d\sigma/dT < 0$) above this temperature.   There
appears to be a very sharp transition between the two types of behavior
at 339°K corresponding to a first-order phase transition from monoclinic
to tetragonal with increasing temperature.   The conductivity appears to
jump by five orders of magnitude at the transition.   Figure 4(b) shows a
semiconductor-to-metal transition at 180°K in $K_{0.30}MoO_3$, the blue
potassium molybdenum bronze.   In this case there is no discontinuity

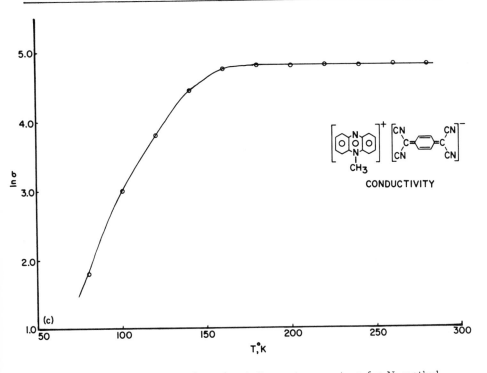

FIG. 4c.   Logarithm of conductivity vs temperature for N-methyl-phenazinium tetracyanoquinodimethane.

at the transition point although it is known that there is a change from electron conduction below 180°K to hole conduction above this tempera-ture.   Finally, Fig. 4(c) shows the peculiar behavior of a class of highly conducting organic salts.   Tetracyanoquinodimethane, TCNQ for short, is a good electron acceptor.   When combined with a good electron donor, R, salts of TCNQ are produced, $R^+[TCNQ]^-$.   The conductivity of one of these salts is shown in Fig. 4(c) with

R^+ = N-methylphenazinium =

and

$$\text{TCNQ}^- = \left[\begin{array}{c} \text{CN} \\ \text{CN} \end{array} \text{C=} \text{=} \text{C} \begin{array}{c} \text{CN} \\ \text{CN} \end{array} \right]^-$$

From Fig. 4(c) it can be seen that below 149°K the salt shows semiconductor properties whereas above 149°K the conductivity appears to
be temperature independent, which is similar to the low temperature
behavior of most metals; this will be discussed later on.

II.   CLASSICAL FREE ELECTRON GAS MODEL

With the above introduction to the general electrical properties of
metals and semiconductors, we should now try to see how the properties
of these materials can be understood in terms of the microscopic ideas
about atomic structure.

The first attempt at understanding the electrical properties of
metals was presented by Drude in 1900. The essentials of the theory
are that the valence electrons of the atoms in a solid are free and are
able to move independently of one another like an ideal gas in a container. In the absence of an electric field the electrons are darting
about with random thermal energy and random velocity, so that the
average thermal velocity $\langle v \rangle$ of all the electrons is zero. In the absence
of an electric field there is thus no current flow. However, under the
influence of an electric field, the electrons tend to drift more in one
direction than in any other, producing a new average velocity $\langle v_d \rangle$
of the electrons which is different from zero. Thus, the electrons flow
or drift preferentially one way, producing an electric current.

Several questions may be raised at this juncture about this model.
First, if the electrons are free, why doesn't the coulomb force of repulsion between electrons dominate the situation so that the gas simply
explodes? Secondly, if the electrons are free, then in the presence of

an electric field, they will constantly be accelerated so that their velocity will increase indefinitely, producing an infinite current. What limits
the velocities so that the average drift velocity and, hence, the current
is constant?

The first question can be answered by realizing that this gas of
electrons has within it a periodically distributed positive charge. Any
one electron in the gas is partially screened from this positive charge
by the surrounding electrons. Thus, although any given electron in the
gas will repel those electrons within its vicinity, it does so at the expense of decreasing the screening from the positive charge that is provided by the surrounding electrons. Hence, the repulsion will stop
when the screening decreases to the point when a given electron "sees"
a positive charge equal to its own negative charge. The electron gas
is thus in a state of equilibrium in which the forces of repulsion between
electrons are just balanced by the forces of attraction between electrons
and the positive ion cores.

The second question is answered by realizing that the "free" electrons must, in fact, be scattered by the lattice atoms. Each time an
electron changes its velocity by scattering off a lattice atom, it loses
almost all of its previous "memory" as to the effects of the electric
field. Thus, after an electron collides with a lattice atom, it loses the
energy it gained from the electric field. We know that this energy is
transferred to the atoms by the fact that, when a current passes through
a conductor, the conductor warms up. Without this lattice scattering
the electron's velocity would increase indefinitely. There would thus
be no resistance to the flow of current and Ohm's law could therefore
not be obeyed. The effect of the scattering process, then, is to limit
the increase of the electron's velocity over the random thermal velocity
which it normally possesses. Thus, in an electric field, an electron
increases its velocity until it is scattered at which time its velocity
returns to the thermal equilibrium value. The process of acceleration
and scattering then repeats itself.

Suppose the average time between scattering events is τ. In an electric field, then, the average increase in velocity of the electrons, which is simply the drift velocity $\langle v_d \rangle$, is simply

$$\langle v_d \rangle = a\tau \quad,$$

where a is the acceleration of the electron between collisions. From Newton's second law,

$$a = \frac{force}{mass} = \frac{|e| E}{m} \quad,$$

where E is the electric field, e the electron's charge, and m its mass. Thus,

$$\langle v_d \rangle = \frac{|e| E\tau}{m} \quad. \tag{10}$$

If we multiply Eq. (10) by the total charge per unit volume, ne, where n is the number of electrons/cm^3 in the gas and e the charge per electron, we have

$$ne \langle v_d \rangle = \frac{ne^2 E\tau}{m} \quad.$$

The quantity on the left is simply the current per unit area or current density. This can be seen by writing out the units; n is in electrons per cm^3, e in coulombs per electron, $\langle v_d \rangle$ in cm/sec. Thus, the units of $ne \langle v_d \rangle$ are

$$\left(\frac{electrons}{cm^3} \right) \left(\frac{coulombs}{electron} \right) \frac{cm}{sec} = \left(\frac{coulombs}{sec} \right) \Big/ area \quad.$$

Calling the current density J, we then have

$$J = \frac{ne^2 \tau}{m} E \quad. \tag{11}$$

Equation (11) is just Ohm's law expressed in terms of current density and electric field rather than current and voltage. If A is the cross-sectional area of the electron gas and L is its length in the electric field direction, then the current I equals JA and the voltage V

equals EL. Substituting into Eq. (11) we get

$$\frac{I}{A} = \frac{ne^2 \tau}{m} \frac{V}{L}$$

or

$$\frac{V}{I} = \frac{m}{ne^2 \tau} \frac{L}{A} \quad .$$

Comparison with Eqs. (1) and (8) shows that the resistivity ρ is

$$\rho = \frac{m}{ne^2 \tau} \tag{12}$$

and the conductivity σ is

$$\sigma = \frac{1}{\rho} = \frac{ne^2 \tau}{m} \quad . \tag{13}$$

Thus, the Drude free electron gas model is able to account for Ohm's law and gives an explicit expression for the conductivity in terms of the density of electrons and the mean time between collisions.

We can get some idea about the mean time between collisions by putting some numbers into Eq. (13). Copper has an electrical conductivity of $5.98 \times 10^5 \ \Omega^{-1} \ cm^{-1}$. In order to use this value in Eq. (13) we have to convert it from practical units into electrostatic units (the cgs units of electricity). The conversion is 1 esu of resistance = $8.99 \times 10^{11} \ \Omega$. The density of copper is $8.92 \ g/cm^3$ and its atomic weight is 63.5. Hence the number of electrons/cm^3, assuming each copper atom donates one electron to the free electron gas, is

$$n = \frac{density}{atomic \ weight} \times Avogadro's \ number$$

$$n = \left(\frac{8.92}{63.5}\right) 6.023 \times 10^{23} = 8.46 \times 10^{22} \ \frac{electrons}{cm^3}$$

The value of e is 4.8×10^{-10} esu and of m is 9.11×10^{-28} g. Therefore,

$$\tau = \frac{\sigma m}{ne^2} = \frac{(5.98 \times 10^5)(9.11 \times 10^{-28})(8.99 \times 10^{11})}{(8.46 \times 10^{22})(4.8 \times 10^{-10})^2} = 2.5 \times 10^{-14} \ sec \ .$$

We can also get some idea how far the electron travels between collisions by defining the mean free path ℓ.

$$\ell = v\tau \quad . \tag{14}$$

Here v will be some sort of average speed of the electrons like
the root mean square velocity $(\overline{v^2})^{\frac{1}{2}}$. From the kinetic theory of gases,
the root mean square thermal velocity of a particle is related to temper-
ature by the relation

$$\tfrac{1}{2}m\overline{v^2} = \tfrac{3}{2}kT \quad .$$

Here k is Boltzmann's constant $= 1.38 \times 10^{-16}$ erg/°K. Thus, taking T
to be 298 °K we find

$$(\overline{v^2})^{\frac{1}{2}} = \left(\frac{3kT}{m}\right)^{\frac{1}{2}} = \left(\frac{3(1.38 \times 10^{-16})(298)}{9.11 \times 10^{-28}}\right)^{\frac{1}{2}} \tag{15}$$

$$= 1.16 \times 10^7 \text{ cm/sec} \quad ,$$

which is a very high velocity – faster than a bullet shot from a gun. We
thus get for the mean free path

$$\ell = (1.16 \times 10^7)(2.5 \times 10^{-14}) = 2.9 \times 10^{-7} \text{ cm}$$

or

$$\ell = 29\,\text{Å} \quad .$$

Two points should be mentioned concerning the calculation of the
distance the electron travels between collisions. In Eq. (14) the speed
used to calculate ℓ comes from the random thermal velocity, not the
drift velocity v_d. On the other hand, the mean time between collisions,
τ, in Eq. (14) was introduced to calculate the average drift velocity $\langle v_d \rangle$
in Eq. (10). It is important to understand that even in the absence of an
electric field, the time it takes the electron to travel from one collision
site to another will depend on how rapidly it is moving, that is to say,
on its random thermal velocity which it normally possesses by being in
equilibrium with its environment. The increase in this thermal velocity,

namely, the drift velocity, when an electric field is turned on will thus depend on how much time the electron has to accelerate in the field before it collides with a lattice atom. Equation (10) relates these facts. In the presence of an electric field, Eq. (14) should properly use the sum of the thermal velocity and the drift velocity. The drift velocity, however, is many orders of magnitude smaller than the thermal velocity, as can be seen by putting some numbers into Eq. (10). Taking $E = 100$ V/cm, converting to esu units, and using $\tau = 2.5 \times 10^{-14}$ sec, we get

$$\langle v_d \rangle = \frac{(4.8 \times 10^{-10})(100)(2.5 \times 10^{-14})}{(9.11 \times 10^{-28})(300)}$$

$$= 4.39 \times 10^3 \text{ cm/sec} \quad ,$$

as compared with the thermal velocity of 1.16×10^7 cm/sec.

The other point to notice about the mean free path is its size. At first thought one would expect that the mean free path should be close to the distance between atoms in a copper lattice which is 2.54 Å. The calculated mean free path of 29 Å is much larger than this and implies that the electron has traveled about 10 unit cells before being scattered. This is not understandable in terms of the free electron gas model. We will have to wait until we discuss the band theory of solids before we can see the reason for this large ℓ.

How well does the Drude theory predict the temperature dependence of the conductivity of metals? In Eq. (13) the only quantity which can vary with temperature is the time between collisions, τ. How does τ change with temperature? If the distance between collisions remains fixed, then as the temperature increases the random thermal velocity will increase according to Eq. (15). Increasing the thermal velocity will thus decrease the time between collisions. Since σ is proportional to τ, the conductivity of metals will decrease with increasing temperature or the resistivity will increase with increasing temperature, which is what is observed experimentally and shown in Fig. 3(a). Thus, the

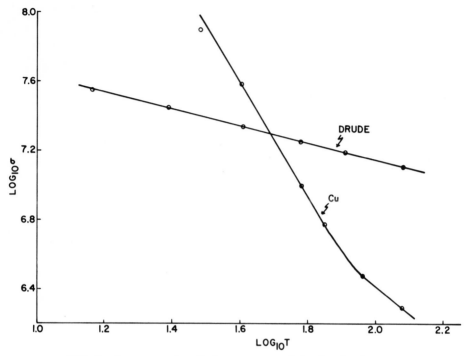

FIG. 5. $\log_{10} \sigma$ vs log T for copper and for Drude theory.

Drude theory qualititively predicts the right direction for the temperature dependence.   However, it is not quantitative.   Substituting Eqs.
(14) and (15) into Eq. (13), we get for the temperature dependence of
metals

$$\sigma = \frac{ne^2 \ell}{mv} = \frac{ne^2 \ell}{(3km)^{\frac{1}{2}} T^{\frac{1}{2}}}$$

or, for the resistivity,

$$\rho = \frac{(3km)^{\frac{1}{2}}}{ne^2 \ell} T^{\frac{1}{2}} \quad . \tag{16}$$

Thus, the resistivity should increase as $T^{\frac{1}{2}}$.   Figure 5 shows a plot of
log σ vs log T   for Eq. (16) and for the experimental data for copper.
The agreement is very poor, indicating that the Drude model while
qualitatively good is incorrect in some essential points.   We will see

that the band theory will give much better results. However, we will
also see that the form of the conductivity equation (13) will still remain
the same. The difference will be in how the time between collisions is
related to temperature. Equations (13) and (10), however, are the essential equations in relating conductivity to the physical and chemical
processes which take place in solids. We can combine these two equations into one statement by defining a new term, the electron mobility
$\mu = \langle v_d \rangle / E$. Equation (10) then becomes

$$\frac{\langle v_d \rangle}{E} = \mu = \frac{|e| \tau}{m} \quad .$$

Substitution of μ into (13) gives

$$\sigma = ne\mu \quad . \tag{17}$$

The units of μ can be determined if we take the units of $\langle v_d \rangle$ to
be cm/sec and that of E to be V/cm, then

$$\mu \equiv \frac{cm/sec}{V/cm} \quad \text{or} \quad \frac{cm^2}{V\text{-}sec} \quad .$$

These are the practical units in which mobility is most often expressed. As an example of the size of μ, consider once more copper
with $\sigma = 5.98 \times 10^5 \, \Omega^{-1} \, cm^{-1}$, $n = 8.46 \times 10^{22}/cm^3$ and $e = 1.6 \times 10^{-19} C$.
Then

$$\mu = \frac{\sigma}{ne} = \frac{5.98 \times 10^5}{(8.46 \times 10^{22})(1.6 \times 10^{-19})}$$

or

$$\mu = 44.2 \, cm^2/V\text{-}sec \quad .$$

Keeping in mind that the mobility is proportional to the time between collisions and, hence, proportional to the mean free path, we
present for observation in Table 2 the mobility of a variety of materials.
It can be seen that mobilities can be extremely small or extremely large.
The Drude model cannot account for such large variations in μ. Once
again we will have to appeal to the band theory of solids for an explanation.

TABLE 2

Room Temperature Mobility μ for Various Materials

Material	μ (cm^2/V-sec)
InSb	65,000
n-Si	240
Cu-phthalocyanine	70
Copper	47
$K_{0.30}MoO_3$	30
CdF_2-0.1 mole % TbF_3	3.5
Hemoglobin	2
$Na_{0.33}V_2O_5$	0.2

Finally, it is to be noted that the Drude model does not predict the temperature coefficient for a semiconductor. Equation (17) indicates that conductivity depends on two quantities; the density of charge carriers and their mobility. If the mobility for a semiconductor varies as some inverse power of T, then, to account for the increasing conductivity with increasing temperature of a semiconductor, n in Eq. (17) must be increasing very rapidly with rising temperature. This could occur if, by some mechanism, the charge carriers were all "frozen" out of the electron gas at 0°K and were released as the temperature was raised. We will again have to appeal to the band theory of solids in order to understand this property.

Before we do, however, it will be useful to describe two other experiments that allow us to independently determine the density of charge carriers, their sign (whether electrons or holes), and their mobility. Earlier, we described how to determine electrical conductivity by measuring the resistance of a specimen. To determine the mobility we then calculated the charge carrier density by assuming each atom donates one electron to the gas. Given σ and n, we then used Eq. (17) to calculate the mobility.

In conjunction with conductivity measurements, the charge carrier density, the sign of the carriers, and their mobility can be determined by an experiment known as the Hall effect, after its discoverer E. H. Hall in 1879 at The Johns Hopkins University. We will again appeal to the free electron gas model to show how the experiment works.

If we place a rectangular parallelepiped of our metal, which contains the free electron gas in a magnetic field, as shown in Figs. 6(a) and 6(b), then the charge carriers in any electric current, which flows perpendicular to the direction of the magnetic field, will experience a Lorentz force given by

$$\underline{F}_x = e\underline{v} \times \underline{B} \quad . \tag{18}$$

Here $\underline{v}$ is the drift velocity of the charge carriers due to the electric field and $\underline{B}$ is the magnetic induction. $\underline{v} \times \underline{B}$ is then a vector perpendicular to both $\underline{v}$ and $\underline{B}$. If the charge carriers are positive (as they would be for holes), the velocity vector will be in the same direction as the electric field that produced the current and the force will be in the direction indicated in Fig. 6(a). The effect of the magnetic field is to force an accumulation of positive charge on one side of the crystal. This accumulation does not build up indefinitely, since the positive charge on one side produces an electric field whose force eE_x will eventually just balance the Lorentz force F_x when enough charge has accumulated on one side. At equilibrium,

$$eE_x = F_x = evB \quad .$$

Since v is the drift velocity v_D, this is related to the current density $J = nev_D$. Making the substitution we get

$$eE_x = \frac{JB}{n} \quad . \tag{19}$$

If the width of the crystal in the direction of E_x is w, the voltage V_H that one measures along this direction is $E_x w$. If the cross-sectional area perpendicular to the current is A, then the current $I = JA$. Making these substitutions into Eq. (19) gives

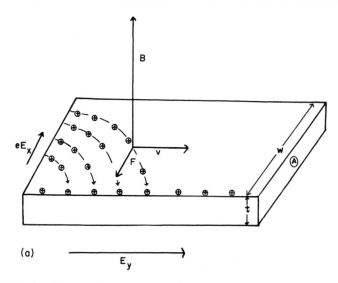

(a)

FIG. 6.   Effect of magnetic induction $\underline{B}$ on charge carriers drifting with velocity $\underline{v}$ in an electric field $\underline{E}_y$.   (a) Effect of Lorentz force $\underline{F}$ on holes.

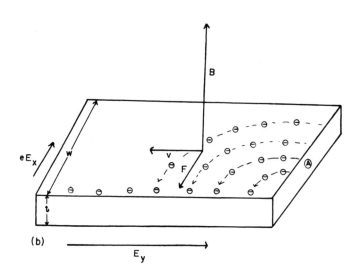

(b)

(b) Effect of Lorentz force on electrons.   Notice that electrons collect on the <u>same</u> side as holes do in (a).

$$\frac{eV_H}{w} = \frac{IB}{nA}$$

or

$$V_H = \frac{wIB}{neA} \quad .$$

The ratio A/w is simply the thickness of the sample, t, in the direction of B:

$$V_H = \frac{IB}{tne} \quad . \tag{20}$$

V_H, I, B, and t can be determined directly from experiment; hence, n can be calculated.

In deriving Eq. (20) it was assumed from the start that the charge carriers were positive. If the charge carriers are negative, then the equation for the Lorentz force changes sign:

$$\underline{F} = -e\underline{v} \times \underline{B} \quad .$$

Also the velocity vector of the carriers will now be in the opposite direction to the applied electric field, as indicated in Fig. 6(b). (This is a general result; positive particles travel in the same direction as the electric field, negative particles travel in the opposite direction to the electric field.) The effect of changing the sign of Eq. (18) and changing the direction of the velocity vector is to leave the force vector pointing in the same direction as before except now the force is on negative charges which therefore accumulate on the same side that the positive charges did before. This is shown in Fig. 6(b). Again electrons will accumulate on one side until the force produced by the electric field $-eE_x$ just balances the Lorentz force:

$$-eE_x = -e\underline{v} \times \underline{B} = -ev_D B \quad .$$

For electrons the current density is

$$J = -nev_D \quad ,$$

the minus sign because e is now negative. Hence

$$E_x = -\frac{JB}{ne} \; .$$

Going through the same substitutions for E_x and J as before gives

$$V_H = -\frac{IB}{net} \; . \tag{21}$$

As can be seen from Eqs. (20) and (21) as well as in Figs. 6(a) and 6(b), the Hall voltage is not only a measure of the charge carrier density but also gives the sign of the charge carriers as well.

In order to apply Eq. (20) or (21) we have to be careful about the units that are used. The derivation was carried out in the mks system. However, for practical purposes one experimentally measures V_H in volts, I in amperes, B in gauss, t in centimeters, and expresses e in coulombs and n in electrons/cm^3. Equations (20) and (21) are then correct for these mixed units only if the right side is multiplied by 10^{-8}. Thus, the Hall voltage in mixed units is

$$V_H = \pm \frac{IB}{net} \times 10^{-8} \; . \tag{22}$$

The quantity $1/ne$ is called the Hall coefficient R and is expressed in units of cm^3/coulomb.

In order to get a feeling for the size of the Hall voltage, let us consider what it would be for a copper crystal whose thickness is 0.01 cm. We will assume a current of 20 mA, a magnetic field of 10,000 G, and a charge carrier density as calculated earlier of 8.46×10^{22} electrons/cm^3. Then the Hall voltage will be

$$V_H = \frac{(20 \times 10^{-3})(10^4)}{(8.46 \times 10^{22})(1.6 \times 10^{-19})10^{-2}} \times 10^{-8}$$

$$= 14.7 \times 10^{-9} \; V \; .$$

This is a rather small voltage and is the kind of result one usually experiences in solid state chemical systems. In samples like germanium or silicon where the carrier density can be reduced to values below $10^{15}/cm^3$, Hall voltages on the order of millivolts can be observed.

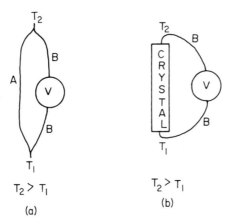

FIG. 7.   (a) Seebeck voltage V for two dissimilar metals A and B connected at their ends and held at different temperatures T_1 and T_2.

(b) Wire A replaced by crystal of material.   Seebeck voltage V measured by keeping the ends of the crystal at different temperatures.

For reasons to be discussed later, the Hall effect is one of the more difficult measurements to make, but not only because of its smallness.

Another method for getting the sign of the charge carriers, which is easier than measuring the Hall effect but does not yield as much immediately useful information, is the Seebeck effect measurement (also called the thermoelectric power).   This experiment makes use of the fact that, when two dissimilar metal wires are connected at their ends and these ends are then held at different temperatures, a voltage develops which is proportional to the temperature difference between the ends, as shown in Fig. 7(a).   Such a device is called a thermocouple and can be used as an accurate means to measure unknown temperatures. To use the Seebeck effect to measure the sign of charge carriers in a sample, wire A in Fig. 7(a) is replaced by the sample to be measured, as shown in Fig. 7(b).   The two ends of the sample are kept at different temperatures (which need not be known if only the sign of the charge carriers is wanted), and the voltage due to the Seebeck effect is then measured.   Appealing to the free electron gas model again, if the warmer end of the sample is at T_2, the electrons at this end will tend to drift down to the cooler end, just as molecules in a container tend to drift toward the cooler side of the container if one side is heated. Just as in the Hall effect, the electrons that drift to the cooler end will set up an electric field whose size will oppose any further buildup of charge density at the cooler end. If the voltage is now measured, say, with a potentiometer or sensitive electrometer, the sign of the cold end will be negative. On the other hand, if the sign of the cold end is positive, then the charge carriers consist of holes. Thus, by simply measuring the polarity of the Seebeck effect the kind of charge carrier in a material can be determined. This argument, of course, depends on the fact that the carriers that drift from the hot end to the cold end in the measuring circuit B produce a much smaller Seebeck voltage than in A. This is usually true when A is a semiconductor where the carrier density is much less than in metals but is not true if A is a metal. Care must be exercised in interpreting the Seebeck effect if A is a metal. Corrections for the measuring circuit must be taken into account.

The Seebeck coefficient S is defined as

$$S = \frac{\Delta V}{\Delta T} \tag{23}$$

in the limit as ΔT goes to zero. Here ΔV is the difference in potential between the hot and cold junctions and ΔT is the temperature difference between the two junctions. After we discuss electron and hole statistics, we will see that the Seebeck coefficient for a semiconductor is proportional to the reciprocal of the charge carrier density. Thus, as the charge carrier density goes up, the Seebeck coefficient goes down. Typical values for S vary from a few $\mu V/°K$ in the case of metals like copper to several $mV/°K$ for semiconductors like germanium.

Up to this point we have introduced much of the experimental measurements that are used to measure electrical properties of solids and have discussed them in a very oversimplified fashion in terms of the classical free electron gas. Nevertheless, we have been able to derive most of the equations that are essential in understanding electrical properties, results which are still correct even in the band theory. We must, however, furnish arguments for the existence of semiconductors and introduce chemical models that will allow us to interpret the electrical properties of solids in terms of the chemical bonds which make up all materials. We shall see that the sophisticated band theory approach of the physicists can be interpreted in terms of molecular orbital concepts familiar to most chemists.

III. QUANTUM MECHANICAL FREE ELECTRON GAS

The difficulty with the classical free electron gas model is its inability to explain the long mean free path that is predicted as well as the temperature dependence of metals. The classical theory predicts a $T^{-\frac{1}{2}}$ dependence. Experiment shows a T^{-1} dependence for temperatures greater than the Debye temperature. It also offers no explanation for the difference between metals, semiconductors, and insulators nor for the existence of holes as charge carriers. It can, however, explain the existence of Ohm's law and allows us to introduce concepts like mean free path, collision frequency $(1/\tau)$, drift velocity, and thermal velocity. It also shows qualitatively that the conductivity of a specimen is proportional to the product of the charge carrier density and the charge carrier mobility.

We might improve on this picture by applying quantum mechanics rather than classical mechanics to the gas. We know that the energy levels of an atom are quantized when they are constrained to move in a coulomb field producing s, p, d orbitals, etc., and that no more than two electrons can be in any one orbital, one with spin up and the other with spin down. Similarly in a box, we would expect that the electron energies are also quantized due to the constraints of the walls of the box, so that, applying Pauli's exclusion principle, no more than two

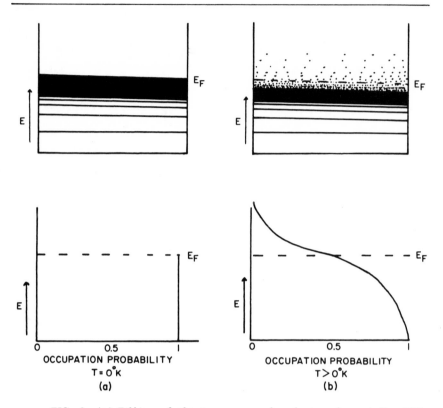

FIG. 8. (a) Filling of electron energy levels in a box at T = 0 °K according to quantum mechanics. The highest filled level E_F is called the Fermi level. Below E_F, all states are filled; above E_F, the occupation probability is zero.

(b) Electron distribution in a box for T > 0 °K; states just above and below the Fermi level are partially occupied.

electrons can be in any one state of the box, one with spin up and the other with spin down. Thus, to make a quantum mechanical free electron gas, we place electrons in a box, filling the lowest energy levels first and at the same time obeying the exclusion principle until we have added as many electrons as desired. This is to be compared with the classical free electron gas in which the electrons can assume continuous energy values with many electrons having the same energy. Thus, the statistical distribution of the electrons in a quantum mechanical gas will be quite different from that in a classical gas. The classical gas is said to follow Boltzmann statistics, the quantum mechanical gas follows Fermi-Dirac statistics.

Figure 8(a) shows the filling of the box for a quantum mechanical free electron gas, taking the exclusion principle into account. At absolute zero all the energy levels below E_F must be filled, all those above E_F must be empty. The energy E_F, below which the probability of electron occupation is unity and above which the electron occupation is zero, is called the Fermi level.

It would be expected that this picture of an electron gas would have different physical consequences than the classical picture. In many respects this is true, especially when the model is applied to specific heat measurements and magnetic susceptibility (see Chapter 5). In the case of electrical conductivity, the model leads to the same result as the classical gas. Resistance is still determined by the scattering of electrons with lattice atoms just as in the classical case with the one difference that now only those electrons near the Fermi level can be scattered. This can be seen by realizing that, when an electron is placed in an electric field, it increases its energy due to the field but loses this energy upon being scattered. If we look at a quantum mechanical gas above $0\,^\circ K$, we see as in Fig. 8(b) that some of the electrons near the Fermi energy have been thermally excited across the Fermi level, so that now there will be empty states both above and below the Fermi energy. Thus, electrons near the Fermi level can be scattered into slightly lower energy levels. Deep down in the "Fermi sea," however, all the energy levels are still occupied and, thus, there is no possibility for an electron to lose energy by scattering since there are no empty energy levels available to it. Since only those electrons near the Fermi energy can be scattered, the time between collisions will thus refer to these electrons. Thus, τ becomes τ_F. The thermal velocity used to calculate the mean free path

will be for those electrons near the Fermi energy. Hence, Eqs. (13) and
(14) will be modified to read

$$\sigma = \frac{ne^2\tau_F}{m} \tag{24}$$

$$\ell = v_F\tau_F \, . \tag{25}$$

One may ask why doesn't the density of charge carriers in Eq. (24)
become the density of carriers only at the Fermi level since these are
the only ones that are scattered and, hence, contribute to the resistance. The reason is that, when the gas is placed in an electric field,
all the electrons, even those deep down below the Fermi level, increase
their energy, but they do so coherently without violating the exclusion
principle, and, hence, have a net drift velocity which contributes to the
total current. It is those electrons near the Fermi level which, upon
being scattered, prevent the dense mass of electrons below them from
increasing their drift velocity indefinitely. It is analogous to people
moving out of a crowded theatre. The people moving at the doors (the
electrons at the Fermi level) govern the rate (the drift velocity) at which
the people in the rest of the theatre can move.

The calculation of τ_F from σ is thus the same as before. However,
the value of v_F to be used in Eq. (25) must now be determined from E_F
and not from the Boltzmann distribution of Eq. (15).

It should be mentioned parenthetically that, if one applies the
Boltzmann distribution to the quantum mechanical gas by neglecting the
exclusion principle, one gets for the average thermal energy $E = (3/2)kT$,
just as in the classical gas. Thus, it is not simply the introduction of
quantized energy levels that changes the average electron energy but
the explicit exclusion of more than two electrons per quantum state.

For periodic boundary conditions, the quantized energy levels for
a particle in a three-dimensional box whose side has length L has been
shown in Chapter 3 to be

$$E = \frac{h^2}{2mL^2} (n_x^2 + n_y^2 + n_z^2) \quad , \tag{26}$$

where n_x, n_y, and n_z are integer quantum numbers with values 0, ±1, ±2, ... Thus, a set of three numbers specifies an energy level. Together with the spin quantum number $m_s = ±\frac{1}{2}$ there will be four quantum numbers which specify the state of the electron. Just as in an atom, no two electrons in the box can have the same four quantum numbers, which is a simple statement of Pauli's exclusion principle. Notice, however, that different sets of quantum numbers can give the same energy, which means that there can be many states with the same value of E, each state will have two electrons, one with spin up ($m_s = +\frac{1}{2}$) and one with spin down ($m_s = -\frac{1}{2}$). For example, if we use the quantum numbers $n_x = 1$, $n_y = 0$, $n_z = 0$ or abbreviating (100), this state has the same energy as (010), (001), (-1, 00) (0, -1, 0), (0, 0-1). Thus, there are six states with the energy $h^2/2mL^2$; each of these states can hold two electrons. Obviously, as the quantum numbers get large, the number of states goes up with increasing energy. When we put electrons in the box, the first electron goes into the state represented by the three integers (0, 0, 0). The second electron also goes into this state but with opposite spin. The third electron goes into any one of the six states mentioned and so do the fourth through twelfth electrons. Filling is continued in this manner until all the electrons have been placed in states. The top state will have the Fermi energy E_F so that

$$E_F = \frac{h^2}{2mL^2} (n_{x_F}^2 + n_{y_F}^2 + n_{z_F}^2) = \frac{h^2 n_F^2}{2mL^2} \tag{27}$$

where n_{x_F}, n_{y_F}, and n_{z_F} are the quantum numbers of the electrons at the Fermi level and $n_F^2 = n_{x_F}^2 + n_{y_F}^2 + n_{z_F}^2$. An equation like (27) has the same form as that for a sphere in Cartesian coordinates $r^2 = x^2 + y^2 + z^2$, with r equal to the radius of the sphere. The radius of our Fermi sphere is thus

$$r = \left(\frac{2mL^2 E_F}{h^2} \right)^{\frac{1}{2}} = n_F \quad ,$$

and the volume of this sphere is

$$\frac{4}{3}\,\pi\,n_F^3 \;=\; \frac{4}{3}\,\pi\left(\frac{2mL^2 E_F}{h^2}\right)^{\frac{3}{2}} \;.$$

Since the values of n_x, n_y, and n_z change by one unit at a time, the volume occupied by any one state is just one. Hence, the total number of electrons located within the Fermi sphere is twice the volume since each state can have two electrons, one with spin up and one with spin down.

The total number of electrons occupying states up to the Fermi level is then

$$2\cdot\frac{4}{3}\,\pi\left(\frac{2mL^2 E_F}{h^2}\right)^{\frac{3}{2}} \;.$$

This many electrons must therefore equal the density of electrons, n, in the gas times the volume of the gas, L^3:

$$nL^3 \;=\; \frac{8\pi}{3}\left(\frac{2mL^2 E_F}{h^2}\right)^{\frac{3}{2}}$$

or

$$E_F \;=\; \frac{h^2}{8m}\left(\frac{3n}{\pi}\right)^{\frac{2}{3}} \;. \tag{28}$$

For copper, $n = 8.46 \times 10^{22}$ electrons/cm^3, $h = 6.63 \times 10^{-27}$ erg-sec, and $m = 9.11 \times 10^{-28}$ g. Therefore,

$$E_F = 11.27 \times 10^{-12}\ \text{erg} = 7.05\ \text{eV} \;.$$

Compare this to kT at a room temperature of 300 °K, which is 0.025 eV. We see that the electrons in a Fermi gas have considerably more energy than those in a classical gas. The velocity to be used to calculate the mean free path is then

$$v_F = \left(E_F \cdot \frac{2}{m}\right)^{\frac{1}{2}} = 1.57 \times 10^8\,\frac{\text{cm}}{\text{sec}} \;,$$

which is considerably larger than that calculated from the Boltzmann distribution in Eq. (15).

Using this value of v_F and $\tau = 2.5 \times 10^{-14}$ sec, we get for the mean free path

$$\ell_F = 393 \,\text{Å} \quad ,$$

which is even longer than that calculated by classical statistics. The quantum mechanical gas model can offer no explanation for this long mean free path nor, like the classical gas model, does it offer an explanation for the existence of holes and semiconductors.

IV. BAND THEORY OF SOLIDS

As long as we insist that the scattering of the free electrons in a crystal occurs by collision with the positive ions, we will always be led to a result for σ which has the wrong temperature dependence and which cannot account for the long mean free paths. The long mean free paths hint at the fact that the previously proposed scattering mechanisms must be incorrect. We can improve on the free electron gas model and at the same time account for the existence of semiconductors in a qualitative way by realizing that electrons have a wave-like property as well as a particle-like property connected by the de Broglie relation:

$$p = \frac{\hbar}{\lambda} = \hbar k \quad , \tag{29}$$

where k, called the wave vector, is defined as $2\pi/\lambda$.

If we go back and look at the equation for the quantized energy levels in a box,

$$E = \frac{h^2}{2mL^2} (n_x^2 + n_y^2 + n_z^2) \quad , \tag{30}$$

we see that this equation has the same form as that for a classical free electron with momentum p:

$$E = \frac{p_x^2 + p_y^2 + p_z^2}{2m}$$

provided we equate n_x^2 with $L^2 p_x^2 / h^2$ and similarly with y and z. Then

$$n_x = \frac{L p_x}{h} = \frac{L k_x}{2\pi}$$

or

$$k_x = \frac{2 n_x \pi}{L}$$

$$k_y = \frac{2 n_y \pi}{L} \qquad\qquad (31)$$

$$k_z = \frac{2 n_z \pi}{L} \; .$$

Thus, in terms of the components of the wave vector, the energy levels in a box become, by substitution of Eqs. (31) into (30),

$$E = \frac{h^2}{8\pi^2 m}(k_x^2 + k_y^2 + k_z^2) = \frac{h^2 k^2}{8\pi^2 m} \; . \qquad\qquad (32)$$

Thus, as we add electrons to our box, the values of k_x, k_y, and k_z will increase, meaning that the wavelength of the electrons will be getting shorter and shorter. For example, we can calculate what the wavelength for an electron will be at the Fermi level in copper:

$$E_F = \frac{h^2 k_F^2}{8\pi^2 m} \; . \qquad\qquad (33)$$

Using the value $E_F = 11.27 \times 10^{-12}$ erg obtained from Eq. (28), we get $k_F = 13.5 \times 10^7$ cm^{-1}, and therefore

$$\lambda = \frac{2\pi}{k_F} = 4.65 \times 10^{-8} \text{ cm} = 4.65 \, \text{Å} \; .$$

This value of λ is considerably longer than the spacing of 2.09 Å between planes of copper atoms. From the Bragg theory of x-ray scattering in periodic structures, we know that an x-ray will pass unimpeded through a crystal without scattering provided $n\lambda \neq 2d \sin \theta$, where θ is the angle of incidence and d is the spacing between the atomic planes. If the angle of incidence is 90°, then no scattering takes place for $n = 1$ if

$$\lambda > 2d .$$

The conclusion we come to for free electrons in a metal like copper is that the wavelength of the electrons at the Fermi surface is much longer than twice the copper planar spacing of $2(2.09 \text{ Å}) = 4.18 \text{ Å}$ and, thus, no scattering takes place. Thus, to a first approximation, metals should have zero resistance. (This, however, is not the origin of superconductivity.) Although this is a rather surprising result, it is a consequence of the wave-like nature of the electrons. The origin of the resistance comes from the deviation from strict periodicity due to thermal vibrations of the atoms, presence of impurities, defects, and dislocations. Thus, the electrons at the Fermi level are scattered not by the presence of the atomic lattice but by the deviations of this lattice from strict periodicity. This will be true provided the electron's wavelength λ is greater than 2d.

If we continue to add more electrons to our gas (say, by increasing the density n), a strange thing happens. As the electron density increases, the Fermi level rises according to Eq. (28) and, thus, larger k vectors and, hence, smaller electron wavelengths appear. As the wavelength gets shorter, it will approach 2d, and at the value $\lambda = 2d$, corresponding to the first Bragg reflection $k = \pi/d$, the electron wave will be totally reflected and, thus, will not be transmitted through the crystal. Thus, Eq. (33), which represents the kinetic energy of a traveling wave, becomes invalid when k approaches the value π/d. At the point of total reflection, the reflected wave will be traveling in

a direction opposite to that of the initial wave. The combination of the
two waves will produce a standing wave which has no net velocity in
any direction. It should be clear that, depending on the value of θ and
n , there will be a whole series of k values for which the Bragg condition

$$n\lambda = 2d \sin \theta \tag{34}$$

will hold. The actual values of λ (and hence k) for which reflection
occurs will, of course, depend on the crystal structure. Without going
into the details, suffice it to say that those k values for which the
Bragg conditions holds form surfaces in k space, called Brillouin zones.

The effect of these Bragg reflections is to cause a discontinuity
in the energy of the electron at the surface of the Brillouin zone. If the
electron density is high enough to just fill all the states in the first
Brillouin zone (n = 1), there will be no higher empty states available
for conduction of sufficiently low energy at the Brillouin zone boundary
when an electric field is turned on. The states inside the Brillouin
zone boundary form an energy band. Figure 9(b) shows a sketch of the
filled energy band separated by an energy gap from the states that occur
in the second Brillouin zone (n = 2).

If the energy gap is large, only a few electrons will be thermally
excited across the gap and, hence, the conductivity will be low.
[Remember from Eq. (17) that the conductivity depends on the mobility
and the electron density.] That is to say, n in Eq. (17) will be small.
If the gap is small, then thermal excitation will be easier and, hence,
n can become large. As we raise the temperature, more and more electrons will be thermally excited across the gap and, thus, n will increase
with temperature. In fact, n follows something like the Boltzmann distribution $e^{-\Delta E/2kT}$, where ΔE is the energy gap and, thus, as T
increases, n increases exponentially. We can now appreciate the distinction between pure semiconductors and metals. In a metal, the k
states in the Brillouin zone are only partially filled with many empty

states just above the Fermi level available for conduction. The density of charge carriers is therefore constant and independent of temperature. The temperature dependence of the conductivity is therefore governed by the temperature dependence of the mobility which decreases with increasing temperature like T^{-1} for temperatures greater than the Debye temperature. For a semiconductor (or insulator), the electron k states are filled right up to the Brillouin zone boundaries, that is, up to the point where the Bragg condition for reflection produces standing waves. There is an energy gap at this boundary so that there are no available energy states just above it and, hence, no conduction. The conductivity is therefore governed by n rather than μ. Since n increases exponentially with temperature, so will σ. Table 3 lists the energy gaps for a number of semiconductors. Experimentally, these energy gaps can be obtained in at least two ways. If n is proportional to $e^{-\Delta E/2kT}$ and μ varies only as T^{-n}, say, then

$$\sigma \propto T^{-n} e^{-\Delta E/2kT} \quad . \tag{35}$$

The exponential term will dominate this expression if ΔE is large compared to $2kT$, so that a graphical plot of $\ln \sigma$ vs $1/T$ will look like

TABLE 3

Intrinsic Energy Gaps ΔE for Various Semiconductors

Material	ΔE (eV)
Diamond	5.3
V_2O_5	2.17
Si	1.2
Ge	0.79
InSb	0.27
$Na_{0.9}WO_3$ (metallic)	0

a straight line with slope $\Delta E/2k$. ΔE can also be determined by optical excitation, to be described in Chapter 7.

The presence of an energy band gap also explains why the mobility of electrons in a semiconductor is much higher than in a metal. In a semiconductor, electrons that have been excited across the gap occupy those states of Eq. (32) with low value of k (k near zero). The energy and, hence, the thermal velocity of these electrons are very low. In a metal, however, the electrons near the Fermi level have large k values and, thus, very high thermal velocities. Since

$$\mu = \frac{e\tau}{m} = \frac{e\ell}{mv} \quad,$$

the lower thermal velocities v in semiconductors implies a higher mobility than that for a metal.

V. SCATTERING MECHANISMS IN METALS

We have shown that, for a metal, the change in conductivity with temperature is governed by what happens to the mobility as a function of temperature. The mobility will be limited by how the electron waves interact with the departures from perfect periodicity. Departures from perfect periodicity, as has already been mentioned, can be the result of a number of factors, some of which have already been discussed in Chapter 2. One of these factors is atoms vibrating about their equilibrium points. From the classical theory of atomic vibration (T > Debye temperature Θ), the square of the amplitude of the vibrations and, hence, their energy increase as $(3/2)kT$. Simple kinetic theory arguments predict that the mean free path for scattering will be inversely proportional to the area occupied by the scatterer. The larger the area, the shorter the mean free path for a given density of scatterers. Now if the square of the vibrational amplitude of an atom increases proportional to T, then the effective scattering area for the electron wave will increase proportional to T. According to Eqs. (24) and (25),

$$\sigma = \frac{ne^2 \ell_F}{mv_F} \tag{36}$$

so that σ is proportional to the mean free path. We therefore expect the mean free path to decrease with increasing temperature as T^{-1}, and thus σ decreases as T^{-1} for $T > \Theta$, which is what is observed experimentally. This argument is based on classical ideas about scattering and is valid provided the temperature is greater than the Debye temperature. Notice that, in determining the temperature dependence of σ, we assumed that the velocity of the electrons v_F in Eq. (36) does not change with increasing temperature. This is the case because the velocity at the Fermi level is determined by the Fermi energy E_F, which is already very large as shown by the calculations in Eq. (28) and does not change very much with temperature. Thus, v_F will be essentially constant and the temperature dependence of σ will be governed by the mean free path ℓ_F. On the other hand, for semiconductors we shall see that v_F is proportional to $T^{-\frac{1}{2}}$ with ℓ_F still proportional to T^{-1}, so that σ will be proportional to $T^{-\frac{3}{2}}$.

For temperatures less than the Debye temperature, the conductivity of metals goes as T^{-5}, and at very low temperatures, the conductivity is temperature independent.

If we try to apply the same argument for $T < \Theta$ as was used for high temperatures, we run into difficulty. For $T < \Theta$, Debye has shown that the vibrational specific heat is proportional to T^3. Thus, we might expect that the mean free path for electrons at the Fermi level will go as T^{-3} and hence $\sigma \propto T^{-3}$. However, this argument assumes that the vibrations act as perfect scatterers so that the electron loses all "memory" of its previous direction before the scattering takes place. This appears not to be the case. At low temperatures, the lattice vibrations are not as effective in scattering electrons as they were at high temperatures, so that it takes many collisions before the electron loses all memory of its forward momentum. It can be shown that this multiplies the mean time between collisions by T^{-2}, so that

$$\sigma = \frac{ne^2 \tau_F T^{-2}}{m} .$$

Since τ_F is still ℓ_F/v_F, and $\ell_F \propto T^{-3}$ for $T < \Theta$, we thus have $\sigma \propto T^{-5}$. At still lower temperatures, the lattice scattering is so weak that the conductivity is governed by scattering from impurities within the crystal. For a metal, scattering by impurities is constant and independent of temperature. Thus, the conductivity of metals at low temperatures is temperature independent. (This is not true for semiconductors, where scattering by impurities goes as $T^{\frac{3}{2}}$.) In fact, if we put a large amount of impurity in a metal, for instance, by making an alloy of two metals, then it is possible for the impurity scattering to totally dominate in the conductivity expression. β-Brass, for instance, which is an alloy of 32% Zn – 68% Cu, has a room temperature conductivity of approximately $14 \times 10^4 \, \Omega^{-1} \, cm^{-1}$, which remains essentially constant down to 4.2 °K.

VI. DYNAMICS OF ELECTRON MOTION

In discussing the quantum mechanical gas we showed that the energy was related to the electron's wavelength by Eq. (33):

$$E = \frac{h^2 k^2}{8\pi^2 m} ,$$

with $k = 2\pi/\lambda$. A plot of E vs k will thus have the shape of a parabola, as shown in Fig. 9(a). As the value of k increases, we know, however, that the above equation will break down because the electrons are not entirely free but are reflected when the Bragg condition is satisfied. This produces an energy gap at certain values of k, shown in Fig. 9(b). Notice in 9(b) that E vs k flattens out near the Brillouin zone boundary.

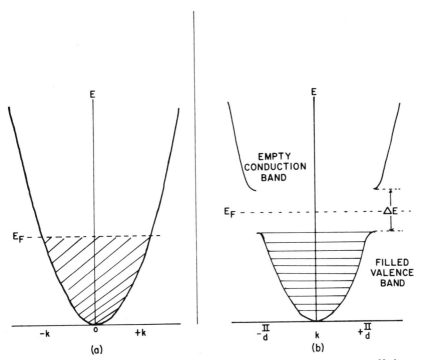

FIG. 9. (a) Parabolic form of E vs k. Electrons occupy all k states up to the Fermi energy E_F, as they would in a metal.

(b) Effects of Bragg reflection near Brillouin zone boundary showing splitting of the parabolic band and the creation of an energy gap ΔE. Conduction occurs only when electrons are thermally excited from the filled valence band across the energy gap into the conduction band. Note position of Fermi level.

This fact is related to the zero electron velocity that must occur at the boundary (a standing wave has no velocity). This can be shown by appealing to Eq. (33) for values of k less than the energy gap (which is identical to a classical electron with $p = \hbar k$).

For a classical electron, $E = p^2/2m$, so that its velocity is simply

$$v = \frac{dE}{dp} \quad .$$

For the quantum mechanical free electron, we know $p = \hbar k$. Therefore, the preceding equation becomes

$$v = \frac{1}{\hbar}\frac{dE}{dk} \quad . \tag{37}$$

We now see how the electron's group velocity is related to the E vs k curve. Thus, at values of k near the energy gap the velocity must be zero and, therefore, the slope of E vs k, dE/dk, must be zero. It should be emphasized that Eq. (37) was derived for a free electron in which $p = \hbar k$. However, it can be shown that (37) is valid regardless of whether an electron is free or not. [Near the energy gap $p = \hbar k$ is <u>not</u> valid. However, Eq. (37) is much more general than its derivation implies and still holds at the gap.]

For a free electron, the acceleration in an applied electric field is, by Newton's second law,

$$a = \frac{force}{mass} = \frac{eE}{m} \quad .$$

How does this acceleration relate to E and k in quantum mechanics? From Eq. (37) we can get the acceleration by differentiating with respect to time:

$$\frac{dv}{dt} = \frac{1}{\hbar}\frac{d}{dt}\left(\frac{dE}{dk}\right) = \frac{1}{\hbar}\frac{d^2E}{dk^2}\frac{dk}{dt} = a \quad . \tag{38}$$

Once again it should be emphasized that this equation holds regardless of whether an electron is free or not. If it <u>is</u> free, then we know that $p = \hbar k$ and, hence, we obtain

$$\frac{dk}{dt} = \frac{1}{\hbar}\frac{dp}{dt} = \frac{force}{\hbar} = \frac{eE}{\hbar} \quad .$$

Thus, $\hbar(dk/dt)$ can be equated with a force which means that, in Eq. (38), the quantity $(1/\hbar^2)d^2E/dk^2$ can be equated with $1/mass = 1/m^*$. The quantity

$$m^* = \frac{\hbar^2}{d^2E/dk^2}$$

is called the effective mass. It equals the true mass only for a completely free electron. Otherwise, m^* will be different from the true electron mass. We equate $\hbar^2/(d^2E/dk^2)$ with a mass because it allows us to use Eq. (33) as an approximation in regions of the E vs k graph where it begins to fail by simply replacing m by m^*, thus

$$E = \frac{h^2 k^2}{8\pi^2 m^*} \quad .$$

The effective mass is therefore a fudge factor which allows us to keep using our ideal equations even in those regions where the ideal equations would break down. (It is equivalent to the use of activities and activity coefficients in solution thermodynamics.) The reason for m^* deviating from m is due to the interaction of the electron with the periodic lattice of positive ions. This interaction increases as we get closer to the Brillouin zone boundaries. All the problems of describing this interaction (calculating wave functions and energies for various periodic potentials) are lumped into the effective mass (just as all interactions of ions in solution are lumped into the activity coefficients). We need not concern ourselves with calculating the effective mass in all its details, but we should realize that, in order to use the results derived from the free electron gas model, we must modify equations like (24), (28), and (32) by substituting m^* for m. Thus

$$\sigma = \frac{ne^2 \tau_F}{m^*} \quad , \tag{39}$$

$$E_F = \frac{h^2}{8m^*} \left(\frac{3n}{\pi}\right)^{\frac{2}{3}} \quad , \tag{40}$$

$$E = \frac{h^2 k^2}{8\pi^2 m^*} \quad . \tag{41}$$

VII. PROPERTIES OF THE EFFECTIVE MASS

Since the effective mass is inversely proportional to the second derivative of the E vs k curve, we would expect that large effective

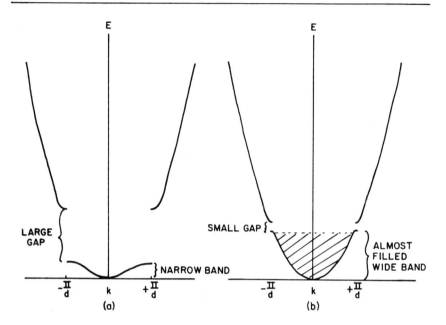

FIG. 10. (a) Narrow band with small band curvature and therefore large m^*. Such narrow bands usually have large energy gaps, as shown.

(b) Almost filled wide band with sharp curvature occurring near Brillouin zone boundary. Therefore m^* is very small and so is the energy gap (e. g., InSb, $m^* = 0.03\, m_0$).

masses ($m^* > m$) occur when the curvature of E vs k, that is, the curvature of the electron band, is very gradual (corresponding to a small value of $d^2 E/dk^2$). This situation usually occurs in materials that have narrow bands, as illustrated in Fig. 10(a). In Fig. 10(b), the band is very wide, so that most of the electrons have $m = m^*$ except near the top of the band where the curvature is very sharp, producing a large second derivative so that m^* is less than the free electron mass. As has already been shown in Chapter 3, the width of the band is related to the amount of overlap between the wave functions on neighboring atoms — the less the overlap, the narrower the bands. Narrow band materials with large m^* will have lower electrical conductivity than wide band materials with small m^*, according to Eq. (39). Thus, the electrical conductivity and related measurements are a measure of the amount and the kind of overlap of neighboring wave functions and thus become a means for determining the kinds of chemical bonds that can occur in solids. Examples of this will be provided at the end of this chapter along with more detailed results in Chapters 13 and 14.

VIII. NEGATIVE EFFECTIVE MASS

Looking at Fig. 10(b), we see that, near the top of the band, d^2E/dk^2 is negative, viz., the band is concave down. Since

$$m^* = \frac{\hbar^2}{d^2E/dk^2} \ , \tag{42}$$

we get the peculiar result that m^* is negative. We can understand this by realizing that, in an applied electric field, the electron waves near the top of the band tend to be Bragg reflected as they approach the band edge. Because of this Bragg condition, the reflected electron waves have opposite momentum of nonreflected electrons. Thus, in an applied electric field the electron changes its momentum by a negative amount on being reflected, therefore giving the appearance of having a negative mass. The negative effective mass simply takes into account the very strong interaction between electron and lattice upon reflection.

IX. HOLE CARRIERS

The negative effective mass concept also allows us to understand the idea of a hole, that is to say, the existence of positive charge carriers in solids as demonstrated by the Hall effect. In its simplest form, a hole is the absence of a negative effective mass electron. For example, in Fig. 10(b), the band is almost filled. Suppose there are only three empty negative effective mass states near the top. Then the Hall effect will show a _positive_ Hall coefficient corresponding to three positive charge carriers. Every empty negative effective mass state is equivalent to a hole of positive charge.

How do negative effective mass electron states act as positive charges? If the band is completely filled, then in an applied electric field, the current density is zero:

$J = 0$ for a filled band.

If we remove one negative effective mass electron at the top of the band, all the remaining electrons will have a current density equal and opposite to what the one removed electron would have since the sum of the current densities must be zero. The removed electron of charge e' would have a current density

$$\underline{J}' = -e'\underline{v}' = -e'\underline{a}'\tau \quad ,$$

where $\underline{a}'$ is the acceleration of this one electron in an applied field; and the minus sign is present because e' is negative. The remaining electrons in the band will, thus, have a current density $\underline{J}_{remaining}$ equal and opposite to J':

$$\underline{J}_{remaining} = +e'a'\tau \quad . \tag{43a}$$

Thus, the remaining electrons in the band have a current density equivalent to a single charged particle. If we removed two electrons from the band, the current density of the remaining electrons would be equivalent to two charged particles.

What is the direction of the acceleration a'?

$$\underline{a}' = \frac{e'E}{m^*} \quad . \tag{43b}$$

In Eq. 43(b) both e' and m^* are negative, hence, the acceleration in Eq. (43a) is in the _same_ direction as the electric field. Thus, Eq. (43a) says that the current density of all the remaining electrons left after one negative effective mass electron is removed is due to a single

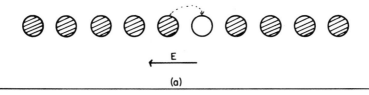

(a)

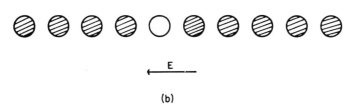

(b)

FIG. 11. (a) Row of atoms in an electric field E with one atom missing an electron (unshaded). This empty state is often called a hole.

(b) Electron from a neighboring filled state jumps into the empty state giving the appearance that a "hole" has moved under the influence of the electric field.

positive charge or hole of value $+e'$ whose acceleration a' is in the same direction as the electric field. This is just what one would expect for a positive charge with positive effective mass. We can thus call all empty negative effective mass electron states at the top of a band hole states with positive charge and positive effective mass.

It should be made clear that not all empty electron states act as holes. Only those are holes which would normally have negative effective mass electrons in them when occupied. The idea of a hole has, in many instances, been pushed beyond its definition. For example, if we have a row of atoms each having one valence electron except one atom, as depicted in Fig. 11(a), the empty state is often called a hole. If an electron from a neighbor jumps into the empty state, then the "hole" state appears to move, as seen in Fig. 11(b). It is then concluded that the row of atoms is a hole carrier. Such an analogy is very useful for describing a solid but can lead to confusion if used indiscriminately. For example, if the electron states in Fig. 11 are localized (viz., are not free as in a gas), then they are usually described as having positive mass. The empty site would normally be occupied by a positive mass electron. Thus, electrons jumping or hopping from site to site, although giving the appearance that the "hole" moves, would not show a positive Hall coefficient but a negative one since there are no empty negative effective mass states. The analogy of Fig. 11 also indicates that the "hole" moves or is accelerated in the opposite direction to the acceleration of the electrons. This is also not the case for a real hole conductor. As indicated by Eq. (43b), the electrons at the top of a band as well as the holes are accelerated in the same direction. This is because the electrons near the top of an almost filled band have negative effective mass.

To summarize, a positive Hall coefficient indicates the presence of empty electron states, which have negative effective mass. Negative effective mass is associated with the negative curvature of an electron band, and, thus, the existence of holes is intimately related to the presence of electron bands in solids. In the absence of negative effective mass states (such as in hopping conduction, examples of which will be given later), no positive Hall coefficient is expected.

X. THE ROLE OF IMPURITIES

The presence of holes as current carriers and the ability to control their number in solids, in addition to controlling the number of electrons, is at the heart of modern solid state electronic technology. As shown in Chapter 8, the controlled addition of impurities allows one to vary the hole content and electron content of a semiconductor. Figure 12(a) and 12(b) is a diagram showing the energy levels of electron donor impurities and electron acceptor impurities, respectively, in silicon. The addition of, say, $10^{16}/cm^3$ impurity atoms to pure silicon

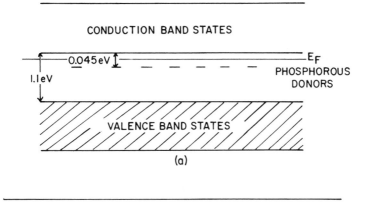

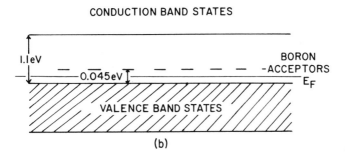

FIG. 12. (a) Position of phosphorus donor energy levels in the energy gap of silicon. Phosphorus has one more valence electron than silicon and, thus, can donate an electron to an empty band.

(b) Position of boron acceptors in the energy gap of silicon. Boron has one less electron than silicon and, thus, can accept an electron from a filled band. Note position of Fermi level for donors and acceptors.

produces extra energy levels in the energy gap, which lie near a filled band (called the valence band) if the impurity is boron (with one less electron than silicon) or near an empty band (called the conduction band) if the impurity is phosphorus (with one more electron than silicon). In the case of boron, the difference in energy between the top of the filled valence band and the boron energy levels is small enough so that, at room temperature, electrons from the filled valence band can be thermally excited into the localized boron energy states. This leaves some empty states, in the otherwise filled valence band, which can now act as holes. Similarly, for the case of phosphorus, the difference in energy between the bottom of the empty conduction band and phosphorus levels is small enough so that thermal excitation of electrons from the phosphorus donors into the empty conduction band may occur producing electron carriers in the solid. The phosphorus-doped silicon is called an n-type semiconductor because the majority of the current carriers are negatively charged electrons. The boron-doped semiconductor is called a p-type semiconductor because the majority carriers are positively charged holes.

XI. DIODES AND TRANSISTORS

What happens if we make a material such that one side is p-type and the other side is n-type? In principle, what we wish to do is to take a p-type semiconductor and place it up against an n-type semiconductor, producing an n-p junction. (The practical way of making junctions is described in Chapter 11.) Figure 13 shows what happens to the energy levels when such a junction is made. Because the Fermi levels of the two regions must coincide, the bands are distorted in the vicinity of the junction as shown. The effect of making the junction is that some holes from the p region will diffuse down to higher energy into the n region and some electrons will diffuse up to higher energy into the p region. At equilibrium, an electric field is established across the junction due to this space charge diffusion, which prevents further diffusion from taking place. This is shown in Fig. 14(a). Suppose we apply a battery voltage across the junction with the positive terminal of the battery connected to the n region, as shown in Fig. 14(b). The effect of the battery voltage is to increase the size of the hill at the

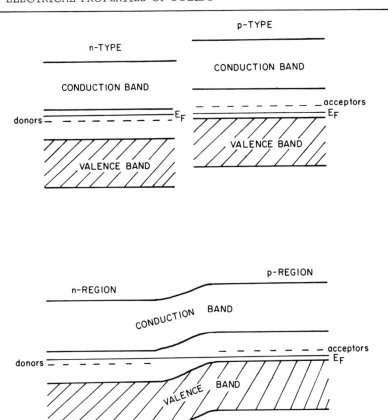

FIG. 13. n-type and p-type semiconductor regions before and after contact. In the region of contact the bands are distorted as shown in order for the Fermi level (the chemical potential) to remain the same on both sides of the junction.

junction still further, which thus reduces the diffusion and prevents any holes from drifting from p to n or electrons from drifting from n to p. The junction is then said to be reverse biased. If we reverse the battery voltage by connecting the positive end of the battery to the p region as in Fig. 14(c), the battery potential will produce an electric field at the junction which will offset the height of the energy barrier for diffusion, thus allowing holes to diffuse rapidly from p to n and electrons from n to p, producing a large current. The junction is

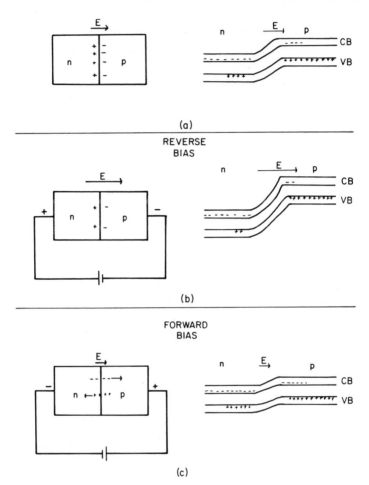

FIG. 14. (a) n-p junction showing diffusion of some electrons into the p region and some holes into the n region producing an electric field and a consequent band distortion at the junction.

(b) Reverse bias increases $\underline{E}$ and thus causes further distortion in the bands, raising the barrier for diffusion still further.

(c) Forward bias reduces $\underline{E}$ and thus decreases the barrier height for diffusion.

said to be forward biased. Such a device is called a diode or rectifier since it can transform an alternating current into a direct current. The reverse-biased junction acts as a large resistance to the applied voltage

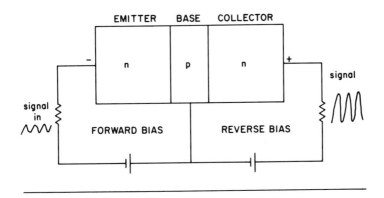

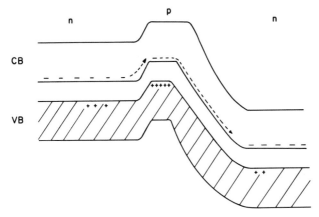

FIG. 15. Schematic and energy level diagram of an n-p-n junction transistor. Electrons diffusing from the emitter through the narrow base find themselves at the top of a high potential hill down which they accelerate, increasing their energy.

while the forward-biased junction acts as a low resistance to the applied voltage. We can use this idea of forward and reverse bias to construct a solid state amplifier called a transistor by putting two junctions together, as shown in Fig. 15, with opposite biases. The n region in the forward-bias circuit is called the emitter. The central p region is called the base and the n region in the reverse-bias circuit is called the collector. Now electrons easily flow in the forward-bias circuit from n to p since the resistance is low (low barrier height for diffusion).

However, when they travel across the p region (whose length is kept
very short to prevent recombination with holes), they very rapidly
accelerate down the potential hill of the reverse-bias circuit, producing
a very large voltage change. Thus, low voltage electrons in the emitter
are transformed into high voltage electrons at the collector, producing
a power gain.

XII.   STATISTICS OF ELECTRONS AND HOLES

In Fig. 8(a) we showed that the Fermi level for a metal lay at the
top of the electron distribution, whereas for a semiconductor with no
impurities as in Fig. 9, we showed the Fermi level lying half-way between the conduction and valence bands. In Fig. 12 we showed the
Fermi level lying between the donor states and the conduction band in
the case of an n-type semiconductor or lying between the acceptor
levels and valence band for a p-type semiconductor. We would like
to look at these situations a little bit more closely to see how they
arise.

As a consequence of the Pauli exclusion principle, electrons (and
holes) are said to obey Fermi-Dirac statistics. In statistical mechanics
one can derive a distribution function which represents the probability
of an energy level E being occupied by an electron. This Fermi-Dirac
distribution function is given by

$$p(E) = \frac{1}{\exp[(E - E_F)/kT] + 1} \quad , \tag{44}$$

where E_F is the Fermi level.

Let us look at this function when T is tending toward zero. For
any energy level E, which is less than the Fermi level E_F, the exponential in the denominator will be tending to zero as T goes to zero.
Therefore, $p(E)$, the probability for any energy level E less than E_F
being occupied by an electron, will be unity. Thus, at absolute zero

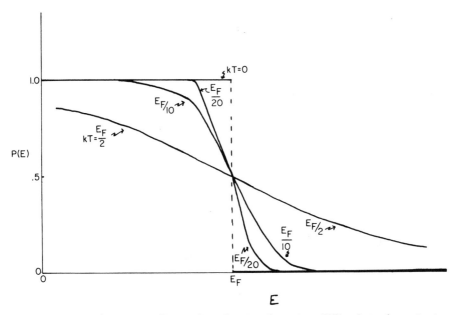

FIG. 16. Fermi-Dirac distribution function P(E) plotted against energy for various temperatures. At T = 0 °K all states below E_F are filled; all states above E_F are empty. For T > 0 °K states above and below the Fermi level will be partially occupied.

all levels below the Fermi level will be filled. For $E > E_F$, the exponential in the denominator will become very large as T goes to zero so that p(E) will tend to zero. Thus, all energy levels above the Fermi level will be empty at absolute zero. Hence, for electrons in a partially filled band, as depicted in Fig. 9(a) for a metal, the Fermi level at absolute zero must be right at the top of the highest filled level. If it were any higher, there would be empty states below E_F at T = 0, in violation of Eq. (44).

Figure 16 shows how the probability of occupation changes as T increases from 0 °K. It can be seen that the point between fully occupied and totally unoccupied levels becomes smeared out, so that partial occupation occurs above and below the Fermi level. For all temperatures above 0 °K, p(E) approaches unity for those states well below the

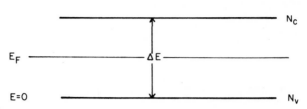

FIG. 17. Simplified picture of a semiconductor with all the empty conduction levels N_c, ΔE above all the filled valence band levels N_v.

Fermi level such that $-(E - E_F) > kT$ (this makes the exponential very small). For states well above the Fermi level, where $E - E_F > kT$, the exponential becomes very large compared to unity, and thus

$$p(E) = \exp[-(E - E_F)/kT] \quad , \tag{45}$$

which has the form of the Boltzmann distribution. Thus, the electrons in the "tail" of the Fermi-Dirac distribution follow classical statistical laws; as we shall see, these are the most important for a semiconductor.

For the case of a pure semiconductor, the Fermi level is half-way between the valence and conduction bands. This is easily seen by considering the oversimplified model of Fig. 17 where all the conduction band states N_c are at an energy ΔE above all the valence band states N_v whose energy is taken to be equal to zero.

According to Eq. (44), the number of electrons, n, thermally excited into the conduction band is equal to the probability of occupation, $p(E)$, times the number of states, N_c.

$$n = \frac{N_c}{\exp[(\Delta E - E_F)/kT] + 1} \approx N_c \exp[(E_F - \Delta E)/kT] \quad \text{if} \quad \Delta E - E_F \gg kT. \tag{46}$$

Similarly for the holes, the number of holes in the valence band, p, is equal to the probability of a valence band state being empty (due to thermal excitation of electrons to the conduction band) times the number of states N_v. The probability of a valence band state being empty is simply 1 - probability of being filled:

$$p = N_V \left(1 - \frac{1}{\exp(-E_F/kT) + 1} \right) = N_V \left(\frac{\exp(-E_f/kT)}{\exp(-E_F/kT) + 1} \right) = N_V \left(\frac{1}{1 + \exp(E_F/kT)} \right)$$

or

$$p = N_V \exp(-E_F/kT) \quad \text{if} \quad E_F \gg kT \quad . \tag{47}$$

For an intrinsic semiconductor (viz., one in which electrons are excited directly from the valence band to the conduction band) we know $n = p$. Combining Eqs. (46) and (47) we have

$$N_C \exp[(E_F - \Delta E)/kT] = N_V \exp(-E_F/kT) \quad .$$

Therefore,

$$E_F = \frac{\Delta E}{2} + kT \ln \frac{N_V}{N_C}$$

or

$$E_F = \frac{\Delta E}{2} \quad \text{if} \quad N_V = N_C \quad . \tag{48}$$

Thus, the Fermi level will be half-way between the valence and conduction bands in this simple model. In the exact treatment where one considers a distribution of energies over a variable density of states in each band, one finds

$$E_F = \frac{\Delta E}{2} + \frac{3}{4} kT \ln \frac{m_h^*}{m_e^*} \quad ,$$

which is the same as the above if the effective mass of holes and electrons are equal.

If we take the product of Eqs. (46) and (47) we find

$$np = N_V N_C e^{-\Delta E/kT} \quad .$$

For an intrinsic semiconductor, $n = p$; therefore,

$$np = n^2 = N_V N_C e^{-\Delta E/kT} \tag{49}$$

or

$$n = (N_V N_C)^{\frac{1}{2}} e^{-\Delta E/2kT} \quad . \tag{50}$$

In the intrinsic region of conduction the conductivity will be proportional to $e^{-\Delta E/2kT}$, so that slopes of $\ln \sigma$ vs $1/T$ will equal $-\Delta E/2$. The appearance of intrinsic conductivity, however, requires that the generation of holes and electrons by impurities be small compared to the thermal excitation from the valence band to the conduction band. This is rarely the case. If the impurity concentration is high, the intrinsic conductivity can be completely masked by the electrons and holes generated from the impurity levels. Intrinsic conductivity has thus been seen in only a few materials where the impurity concentration is low so that most of the impurity sites have been emptied before the intrinsic conductivity appears.

Equation (49) did not depend on the existence of impurity levels and, in fact, is still correct whether there are any impurities or not (although in the presence of impurities $n \neq p$). The result can also be obtained from chemical equilibrium theory by starting with a neutral crystal and generating some holes and electrons in equilibrium by any means available (intrinsically or from impurity levels or defects):

$$\text{neutral crystal} \rightarrow [n] + [p] \quad .$$

We can write an equilibrium constant for this reaction by taking the activity of the crystal to be unity

$$K = [n][p] \quad .$$

K is related to the standard free energy of the reaction

$$\Delta G^\circ = -RT \ln K \quad .$$

Therefore,

$$[n][p] = \exp(-\Delta G^\circ/RT) = \exp\left(-\frac{\Delta H^\circ}{RT} + \frac{\Delta S^\circ}{R}\right) \quad ,$$

where ΔH° and ΔS° are the standard enthalpy and entropy per mole for the reaction. Since R is equal to Avogadro's number times k, then in terms of enthalpy and entropy per particle, Δh° and Δs°, we have

$$np = \exp(-\Delta h°/kT) \exp(\Delta s°/k) \ .$$

But $\Delta s° = k \ln W$, where W is the number of ways that the hole and electron can be created. For our simple model, W is simply the number of available states for holes and electrons:

$$\Delta s° = \Delta s_e + \Delta s_h = k \ln N_c + k \ln N_v$$

$$= k \ln N_c N_v$$

$$\therefore \ np = N_c N_v \exp(-\frac{\Delta h°}{kT}) \ ,$$

which is identical to Eq. (49) with $\Delta h = \Delta E$.

Thus, the energy gap is simply the enthalpy change per electron.

Suppose we now introduce into our simple model a few donor impurities whose number is N_D, as shown in Fig. 18. We now require that, at absolute zero, the Fermi level lie half-way between the donors and the conduction band because, as already shown, all states below the Fermi level must be filled and all those above the Fermi level must be empty at $0°K$. In this picture, however, thermal excitation of electrons only produces negative charge carriers with no holes.

Let us apply the Fermi-Dirac distribution equation (44) to this case to see how the number of electron carriers changes with temperature. The donor energy will be taken equal to zero. The number of electrons N_D in the donor levels at any temperature T is equal to the probability that the donor is occupied, $p(E)$, times the number of donors N_D:

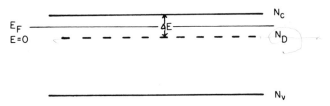

FIG. 18. Semiconductor with N_D donors. The conduction band states N_C are an energy ΔE above the donor levels.

number of electrons on donor sites, $n_D = \dfrac{N_D}{\exp[(0 - E_F)/kT] + 1}$ (51)

The number of electrons n in the conduction states is equal to the probability that a conduction level is occupied, $p(E)$, times the number of conduction levels:

number of electrons in conduction band, $n_c = \dfrac{N_c}{\exp[(\Delta E - E_F)/kT] + 1}$

(52)

where n_c is the number of conduction electrons that we are interested in. This is the number that determines the electrical conductivity. Equation (52), however, contains E_F as well as ΔE so that we must eliminate E_F to obtain an expression for n_c in terms of ΔE alone. We can do this by realizing that the number of electrons/cm^3 in the conduction band plus the number of electrons in the donor levels must add up to the number of donor levels N_D since all the electrons originate from the donor sites (electrons from the valence band are neglected since the thermal energy needed to excite them to the conduction band is too high):

$$n_D + n_c = N_D .$$

Adding Eqs. (51) and (52) we get

$$N_D = \frac{N_D}{\exp(-E_F/kT) + 1} + \frac{N_c}{\exp[(\Delta E - E_F)/kT] + 1} .$$

(53)

As in the intrinsic case we will assume that the Fermi level is below the conduction band by several kT so that $\Delta E - E_F > kT$ and the approximation of Eq. (45) holds. Then rearranging Eq. (53) gives

$$N_D \left(1 - \frac{1}{\exp(-E_F/kT) + 1}\right) = N_c \exp[(E_F - \Delta E)/kT]$$

or

$$\frac{N_D}{\exp(E_F/kT) + 1} = N_c \exp[(E_F - \Delta E)/kT] .$$

(54)

We will also assume that E_F is far enough away from the donor levels so that $E_F > kT$. Then Eq. (54) becomes

$$N_D \exp(-E_F/kT) = N_C \exp[(E_F - \Delta E)/kT]$$

or

$$E_F = \frac{\Delta E}{2} - \frac{kT}{2} \ln \frac{N_C}{N_D} . \qquad (55)$$

Equation (55) shows how the Fermi level changes with temperature. At $T = 0$ the Fermi level is half-way between the conduction band and the donor levels, as already stated. However, notice that, as T increases, E_F begins to fall (the logarithm is positive since $N_C \gg N_D$). Eventually, E_F will fall below the donor levels as more electrons are excited into the conduction band. [Equation (55), however, will break down when E_F approaches the donor level energy.] In the region where our approximations are valid, substitution of Eq. (55) into Eq. (52) gives the number of conduction electrons n_c:

$$n_c = \frac{N_C}{\exp[(\Delta E - E_F)/kT] + 1} \approx N_C \exp[(E_F - \Delta E)/kT]$$

$$= N_C \left[\exp\left(\frac{\Delta E}{2kT} - \frac{1}{2} \ln \frac{N_C}{N_D} \right) \right] \exp\left(-\frac{\Delta E}{kT} \right)$$

$$= (N_C N_D)^{\frac{1}{2}} \exp(-\Delta E/2kT) . \qquad (56)$$

Thus, when the Fermi level lies between the conduction band and the donor levels, which will occur if T is small, the electrical conductivity will vary as $\exp(-\Delta E/2kT)$. Notice the 2 in the denominator of the exponential. The form of Eq. (56) is identical to that for the intrinsic situation, Eq. (50). At higher temperatures when E_F has fallen well below the donor levels compared to kT, $(-E_F > kT)$, the exponential in the first term of Eq. (54) becomes small compared to 1, so that this equation reduces to

$$N_D = N_C \exp[(E_F - \Delta E)/kT]$$

or

$$E_F = \Delta E - kT \ln \frac{N_C}{N_D} \quad . \tag{57}$$

Substitution of this E_F into Eq. (52) for n_c gives

$$n_c = N_C \left[\exp \left(\frac{\Delta E}{kT} - \ln \frac{N_C}{N_D} \right) \right] \exp \left(- \frac{\Delta E}{kT} \right)$$

$$= N_D \quad .$$

Thus, the number of carriers becomes constant at higher temperatures, equal to the number of donors. Since the conductivity σ equals $ne\mu$, if n is constant, then the conductivity will be dominated by the mobility which normally decreases as some power of T. This is the situation at the high temperature end of Fig. 3 and is known as the exhaustion region in which all the donor sites have been emptied. At still higher temperatures the Fermi level drops to the center of the intrinsic energy gap at which point electrons are excited from the valence band into the conduction band producing a large increase in conductivity due to electrons and holes which change with temperature like $e^{-\Delta E/2kT}$. This is seen in Fig. 3 above the exhaustion region.

Equations like (55) or (57) which relate the Fermi level to the energy gap and to the logarithm of the number of states have a simple macroscopic thermodynamic equivalent.

We have already shown that ΔE is simply the standard enthalpy change per electron. In Eq. (57) the quantity $k \ln N_C$ is simply the entropy per electron associated with the conduction band whereas $k \ln N_D$ is the entropy associated with the donor levels. Thus,

$$k \ln (N_C/N_D) \equiv k \ln N_C - k \ln N_D$$

is the entropy change upon thermal excitation from the donor levels to the conduction band. This gives Eq. (57) the form of

$$E_F = \Delta h^\circ - T \Delta s^\circ \equiv \text{chemical potential} \quad ,$$

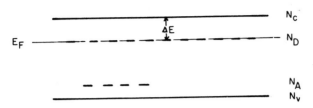

FIG. 19.   Semiconductor with donors N_D and acceptors N_A.
$N_D \gg N_A$.

so that the Fermi level is simply equivalent to the chemical potential.
The fact that the Fermi level falls with increase in temperature is, thus,
understandable as an increase in the contribution of the entropy to the
standard free energy change.

If there happens to be a small number of acceptors N_A present
as well as donors ($N_A \ll N_D$), which is usually the case, then at
absolute zero the donors will be partially empty since some of the
electrons in the donor levels will lose energy and fill the acceptor
levels, as shown in Fig. 19.

Then

$$n_D + n_c = N_D - N_A \tag{58}$$

since some of the electrons that were in the donor sites were used to
"compensate" the N_A acceptor sites.

Since $n_c = N_c \exp[(E_F - \Delta E)/kT]$, we can write the Fermi level
in terms of n_c:

$$E_F = kT \ln \frac{n_c}{N_c} + \Delta E \ . \tag{59}$$

Substituting this expression for E_F into Eq. (51) and then putting
Eq. (51) into Eq. (58), we get

$$\left(\frac{N_A + n_c}{N_D - N_A - n_c} \right) n_c = N_c \, e^{-\Delta E/kT} \ . \tag{60}$$

This equation has two interesting limiting forms.   At very low

temperatures when there are only a few electrons in the conduction band
so that $n_c \ll N_A$ and if $N_A < N_D$, then Eq. (60) becomes

$$n_c = \frac{N_c(N_D - N_A)}{N_A} e^{-\Delta E/kT} .$$

(61)

Thus, at low temperatures the presence of acceptors causes the
electrical conductivity to go as $e^{-\Delta E/kT}$. Notice the 2 is missing
from the denominator of the exponential.

At still higher temperatures where the number of electrons in the
conduction band becomes greater than N_A but still much less than N_D,
Eq. (60) reduces to

$$n_c^2 = N_D N_c e^{-\Delta E/kT} ,$$

$$n_c = (N_D N_c)^{\frac{1}{2}} e^{-\Delta E/2kT} ,$$

(62)

which is equivalent to the model with no acceptors. Thus, the low
temperature behavior is modified by the presence of acceptors. A practical example of this is seen in silicon doped with arsenic (a donor)
(Fig. 20). Below 45 °K the slope is twice that above this temperature.

Let us look at what happens to the Fermi level in this model
where acceptors are present. Without going through the calculation
we can predict that, at absolute zero, the Fermi level must lie exactly
at the donor energy level. This occurs because, as Fig. 19 shows,
some of the donor levels are empty even at 0 °K due to compensation of
the acceptor levels. The Fermi distribution requires all levels below
E_F filled at 0 °K, which is impossible if E_F is above the donor levels.
E_F cannot be below the donor levels since all levels above the donors
must be empty at 0 °K. Thus, $E_F = 0$ at absolute zero. Quantitatively,
we can see what happens in our model by equating the low temperature
approximation for n_c, Eq. (61), with

$$n_c = N_c \exp[(E_F - \Delta E)/kT]$$

(63)

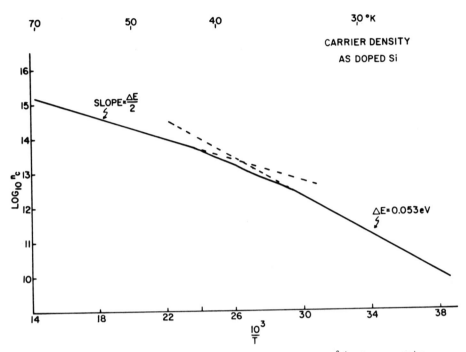

FIG. 20. Logarithm of the carrier density vs $10^3/T$ for arsenic-doped silicon (n-type). Below 45 °K where $n_C \ll N_A$ the slope is twice as great as above 45 °K where $N_D > n_C > N_A$. $N_D = 10^{16}/cm^3$, $N_A = 3 \times 10^{13}/cm^3$.

$$\frac{N_c(N_D - N_A)}{N_A} \exp\left(-\frac{\Delta E}{kT}\right) = N_c \exp\left(\frac{E_F - \Delta E}{kT}\right) \quad ,$$

$$E_F = kT \ln \frac{N_D - N_A}{N_A} \quad .$$

At $T = 0$, $E_F = 0$. Notice, however, that as T increases, the Fermi level rises above the donor levels. It will continue to rise until n_c becomes greater than the number of acceptors N_A, whereupon the approximation of Eq. (62) becomes valid. We can see what happens at this point by combining Eqs. (62) and (63):

$$(N_D N_c)^{\frac{1}{2}} \exp(-\Delta E/2kT) = N_c \exp[(E_F - \Delta E)/kT] \quad ,$$

$$E_F = \frac{\Delta E}{2} - \frac{kT}{2} \ln \frac{N_c}{N_D} \quad .$$

Thus, the Fermi level will rise to half-way between the donor levels and the conduction levels and then begin to fall. This result is equivalent to the model with no acceptors present [Eq. (55)].

The importance of these conclusions is related to how one calculates the activation energy for conduction from plots of $\ln \sigma$ vs $1/T$. The slope can be proportional to $\Delta E/2$ or ΔE, depending on the presence of acceptors and their concentration relative to the density of carriers. Great care must then be taken in evaluating conductivity data alone, especially if it covers only a limited temperature range.

XIII. ENERGY BANDS AND MOLECULAR ORBITALS

In the previous sections we were able to introduce the idea of energy bands with an energy gap by considering the electron as a wave interacting with the periodic lattice array of ions in which the electron finds itself. This concept is equivalent to the electron delocalization picture of molecular orbital theory that was treated in Chapter 3. If we allow the wave functions of a large number of atoms to overlap with one another, the net result is to produce an energy band of states, the width of the band being proportional to the amount of overlap of the atomic wave functions. Each atom will contribute one orbital state to the band. Since each state can hold two electrons, one with spin up and the other with spin down, the band will be full if each atom has two electrons.

As an example of the direct application of this idea, consider the square planar ion $[Pt(CN)_4]^{2-}$ shown in Fig. 21(a). The $5d_{z^2}$ orbital points above and below the plane of the complex and has two

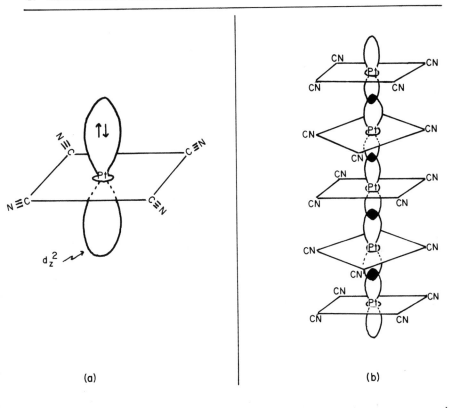

(a)   (b)

FIG. 21. (a) Geometric arrangement of four cyanide groups around a divalent platinum ion. The d_{z^2} orbital is nonbonding and holds two electrons.
  (b) Stacking of $[Pt(CN)_4]^{2-}$ ions in the solid state showing overlap of $5d_{z^2}$ orbitals. The d_{z^2} band is thus filled. Partial oxidation removes some electrons from the filled band producing a hole conductor.

electrons in it. In the solid state, these complex ions stack one on top of the other to form a linear chain, as shown in Fig. 21(b), with the CN^- array staggered from Pt to Pt. (This reduces the coulomb repulsion between CN^- groups.) Since each platinum has two electrons, overlap of all the platinum $5d_{z^2}$ orbitals will produce a filled band separated by a large gap from the next band formed by overlap of the $6p_z$ orbitals. Thus, $K_2Pt(CN)_4$, for example, will be an insulator or semiconductor with a large band gap. We can make it into a conductor by removing a few electrons from the $5d_{z^2}$ band. Chemically, this is equivalent to oxidizing some of the Pt(II) to Pt(IV). Just about any oxidizing agent will do the job (e.g., Cl_2, Br_2, H_2O_2, HNO_3). Addition of Br_2, for example, oxidizes some Pt(II) to Pt(IV), producing Br^- ions which are incorporated between the complex chains. Crystals of this partly oxidized complex or mixed valence complex, as it is called, have the appearance of metallic copper with a conductivity $\sigma = 10\Omega^{-1}\,cm^{-1}$. In fact, if the Br^- ion concentration is not too high, one might predict that this material should be a one-dimensional hole conductor (conductivity perpendicular to the z axis should be negligible since there is no Pt-Pt overlap).

The molecular orbital picture can also be used to explain what happens when the concentration of impurity levels in a semiconductor, such as those in Fig. 19, becomes very large ($> 10^{19}$ phosphorus atoms per cm^3 in Si, for example). In this case the impurities get close enough so that the atomic orbitals on the impurities can overlap sufficiently to form a band which, if wide enough, will overlap the conduction band of the host lattice. As must be the case, the Fermi level will lie in the band at the top of the electron distribution, as in a metal. The material will thus show metallic behavior. With such a large impurity concentration, the electron scattering will be dominated by impurity scattering rather than lattice vibrations, so that the electrical conductivity will be temperature independent. This is shown in Fig. 22.

XIV.   HOPPING CONDUCTION

In all of our previous discussions we have assumed that we can treat the electron as a wave that is delocalized over the entire crystal lattice. The electrons are essentially free, moving independently of one another, the independent motion being guaranteed by the presence of the uniformly distributed background positive charge. It has been

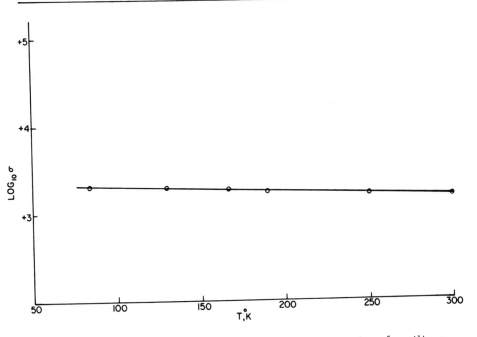

FIG. 22. Logarithm of the conductivity vs temperature for silicon doped with greater than 10^{19} phosphorus atoms (n-type). Conductivity is temperature independent due to impurity scattering.

shown that this is essentially equivalent to the molecular orbital scheme where there is large overlap between atomic wave functions. A new situation arises, however, if we start pulling the atoms apart, reducing the overlap between atomic states, or equivalently, thinning out the background positive charge which will increase the coulomb repulsion between electrons. Eventually, a point will be reached where the electron gas is no longer stable because of the large coulomb repulsions between electrons. The electrons will tend to "condense" out onto the atomic cores and become localized. The material will thus change from a metal to an insulator, in which the electrons are localized on atomic or molecular sites. This is the situation for the impurities in our n-type semiconductor shown in Fig. 18. The impurity levels are far enough apart so that overlap of atomic wave functions is small. Electron transfer from one filled impurity level to another does not take place because

PERIODIC POTENTIAL

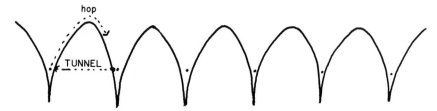

FIG. 23. Simple model of hopping vs band conduction. Electrons in potential wells can either "hop" over the barrier or tunnel through it. Hopping requires an activation energy whereas tunneling through the wells does not. At high temperatures ($T > \Theta/2$) probability for hopping over the barrier is higher than for tunneling through the barrier. At low temperatures ($T < \Theta/2$) just the reverse is true.

of the strong coulomb repulsion which would occur for two electrons on one site. However, if there happens to be some compensating accept- ors, as in Fig. 19, then an electron in one donor impurity can jump to a neighboring site if it happens to be empty. Such hopping conduction, as it is called, occurs among the impurity levels of germanium and silicon at very low temperatures where the effects of conduction in the host band is negligibly small. However, this kind of conduction is more likely to occur in compounds where the metal ions can have more than one oxidation state. An example is V_2O_5 doped with an electro- positive metal $M_xV_2O_5$ where M can be an alkali metal, Li, Na, K, an alkaline earth, Mg, Ca, or a trivalent metal like Al. In all cases M gives up its electrons to the vanadium to form V^{4+} centers so that $M_xV_2O_5$ contains vanadium in two different oxidation states, V^{4+} and V^{5+}. The added electron on V^{4+} (d^1 ion) can now jump to an empty neighboring V^{5+} site:

$$V^{4+} \ldots V^{5+} \;\rightarrow\; V^{5+} \ldots V^{4+} \quad ,$$

producing an effective means of electron transport. As an example, $Na_{0.33}V_2O_5$ has a room temperature conductivity of $\sigma = 100\,\Omega^{-1}\,cm^{-1}$, whereas nominally pure V_2O_5 has a room temperature $\sigma = 10^{-3}\,\Omega^{-1}\,cm^{-1}$.

This is another example of how mixed valence chemistry can produce interesting electrical effects.

In order for hopping conduction to occur, the wave functions on neighboring sites must overlap a small amount but not sufficiently so that the electrons become delocalized in a band. Pictorially, what occurs is shown in Fig. 23. The positive atomic cores form a periodic coulomb potential with the electrons localized within the potential well. The small overlap between adjacent wave functions is equivalent to the electrons tunneling through the potential barriers producing a delocalized state. However, at ordinary temperatures $(T > \Theta)$ the tunneling probability is less than that for jumping over the potential barriers, so that the electron "hops" from site to site. At lower temperatures $(T < \Theta)$, the probability for jumping diminishes below that for tunneling, at which point the electron acts like it is in a narrow conduction band. In the high temperature region, the electron's velocity or mobility will depend on its being activated over the potential barrier, so that

$$\mu \propto e^{-\Delta E/kT} \quad ,$$

where ΔE is the energy of activation. Thus, μ increases with increasing temperature. In a normal band semiconductor, μ decreases with increasing temperature. Thus, the two modes of conduction can be distinguished by measuring the mobility as a function of temperature via the Hall effect. At low temperatures, where the hopping probability is small, μ should decrease with increasing temperature as in a normal semiconductor. As we pictured it, the transition between the two conduction modes should be continuous. This has been observed in $Na_{0.33}V_2O_5$.

XV. SCATTERING MECHANISMS IN SEMICONDUCTORS

We have shown previously that vibrational scattering mechanisms for electrons in a metal lead to an electron mean free path and, hence,

mobility which is proportional to T^{-1}. The derivation of this depends
on the fact that the distribution of electron velocities at the Fermi level
is independent of temperature, so that the average electron speed remains fixed when the temperature is changed. Since

$$\sigma = \frac{ne^2 \ell_F}{m^* v_F} = ne\mu \quad,$$

we have

$$\mu = \frac{e\ell_F}{m^* v_F} \quad . \tag{64}$$

The conductivity will vary with temperature because the mean
free path at the Fermi level, ℓ_F, varies with temperature as T^{-1} while
v_F is constant.

In a semiconductor, however, we have shown that the electrons
follow Boltzmann statistics when the Fermi level is well below the
bottom of the conduction band compared to kT. In this case the electrons in the conduction band will have a distribution of velocities which
changes with temperature producing a mean square velocity which
changes with temperature by the well-known relation

$$\frac{1}{2} m\overline{v^2} = \frac{3}{2} kT$$

or

$$(\overline{v^2})^{\frac{1}{2}} \propto T^{\frac{1}{2}} \quad .$$

This velocity distribution must be taken into account in equations
like (64) in addition to the fact that the mean free path due to lattice
vibrational scattering is proportional to T^{-1}. Hence, for a semiconductor

$$\ell \propto T^{-1} \quad \text{(as for a metal)} \quad,$$

$$v \propto T^{+\frac{1}{2}} \quad,$$

therefore,

$$\mu \propto T^{-\frac{3}{2}} = T^{-1.5} \ .$$

Many semiconductors follow this relationship in only a very crude way. For example, n-type germanium has $\mu \propto T^{-1.66}$ whereas n-type silicon has $\mu \propto T^{-2.6}$. Large deviations from the $T^{-\frac{3}{2}}$ relation are indicative of other kinds of vibrational scattering mechanisms, which will not be discussed here. Table 4 shows how the mobility varies with temperature depending on the type of scattering mechanism.

In the case of ionized impurity scattering, we indicated that this led to a mobility that is temperature independent for a metal but that goes as $T^{\frac{3}{2}}$ for a semiconductor. Once again the difference lies in the fact that the root mean square velocity of electrons in a semiconductor varies as $T^{+\frac{1}{2}}$, whereas in a metal the root mean square velocity of electrons at the Fermi level is independent of temperature. As a model for ionized impurity scattering, we can use the result derived by Rutherford

TABLE 4

Temperature Dependence of Mobility for Different Scattering
Mechanisms in Metals and Semiconductors

Scattering mechanism	Temperature dependence of
Nonpolar acoustical mode vibrations	T^{-1} (metals) $T^{-\frac{3}{2}}$ (semiconductors)
Optical-mode lattice vibrations (large polaron)	$(T/\Theta)^{\frac{1}{2}} [e^{\Theta/T} - 1]$ (semiconductors) $\sinh^2(\Theta/2T)$ (metals)
Ionized impurities	$T^{\frac{3}{2}}$ (semiconductors) Temperature independent (metals)
Neutral impurities	Temperature independent
Small polaron (hopping)	$T^{-\frac{3}{2}} e^{-\Delta E/kT}$

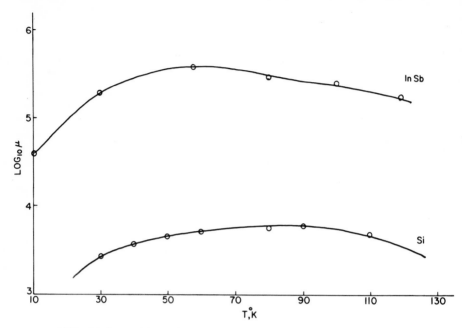

FIG. 24. Logarithm of mobility vs temperature for silicon and
indium antimonide, showing maxima due to appearance of ionized
impurity scattering.

for electron scattering off nuclei. Rutherford showed that the mean free
path for scattering is proportional to the square of the kinetic energy of
the scattered electrons:

$$\ell \propto E^2 \propto v^4 \quad ,$$

therefore,

$$\mu \propto \frac{\ell}{v} \propto v^3 \quad .$$

For a metal $v = v_F$, the velocity of electrons at the Fermi level,
which is independent of T. For a semiconductor, however, since
$v \propto T^{\frac{1}{2}}$, we have

$$\mu \propto T^{\frac{3}{2}} \quad .$$

Thus, it is expected that, at low temperatures where lattice vibrational scattering is small, the mobility of a semiconductor with impurities

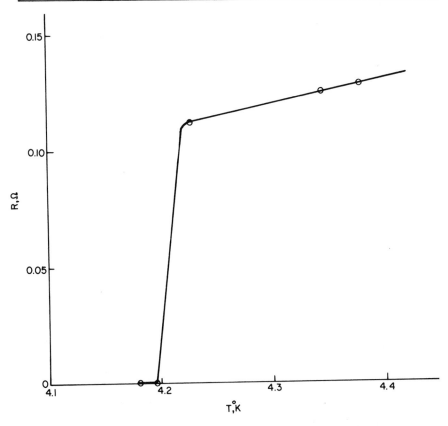

FIG. 25. Resistance vs temperature for mercury, showing superconducting phase transition near 4.2 °K.

should increase with increasing temperature, reach a maximum, and
then begin to fall with increasing temperature as lattice vibrational
scattering begins to dominate. This is shown to be the case for n-type
silicon and n-type InSb in Fig. 24.

XVI. SUPERCONDUCTIVITY

The phenomenon of superconduction was discovered in 1908 by
Kamerlingh Onnes, who found that the resistance of a sample of mercury
metal fell sharply to zero when the temperature fell below 4.2 °K, as is
shown in Fig. 25. At first one might surmise that this effect is simply

the disappearance of the lattice vibrational scattering at low temperatures. As we have shown, a rigid periodic lattice should offer no
resistance to electron motion, and, thus, at low enough temperature
the resistance should approach zero. (Parenthetically, we should
remark that the last statement is true provided we neglect Bragg reflection. In an applied electric field and with no lattice scattering, an
electron will increase its velocity and thus decrease its wavelength
until the Bragg condition for reflection $\lambda = 2d$ occurs. At this point
the electron wave will be reflected and combined with the incident wave
forming a standing wave pattern, so that the net contribution to the conductivity will be zero. Thus, in the absence of any kind of scattering,
a partially filled band would have infinite resistance. Practically,
there are always impurities present which will cause scattering at low
temperatures and thus prevent the Bragg condition from occurring.) If
this were true, then if one adds impurities to the mercury, the low temperature resistance should be dominated by impurity scattering. Nevertheless, the transition to zero resistance at 4.2 °K is still observed even
with impurities present. The mercury must therefore be in some kind of
new state called the superconducting state, with a critical transition
temperature T_c of about 4.2 °K.

Since 1908 there have been reported more than 1000 compounds
which become superconducting, with transition temperatures as low as
a few millidegrees to as high as 20.98 °K for the nonstoichiometric alloy
$Nb_{0.79}(Al_{0.75}Ge_{0.25})_{0.21}$. Superconductivity appears as a property in
at least 25 pure elements of the periodic table but is not limited to only
compounds of those elements, as demonstrated by Ag_2F and $[Ag_7O_8]^+NO_3^-$
with transition temperatures of 0.066 and 1.04 °K, respectively. Strangely
enough, Ag, which is the best metallic conductor at room temperature,
does not appear to be superconducting. In fact, most of the good
metallic conductors like the alkali metals are not superconducting. At
present there do not appear to be good criteria for predicting whether a
compound will be a superconductor nor is there any theoretical upper

limit for the transition temperature. The search for new superconductors with higher transition temperatures, therefore, goes on unabated.
Whether or not superconductivity exists in two-dimensional metals or
possibly one-dimensional metals like $K_2Pt(CN)_4Br_{0.3}$ described previously remains to be seen. Most theoretical arguments support the notion
that superconductivity must be a three-dimensional cooperative effect.
This is supported by the observation that, in anisotropic solids, the
superconducting transition temperature is independent of direction.
Nevertheless, a chemist's approach to this problem should offer syn-
thetic routes to new materials which may possess superconducting
properties with higher transition temperatures presently not available
in alloys.

Associated with the electrical effects of the superconducting
state is a magnetic phenomenon known as the Meissner effect. If a
sample of material, which has a transition temperature T_c, is placed
in a magnetic field, then above T_c the magnetic flux will pass through
the sample as shown in Fig. 26(a). In the superconducting state below
T_c, the magnetic flux is expelled from the interior of the sample provided the externally applied field is below a critical value, as shown
in Fig. 26(b). The Meissner effect is another fundamental property of
the superconducting state and is often used as the means for detecting
the transition from the normal state to the superconducting state. For
any material in a magnetic field,

$$B = H + 4\pi M = (1 + 4\pi\chi)H \quad ,$$

where B is the total magnetic induction, H is the applied magnetic
field, M the magnetic moment per unit volume, and $\chi = M/H$ is the
magnetic susceptibility of the sample (see Chapter 5). For $T < T_c$,
we know $B = 0$ in the interior of the sample. Therefore,

$$H = -4\pi M$$

or

$$\chi = \frac{M}{H} = -\frac{1}{4\pi} \quad .$$

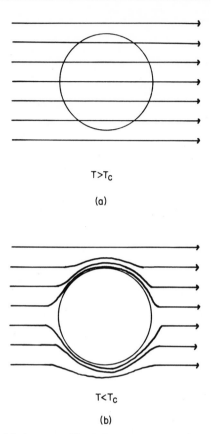

T>T$_c$

(a)

T<T$_c$

(b)

FIG. 26.   Meissner effect for a superconducting sphere.

(a) Above the critical temperature T$_c$, the magnetic flux penetrates the sphere.

(b) Below T$_c$, the magnetic flux cannot penetrate the sphere.

For a normal metal, $T > T_c$, $M \approx 0$, so that in passing through the transition temperature the susceptibility abruptly changes from 0 to $-1/4\pi$. The inductance of a coil of wire wrapped around the sample depends on the susceptibility of the material occupying the core of the coil, so that measurement of the inductance of the coil will show a sharp transition at $T = T_c$. The absence of a Meissner effect measured this way is thus taken as indicating the absence of the superconducting

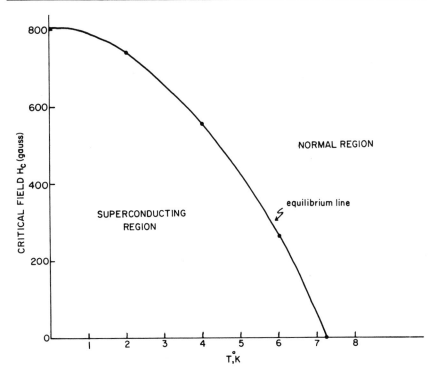

FIG. 27. Phase diagram for lead showing equilibrium between the normal and superconducting states as a function of magnetic field and temperature.

state. (A Meissner effect would not be seen for a fictitious material which is not in the superconducting state but shows zero resistance. M is still approximately zero in this case.) If the applied magnetic field in the superconducting state exceeds a critical value, then the superconducting state disappears and the material returns to the normal state. The magnetic field at which this occurs depends on the temperature. We can thus draw a phase diagram with critical field and critical temperature as axes, as shown for lead in Fig. 27. Points below the curve represent the superconducting region, points above the curve are in the normal region. The curve itself is the equilibrium line representing those values of H and T where the two phases, normal and superconducting, are in equilibrium.

Given this type of phase diagram, we can learn something about the thermodynamic properties of superconductors. If a magnetic field H is applied to a superconductor below T_c, the free energy of the superconductor will change by an amount equal to the work done by the field:

$$G_{H_S} = G_S - \int_0^H \mu\, dH \quad .$$

Here G_S is the free energy of the superconductor in the absence of a field, the integral is the work done by the field, and μ is the total magnetic moment of the specimen $= MV$, where V is the volume of the sample. Since $M/H = -1/4\pi$ for a superconductor, then

$$G_{H_S} = G_S + \frac{V}{4\pi} \int_0^H H\, dH$$

$$= G_S + \frac{H^2 V}{8\pi} \quad .$$

For a normal metal the integral will be approximately zero since $M \approx 0$:

$$G_{H_N} = G_N \quad ,$$

where G_N is the free energy of the fictitious normal state below T_c in the absence of a magnetic field and G_{H_N} is the free energy in the presence of a field. At equilibrium, for which $H = H_c$ (viz., along the curve shown in the phase diagram),

$$G_{H_S} = G_{H_N} \quad ,$$

$$G_S - G_N = \Delta G° = -\frac{H_c^2 V}{8\pi} \quad . \tag{65}$$

This equation gives the standard free energy change that would take place at some fixed temperature less than or equal to T_c when a normal metal is transformed into the superconducting state in the absence of a magnetic field. Notice that this free energy change can be calculated by knowing what the critical magnetic field is at the temperature in question.

Below T_c the superconducting state is the thermodynamically stable state, so that $G_S - G_N$ is always negative. Equations like (65), as all such equations for standard free energy changes of chemical reactions, demonstrate the subtlety of thermodynamic arguments. The left side of the equation, $\Delta G°$, is the free energy lowering in the absence of a magnetic field, and this is equal to the right side which has in it a quantity for the magnetic field necessary to achieve equilibrium, H_c.

The entropy change for the transformation can be found by differentiating with respect to T at constant pressure:

$$\left(\frac{d\Delta G°}{dT}\right)_P = -\Delta S°$$

or

$$S_S° - S_N° = +\frac{H_c V}{4\pi}\frac{dH_c}{dT} \quad , \tag{66}$$

assuming the volume is independent of T. Differentiating again will give us the change in heat capacity for the transition:

$$\frac{d\Delta S°}{dT} = \frac{\Delta C}{T}$$

or

$$C_S - C_N = +T\frac{d}{dT}\left(\frac{H_c V}{4\pi}\frac{dH_c}{dT}\right)$$

$$= +\frac{VT}{4\pi}\left[\left(\frac{dH_c}{dT}\right)^2 + H_c\frac{d^2H_c}{dT^2}\right] \quad . \tag{67}$$

Looking back at the phase diagram, Fig. 27, we see that, at $T = T_c$, $H_c = 0$ but dH_c/dT is still finite. Equations (65), (66), and (67) then tell us that at $T = T_c$ the entropy change and, hence, the enthalpy change are zero, whereas the heat capacity changes abruptly. Thus, at T_c the transition to the superconducting state is a second-order one. For $T < T_c$, H_c is finite and dH_c/dT is always negative; thus, below the transition temperature the entropy of the superconducting state is always less than the entropy of the normal state. The superconducting state is therefore a more ordered state than the normal state below T_c. Approaching absolute zero, we see from the curve that $dH_c/dT \rightarrow 0$, so that the difference in entropy between the normal and superconducting states approaches zero, as required by the third law of thermodynamics. At both ends of the equilibrium line, then, the entropy change is zero. In between, the entropy must pass through a minimum.

Equation (66) is simply the Clausius-Clapeyron equation for the change in the equilibrium magnetic field with temperature analogous to the change in equilibrium pressure with temperature:

$$\frac{dP}{dT} = \frac{\Delta S}{\Delta V} \quad ,$$

$$\frac{dH_c}{dT} = -\frac{\Delta S}{MV} \quad . \tag{68}$$

In Eq. (68) H_c replaces P and total magnetic moment MV replaces volume. The magnetic moment of the normal phase has been left out since it is small compared to the superconducting moment, but we could just as well include it so that Eq. (68) becomes

$$\frac{dH_c}{dT} = -\frac{\Delta S}{\Delta(MV)} \quad , \tag{69}$$

where $\Delta(MV)$ is the difference in magnetic moment between the super-conducting and normal phases.

We can summarize the thermodynamic results by considering the transformation from the normal state to the superconducting state as a chemical equation with associated $\Delta G°$, $\Delta H°$, and $\Delta S°$:

Normal state $\rightarrow$ Superconducting state

$$\Delta G° = -\frac{H_c^2 V}{8\pi} \quad , \tag{70}$$

$$\Delta S° = \frac{H_c V}{4\pi} \frac{dH_c}{dT} \quad , \tag{71}$$

$$\Delta H° = \Delta G° + T\Delta S° = -\frac{H_c^2 V}{8\pi} + \frac{H_c VT}{4\pi} \frac{dH_c}{dT} \quad \left(\frac{dH_c}{dT} < 0 \right) \quad . \tag{72}$$

At any T less than T_c, this reaction is spontaneous and is accompanied by a decrease in enthalpy and entropy. At T = 0 the reaction is still spontaneous ($\Delta G° < 0$) but the free energy change is due to a change in enthalpy only since $\Delta S° = 0$ at T = 0. As T approaches T_c, the enthalpy, entropy, and free energy approach zero, so that the reaction changes from a first-order phase transition to a second-order phase transition.

A model consistent with these facts is one in which the electrons in the superconducting state are separated from the normal state by an energy gap (incorporated into ΔH for the transition). The superconducting states are, however, not entirely free from interacting with the normal states because, as the temperature increases from absolute zero, the enthalpy according to Eq. (72) tends to zero. Thus, the energy gap must also tend to zero and become zero at $T = T_c$.

As to the nature of the superconducting state, Bardeen, Cooper, and Schrieffer (BCS) have postulated that the electrons near the Fermi surface attract one another forming bound electron pairs whose energy is lower than the free electrons in the normal state of the metal. The electron pair state would, thus, have a lower entropy than the normal state. The pairing occurs in such a way that an electron of wave vector +k and spin up is paired with an electron with wave vector -k and spin down. Since the energy gap is in part determined by the pairing energy of electrons, as the temperature is raised, the paired states begin to break up into normal states, so that the energy gap begins to collapse. At $T = T_c$ it collapses to zero where there are no longer any paired states. As long as there are some paired electrons, there can be a persistent current induced in the specimen. This can be seen by realizing that the total k vector for a paired state is zero (+k and -k are occupied by a pair). Any scattering of this pair by lattice vibrations can only scatter a paired state into another paired set of k vectors whose momentum is still zero. Scattering thus has no effect on the momentum of a pair. If we thus induce an additional momentum into the pair (viz., induce current flow), there is no way that this additional momentum can be damped out so long as the electrons remain paired. Thus, at any temperature between 0 °K and the critical temperature T_c, there will always be some electron pairs whose momentum is unaffected by scattering.

As to the question of what produces the pairing, the BCS theory assumes that an electron distorts a lattice in its neighborhood due to polarization of the surrounding atoms by the coulomb charge on the electron. This surrounding distortion will have an effect on the energy of a second electron in the neighborhood of the distortion such that the total energy of the two electrons is less than without the electron lattice interaction. The better this electron lattice interaction, the lower the energy of the pair that can be produced and, hence, the higher the critical temperature for the superconducting state. A strong electron lattice interaction in the normal state implies a low mobility and, therefore, low conductivity for the normal state. Thus, very good conductors like the alkali metals do not become superconductors.

If an electron interacting with the lattice is the major reason for the occurrence of pair states, then changing the Debye temperature Θ, which is a measure of the lattice stiffness, should change T_c. Θ can be changed by varying the isotope of the superconductor since

$$\Theta \propto (\text{mass})^{-\frac{1}{2}} \ .$$

It has been found for many superconductors that $T_c \propto (\text{mass})^{-0.5}$, although not all superconductors seem to have an isotope effect. Thus, for ruthenium and zirconium there is no isotope effect at all. It would appear that the electron lattice interaction may not be the only kind of interaction which can lead to electron pair formation.

XVII. ELECTRICAL CONDUCTIVITY MEASUREMENTS

Measurement of electrical conductivity as described earlier requires some precautions if reproducible and meaningful results are to be obtained. All measurements should be made on oriented single crystals since polycrystalline samples have grain boundaries which act as high resistance regions and grossly distort the true activation energies of the bulk. Usually, the crystals that are available are only a few millimeters in length, so that the samples must be worked with under a microscope, preferably a binocular or stereo scope which produces a three-dimensional image for ease in working. Samples can be handled with a fine tweezer or a vacuum pick-up pencil.

If measurements are performed as indicated in Fig. 2, it will usually be found that the total voltage across the standard and sample will be less than the battery voltage. The reason for this is the presence of an additional resistance present between the sample and the wire leads making contact to the sample. This is known as contact resistance. The contact resistance can be many orders of magnitude higher than the sample resistance, especially for materials which have high conductivity. Consequently, the measurement of conductivity as illustrated in the block diagram of Fig. 2 in which the potential measuring probes are placed directly on the crystal and separated from the current electrodes is essential for reliable results. Much of the literature is cluttered with electrical measurements made on compressed powders, polycrystalline samples, and electrical potential leads attached directly to the current leads rather than on the sample, so that caution should be taken in evaluating this data.

For example, on single crystals of $K_2Pt(CN)_4Br_{0.3}$, two probe electrical conductivity measurements (potential probes connected

across the current leads attached to the crystal) give $\sigma = 10^{-3} \, \Omega^{-1} \, cm^{-1}$.
Four probe measurements with the potential leads attached directly to
the crystal give $\sigma = 10 \, \Omega^{-1} \, cm^{-1}$ or a change of four orders of magnitude.

In the case of the four probe techniques illustrated in Fig. 2, the
value of L to be used in Eq. (9) to calculate the conductivity is the
separation between the voltage probes. The cross-sectional area can
be determined from the measurement of width and thickness. All three
measurements can be determined using a calibrated microscope eyepiece.

There will sometimes be a Seebeck effect across the potential
leads due to uncontrollable temperature gradients. This can be easily
eliminated by measuring the potential across the sample with current
first traveling in one direction and then reversing the current and
remeasuring the potential. The Seebeck effect does not reverse with
current, so that the average of the two potentials will eliminate it.

Attaching current and potential leads to a sample a few millimeters long is an art which can be learned in a few days work. For
materials with high decomposition points, contacts can be applied
with a fine pointed soldering iron using indium or other suitable low
melting solders. Indium is good because wires can be pressed into it
after it has hardened. Engelhard Co. sells a liquid gold and liquid
platinum which are organic complexes of these metals. When painted
on the crystal and heated to about 500 °C, the complex decomposes
leaving a fine film of gold or platinum on the surface to which wires
can be soldered. Needless to say, the crystals must be able to with-
stand high temperature baking.

For heat sensitive crystals, one must rely on contacts which can
be easily painted onto the sample. These include silver metal based
paints which are normally used on printed circuit boards (duPont Cor-
poration makes several varieties); silver-filled epoxies; an amalgam of
In and Hg which is made by dissolving enough indium in mercury until
the liquid will wet a glass surface; liquid gallium; and aquadag, which
is a commercially made suspension of graphite in 2-propanol used to

coat the inside of television picture tubes. Addition of butyl cellosolve acetate to the aquadag will give it a better consistency for handling and drying. There is no set rule as to which contacting agent to use (sometimes simple pressure contact between the wire and sample is all that is necessary). Finding the best contact (the one with the lowest contact resistance) is usually a hit or miss proposition. Generally, contact resistance will increase with decreasing temperature, so it is essential to get the contact resistance as low as possible at room temperature. If it becomes too high (say, greater than $10^5\Omega$), a potentiometer will be insensitive as a voltage measuring device, so that one will have to resort to using an electrometer. Good electrometers have input impedances greater than $10^{14}\Omega$ and are very versatile instruments in any laboratory studying electrical properties. (One can use an electrometer to measure voltages across samples whose resistances are about $10^{13}\,\Omega$ or less.)

The above method is a standard dc technique which requires four voltage measurements, across the standard resistor and across the sample with forward and reverse current. This can be reduced to two measurements if ac techniques are used. In its simplest form one substitutes a sine wave generator for the battery and uses an ac voltmeter as the measuring device. One difficulty with the ac technique, however, is that the sample has a capacitance as well as a resistance, so that the voltage seen by a voltmeter will be the vector sum of the resistive voltage and capacitive voltage. One is mainly interested in the resistive voltage. Since the voltage due to capacitive effects is 90° out of phase with the voltage due to resistance, one can use an ac measuring device that is phase sensitive. A schematic circuit incorporating one of these instruments called a lock-in detector amplifier is shown in Fig. 28. An audio signal from the lock-in is amplified by an audio amplifier, the output of which is used to drive current through the sample and standard. The standard must be a purely resistive one (available from General Radio Corp.). The voltage across the standard

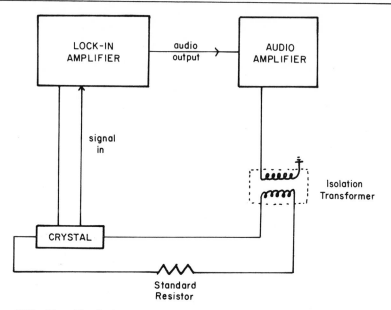

FIG. 28. Block diagram of phase sensitive ac technique for measuring conductivity. Audio signal from the lock-in which supplies current
to the crystal is matched in frequency and phase with the potential
across the crystal. Any spurious outside signals are eliminated.

is first measured with the lock-in, and the phase of this voltage, which
is also detected by the lock-in, is made to coincide with the phase from
the signal generator. The voltage across the sample is then read. The
voltage will be in phase with that of the internal-signal generator as
well as the standard. Any out-of-phase signal (due to capacitance) will
be effectively canceled. Any thermal emf's (and for that matter, any dc
components) will be eliminated.

XVIII. THE SEEBECK EFFECT

In a semiconductor in which conduction occurs in one band, the
Seebeck coefficient is related to the Fermi level by

$$S = \pm \frac{k}{e} \left(\frac{E_F}{kT} + A \right) \quad ,$$

the plus sign taken for hole carriers, the minus sign for electron carriers. A is a constant which can be 0, 1, or 2 depending on the type of scattering mechanism. A appears to be zero if the conduction is by hopping.

To see how S changes with carrier density, let us go back to our simple model of N_D donors below N_c conduction levels. The results for the Fermi level E_F and the number of carriers, n_c, in the conduction levels are given by Eqs. (55) and (56):

$$E_F = \frac{\Delta E}{2} - \frac{kT}{2} \ln \frac{N_c}{N_D} \quad ,$$

$$n_c = (N_c N_D)^{\frac{1}{2}} e^{-\Delta E/2kT} \quad .$$

Combining these two equations by eliminating $\Delta E/2$ we get

$$n_c = N_D \exp(-E_F/kT)$$

so that the Seebeck coefficient in terms of E_F is

$$S = -\frac{k}{e} \left[\ln \left(\frac{N_D}{n_c} \right) + A \right] \qquad (N_D > n_c) \quad .$$

The terms in brackets are both positive. With an increase in temperature, n_c increases and the absolute value of the Seebeck coefficient becomes smaller. Thus, a measurement of S will also give the density of carriers provided A is known. The situation becomes a bit more complicated if there is more than one type of charge carrier or more than one conduction band (see Harman and Honig).

The measurement of S requires that (a) a temperature gradient be established across the sample, (b) some means be provided for measuring the temperature at the two ends, and (c) some means be provided for measuring the Seebeck voltage across the ends. All this can be

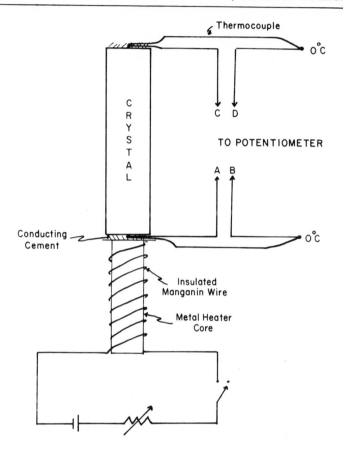

FIG. 29. Method for measuring the Seebeck effect. The heater supplies a temperature gradient across the crystal which can be measured with the thermocouples. The Seebeck voltage is then measured between A and C.

accomplished by using the setup shown in Fig. 29. Two small thermocouples are attached to the ends so that good thermal and electrical contact is maintained. This can usually be accomplished by placing the thermocouples up against the crystal and painting one of the contacting agents described for electrical resistance measurements over the junctions. The thermocouples can be made from 0.001 to 0.005-in. thermocouple wire (available from Sigmund Cohn or Omega Engineering). For low temperature work, copper constantan or the new gold (with

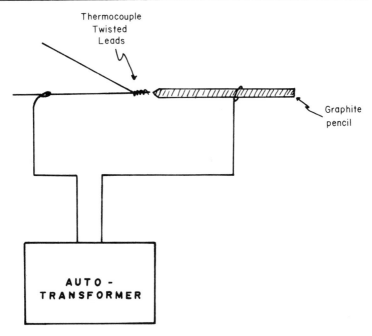

FIG. 30. Simple technique for making small thermocouple junctions. One of the twisted thermocouple leads and a graphite pencil are connected to an autotransformer (Variac). Momentary contact produces a weld at the thermocouple junction.

0.07% iron)-constantan are useful. The thermocouples can be made by using a simple technique, shown in Fig. 30. The thermocouple leads are twisted together. The output of an autotransformer is attached to one of the twisted pair and to a graphite pencil lead. The pencil lead is momentarily touched to the twisted thermocouple leads forming a thermocouple bead. The voltage on the autotransformer is adjusted by trial and error to give the best bead.

A heater can be made by wrapping some insulated manganin or nichrome wire around a support which can be placed up against one end of the crystal. The ends are connected to a small battery and the power to the heater is adjusted with a variable resistor (small 10-turn potentiometers are useful here). With the heater turned on, the temperatures at the two ends are measured with the thermocouples. Then one lead

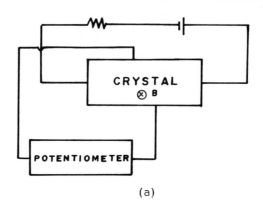

(a)

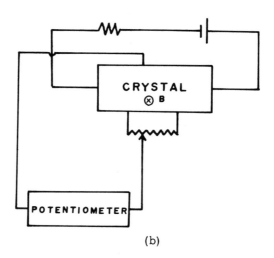

(b)

FIG. 31. (a) Schematic for measuring the Hall voltage showing exaggerated misalignment of potential probes. Magnetic induction B̲ is pointing into the page.

(b) Elimination of misalignment potential using a high resistance slide wire (e. g. , 10-turn potentiometer).

from each thermocouple is used to measure the Seebeck voltage across the crystal. $\Delta V/\Delta T$ is computed and several measurements are taken with decreasing ΔT to see if the ratio is constant. Temperature gradients of less than 10 °C will usually give constant S. Sometimes, however, heat leaks down the thermocouple wires, poor contacts, or poorly made thermocouples will lead to erroneous results. If the heat leaks are substantial, S will not appear to be constant with variable temperature. The heat leaks can be reduced by using smaller thermocouples or by placing additional resistance heaters in contact with the thermocouple wires downstream from the beads. These extra heaters can be turned on to reduce the heat flux away from the thermocouple beads.

XIX. THE HALL EFFECT

The equipment needed to measure the Hall effect is quite variable, depending on the size of the Hall voltage being measured relative to the resistive voltage that would be measured in the absence of a magnetic field. Figure 31(a) shows schematically the arrangement of current and potential leads for determination of the Hall voltage. Ideally, the Hall probes are supposed to be directly opposite one another. In practice this is never achieved. There is always some misalignment between the two probes which for clarity we show in Fig. 31(a). A current flowing through the crystal will therefore produce a voltage drop between the misaligned Hall probes; this misalignment voltage will usually be larger than the Hall voltage itself. If Δy is the amount of misalignment and x is the width of the crystal (distance between Hall probes), then the ability to measure a Hall voltage will be proportional to the mobility of the charge carriers. This can be seen by going back to the definitions for the Hall electric field and mobility, respectively,

$$E_x = vB \quad ,$$

$$\mu = v/E_y \quad ,$$

therefore

$$\frac{E_x}{E_y} = \mu B \quad . \tag{73}$$

Equation (73) gives the ratio between the Hall electric field E_x and the electric field producing the current, E_y. In terms of voltages $(E_x)x = V_x$ and $E_y \Delta y = V_y$,

$$\frac{V_x}{V_y} = \frac{\mu B x}{\Delta y} \quad .$$

For fixed dimensions, the greater the mobility, the larger the ratio of V_x to V_y and, therefore, the easier the measurement. For low mobility specimens (< 1 cm^2/V-sec), the Hall voltage V_x can be orders of magnitude less than the misalignment voltage V_y.

To reduce the misalignment voltage, what is often done is to introduce a third probe, as shown in Fig. 31(b). The two probes are connected by a multiturn potentiometer of high resistance compared to the sample. The slider is adjusted so that the misalignment voltage between probes is close to zero. The magnetic field is then turned on and V_x is measured. In order to be confident that a Hall voltage is being observed and not some other effect (such as magnetoresistance, for example), reversal of the magnetic field should reverse the Hall voltage polarity. In addition, for small magnetic fields, the Hall voltage should be linear with current and field strength. It is important that all these criteria be met if a true Hall signal is to be observed.

Several problems may arise with the above dc technique. Poor contacts to the sample will introduce noise problems which may overwhelm the Hall signal. It is thus necessary to achieve as low a contact resistance as possible. Noisy contacts are probably the greatest obstacle to overcome in this experiment. A Seebeck effect due to a temperature gradient across the probes can interfere with the Hall voltage, especially if it is not constant with time. Associated with this is the general drift of the misalignment potential with temperature in anisotropic solids. If the misalignment potential drifts due to small temperature changes, then the base line voltage, which is measured before the magnetic field is turned on, will be changing with time.

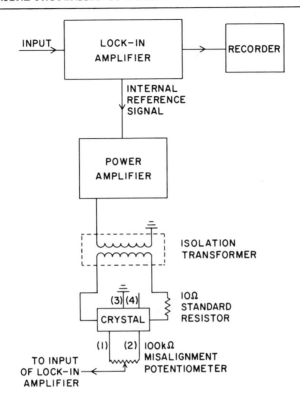

FIG. 32. Phase-sensitive detection of Hall voltage. Internal reference audio signal from the lock-in is amplified and used to feed current to the sample. The Hall voltage is fed back to the lock-in and matched in frequency and phase to the audio reference signal, thus eliminating any outside noise.

During the time it takes to turn on the magnetic field and measure the Hall voltage, the misalignment voltage can change by an amount larger than the Hall signal itself. This drift is often confused with the Hall signal itself. Thus, the importance of measuring the polarity of the voltage with field reversal as well as the linearity of the Hall voltage with current and field is most important.

Oscillating voltages due to noisy contacts as well as Seebeck effects due to temperature gradients can be eliminated by using an ac technique in which alternating current is passed through the sample at a fixed frequency and the ac Hall voltage is detected with a lock-in

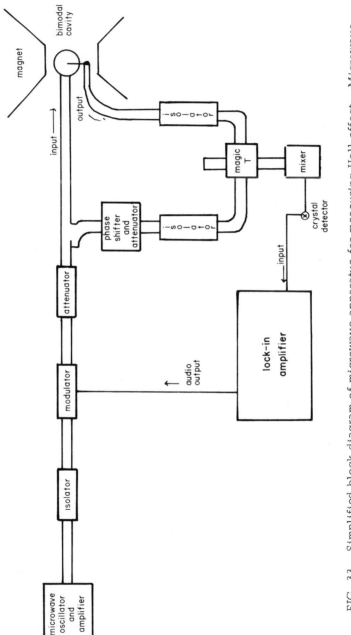

FIG. 33. Simplified block diagram of microwave apparatus for measuring Hall effect. Microwave power from the oscillator is first modulated by an audio frequency from the lock-in and fed to the cavity containing the sample. A small modulated output signal appears even in the absence of $\overline{B}$ because of cavity nonideality. The modulated output and input are compared at the mixer and the phases are adjusted to give a minimum reading on the lock-in. The magnetic field is turned on, producing a larger 90° output because of the Hall effect which is read on the lock-in (after Mohamed Sayed, JHU).

amplifier that is phase sensitive (Fig. 32). The advantages of this technique over the dc one is elimination of all spurious effects whose voltages are not at the same frequency and in phase with the Hall frequency. Thus, any dc effects are eliminated and most of the oscillating potentials from bad contacts can be reduced to tolerable levels. The technique has the advantage of high sensitivity, with signals as small as 10^{-9} V being detectable. It does not, however, eliminate the problems associated with misalignment potentials. The misalignment voltage will now be an ac potential which can still drift with temperature.

A technique that eliminates the misalignment problem is to use an alternating current and an alternating magnetic field whose frequency is different from the current frequency. The Hall voltage will then appear at a frequency equal to the sum and difference of the current and magnetic field frequencies. The misalignment potential is effectively eliminated since it occurs at a different frequency from that of the Hall voltage. This technique suffers from several serious problems and, thus, has not been widely used. Voltages other than the Hall voltage can occur at the sum and difference frequencies due to mixing of the current and field frequencies by nonlinear contacts and nonlinear circuits. It is very difficult to sort out these spurious effects since they will vary with field strength and current just like the Hall voltage.

A new technique which holds promise for the future is one which employs microwave radiation. Looking at Eq. (73) we see that, in order to determine the mobility of the carriers in a material, we have to establish an electric field E_y along the specimen, turn on a magnetic field B perpendicular to E_y, and look at the electric field E_x that is developed perpendicular to E_y and B. If we place a sample in an electromagnetic field, say, of frequency 9×10^9 Hz, we essentially establish E_y in the specimen due to its interaction with the electromagnetic field. By turning on a magnetic field perpendicular to E_y, an electromagnetic wave E_x will be produced which will be perpendicular to both E_y and B. Measurement of the electromagnetic radiation emitted from the specimen

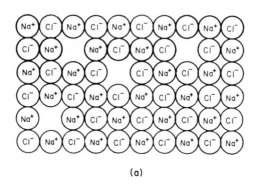

(a)

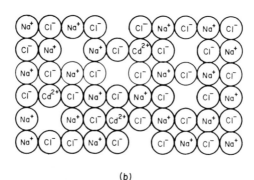

(b)

FIG. 34. (a) Two-dimensional array of sodium and chloride ions showing some Schottky defects. Migration occurs by hopping of sodium or chloride ions into the vacancy.

(b) Substitution of Cd^{2+} for Na^+ requires the formation of additional Na^+ vacancies, one for each Cd^{2+} present.

perpendicular to E_y and B will then give the Hall field. The advantage of a technique such as this is elimination of all contacts to the sample. A block diagram of an apparatus that uses this technique is shown in Fig. 33.

XX. IONIC CONDUCTION

We conclude this chapter with a brief discussion of ionic conductivity. As explained in Chapter 8, all ionic solids contain an equilibrium number of random vacancies of the Schottky type, the presence of
which increases the entropy over that of the perfectly ordered structure
and thus lowers the free energy. The number of such vacancies at any
temperature follows the Boltzmann distribution

$$n = N \exp \left(- \frac{\Delta E}{2kT} \right) \quad , \tag{74}$$

where n is the number of vacancies per cubic centimeter, N the total
number of ions per cubic centimeter, and ΔE the energy necessary to
create the vacancy (viz., the work that must be done in order to remove
an ion from the interior of the solid). The 2 appears in the denominator
of the exponential since for every positive ion vacancy created there
must be a negative ion vacancy in order to conserve charge neutrality.
ΔE is estimated to be about 2.3 eV in NaCl. With the total ion density
of $10^{21}/cm^3$ the number of vacancies at 1000 °K, for example, is

$$7.07 \times 10^{16} \quad .$$

Figure 34(a) shows a two-dimensional array of Na ions and Cl
ions with some vacancies present. The presence of these Scottky defects allows neighboring ions to jump into the empty sites. In the
presence of an electric field, there will be a preferential jumping of
Na^+ in the direction of the field and of Cl^- in the opposite direction.
The effective number of ions that can diffuse through the solid in this
way will be equal to the total number of Schottky defects, n. Figure 35
shows how the conductivity varies with temperature for several specimens of NaCl crystals. The high temperature conductivity is identical
for all specimens, whereas the low temperature conductivity varies
from sample to sample and has a lower activation energy. As explained
in Chapter 8, the presence of divalent impurities like Cd^{2+} substituting

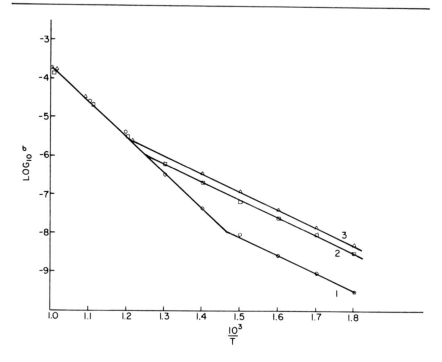

FIG. 35. Logarithm of the conductivity vs $10^3/T$ for several NaCl specimens. The lower temperature part of σ (to the right) varies from sample to sample due to different impurity concentrations. The higher temperature part, due to normal Schottky defects, is the same for all samples.

for Na^+ requires the creation of an extra Na^+ vacancy per Cd^{2+} ion in order to maintain charge neutrality, as shown in Fig. 34(b). These extra vacancies will be present even at low temperatures where the Schottky defect concentration is very low. For example, at 550°K, the number of Schottky defects is only about $10^{11}/cm^3$. The presence of, say, 10^{12} Cd^{2+} ions/cm^3 will create 10^{12} additional Na^+ vacancies, which is an order of magnitude larger than the Schottky defect concentration. Since there are 10^{21} Na^+ and Cl^- ions/cm^3, 10^{12} Cd^{2+}ions/cm^3 represents only a 1-ppb impurity level, a very small amount of impurity but more than enough to dominate the conductivity at low temperatures. The impurity concentration will vary from sample to sample, and, thus, the low temperature conductivity will be different for different samples. With increasing temperature, the Schottky defect concentration increases as described by Eq. (74), eventually becoming larger than the impurity concentration. The high temperature conductivity will therefore be dominated by the Schottky defect concentration which is the same for all samples.

As is the case for electron or hole conductors, ionic conductivity is governed by the equation

$$\sigma = ne\mu \quad ,$$

where n is the defect density and μ their mobility. For a given sample at low temperatures, the defect density is independent of temperature and proportional to the impurity concentration. As Fig. 35 shows for NaCl, the conductivity still has an activation energy at low temperatures. If n is constant, then

$$\mu \propto \exp(-\Delta E^*/kT) \quad ,$$

just as for hopping electron conduction described earlier. Ions jumping from site to site thus require an activation energy. From the slope of the low temperature part of Fig. 35, ΔE^* is calculated to be 0.90 eV.

At high temperature both n and μ have an activation energy, so that

$$\sigma \propto \left[\exp\left(-\frac{\Delta E}{2kT}\right)\right] \exp\left(-\frac{\Delta E^*}{kT}\right) \propto \exp\left(-\frac{\frac{1}{2}\Delta E + \Delta E^*}{kT}\right) \quad .$$

We already know ΔE^*. The sum can be determined from the high temperature slope of Fig. 35. This is

$$\Delta E/2 + \Delta E^* = 2.05 \text{ eV} \quad ,$$

so that $\Delta E = 2.30$, which is the value we used earlier to calculate the Schottky defect concentration.

For sample 1, the high conductivity and low conductivity intersect at $10^3/T = 1.48$ or $T = 675\,°K$. At this point the concentration

TABLE 5

Ionic Conductors with High Conductivity

Material	σ $(\Omega^{-1} cm^{-1})$
$RbAg_4I_5$	0. 26 (25 °C)
α-Ag_2HgI_4	10^{-3} (57 °C)
α-AgI	1. 3 (150 °C)

of Schottky defects and impurities must be equal. We can thus calculate the number of impurities from Eq. (74) by evaluating it at $T = 675\,°K$.
The number of impurities is thus 1.1×10^{14}. Extrapolating the low
temperature curve of sample 1 to room temperature gives $\sigma = 3.4 \times 10^{-17}$
$\Omega^{-1} cm^{-1}$. Hence, the mobility of the ions at room temperature in NaCl
is

$$\mu = \frac{\sigma}{ne} = \frac{3.4 \times 10^{-17}}{(1.1 \times 10^{14})(1.6 \times 10^{-19})} = 1.9 \times 10^{-12} \ cm^2/V\text{-}sec \quad .$$

These values of σ and μ should be compared to the values of σ and μ
for electron and hole conductors given in Tables 1 and 2. In general,
ionic conductors have very low conductivities and mobilities.

There are, however, some unusual ionic conductors with high
conductivities and high mobilities. Some of these are listed in Table 5.
In all cases the high conductivity is due to a high mobility of a large
number of Ag^+ ions in the structure. All these materials have structures in which there is more than one equivalent site available for each
Ag^+ ion present. Thus, all Ag^+ ions migrate among these equivalent
sites with a low activation energy. Since the density of Ag^+ is quite
high (equal to $10^{22}/cm^3$ for $RbAg_4I_5$), even if the mobility of the ions
were the same as in NaCl, the conductivity would still be eight orders
of magnitude higher because of the increased density of the mobile ions.
The mobility, however, is also many orders of magnitude larger than
typical ionic materials. For $RbAg_4I_5$,

$$\mu = \frac{\sigma}{ne} = \frac{0.26}{(10^{22})(1.6 \times 10^{-19})} = 1.6 \times 10^{-4} \, cm^2/V\text{-}sec \quad .$$

Such high conductivity ionic materials open up a new area of research for solid state scientists.

BIBLIOGRAPHY

J. S. Blakemore, Solid State Physics, Saunders, 1969.

W. Hume-Rothery, Electrons, Atoms, Metals and Alloys, Philosophical Library, 1955.

N. Cusack, The Electrical and Magnetic Properties of Solids, Longmans, Green, and Co., 1958.

A. J. Dekker, Solid State Physics, Prentice-Hall, Englewood Cliffs, New Jersey, 1962.

D. Greig, Electrons in Metals and Semiconductors, McGraw-Hill, New York, 1969.

N. B. Hannay, ed., Semiconductors, Reinhold, New York, 1959.

T. C. Harman and J. M. Honig, Thermoelectric and Thermomagnetic Effects and Applications, McGraw-Hill, New York, 1967.

C. Kittel, Elementary Solid State Physics, Wiley, New York, 1962.

C. Kittel, Introduction to Solid State Physics, Wiley, New York, 1968.

R. A. Levy, Principles of Solid State Physics, Academic, New York, 1968.

B. T. Matthias, "Systematics of Superconductivity, " in Superconductivity, (P. R. Wallace, ed.), Vol. 1, Gordon Breach, 1969.

B. Serin, "Superconductivity Experimental Part," Encyclopedia of Physics XV, Springer-Verlag, Berlin, 1956.

D. Shoenberg, Superconductivity, Cambridge Univ. Press, London and New York, 1962.

N. F. Mott and R. W. Gurney, Electronic Processes in Ionic Crystals, Dover, New York, 1964.

B. B. Owens and G. Argue, "High Conductivity Solid Electrolyte System RbI-AgI," J. Electrochem. Soc., 117, 898 (1970).

Chapter 5

MAGNETIC PROPERTIES

J.J. Steger

Department of Chemistry
Cornell University
Ithaca, New York

I.   INTRODUCTION

Ever since the discovery of lodestone (leading stone), magnetism
has intrigued man.   For centuries magnetism was purported to be a
"mystical power."   Today it is known that the cause of magnetism is
the motion, both spin and orbital, of the electrons in the atoms making
up matter.   Nuclear spins also produce magnetic fields, but their magnitude is negligible in comparison to atomic effects.   The purpose of
this chapter is to discuss briefly the highlights and basic physical
features of atomic magnetism.   Special emphasis will be placed on the
information which can be obtained by magnetic measurement.   Specific
examples will be avoided so as not to obscure the general features of
a particular topic.   It is hoped that this chapter will give the reader a
foundation for further study of magnetism.

II.   INTERACTION OF A MAGNETIC DIPOLE
WITH A MAGNETIC FIELD

Consider a rigid magnetic dipole, equal and opposite ends of a
magnet separated by some finite distance (n - s), placed in a uniform
magnetic field $\underline{H}$, as shown in Fig. 1.   From a qualitative point of
view, the dipole will experience a torque, or turning force, which will
tend to align it with the field.   In order to treat the problem quantitatively some basic definitions are required.

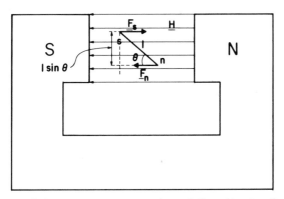

FIG. 1.   Schematic representation of the aligning forces, F_s and F_n, on a magnetic dipole immersed in a uniform magnetic field $\underline{H}$.

First, for convenience, a magnetic unit pole[*] is defined as one which will attract an equal and opposite pole with a force of one dyne when separated by a distance of one centimeter in a vacuum.   There- fore, when an arbitrarily chosen magnet exerts a force of X dynes on a unit pole placed one centimeter away, the magnet is defined to have a pole strength equivalent to X unit poles.   The magnitude, intensity, or strength of a magnetic field $\underline{H}$ at some point in space is just equal to the number of dynes, expressed in Øersteds, acting on a unit pole placed at that point in space, where the Øersted is the conventional unit of magnetic magnitude.   By convention, the direction of the field is that direction in which a unit north pole will move when placed in the field.   Thus, if a unit north pole is placed some distance away from the face of a magnet, and it experiences an attractive force of X dynes, the magnitude of the field at that point is X Øersteds.   Since the conventional unit north pole is attracted, the magnetic pole in question is by definition a south pole.

If, as depicted in Fig. 1, a north-south dipole of length ℓ and pole strengths p interacts with a uniform magnetic field $\underline{H}$, both its

[*] No magnetic unit pole has ever been found experimentally.

north and south poles will experience forces given, respectively, by

$$\underline{F}_n = +p\underline{H} \tag{1}$$

and

$$\underline{F}_s = -p\underline{H} \quad . \tag{2}$$

Taken together, these two equal magnitude forces constitute a couple which tends to align the dipole with the field. The fact that these forces constitute a couple means that the turning moment or torque τ is independent of the origin, and its magnitude can be written simply as

$$\tau = F_s \ell \sin \theta \quad , \tag{3}$$

where the origin is now the north pole. Thus, the turning moment as given above is simply the force $\underline{F}_s$ acting through the distance $\ell \sin \theta$. Substitution of pH for the magnitude of $\underline{F}_s$ yields

$$\tau = pH \ell \sin \theta \quad . \tag{4}$$

Next, the dipole moment μ is defined as a vector quantity with direction north to south along the magnet, with magnitude

$$\mu = p\ell \quad . \tag{5}$$

Substituting into Eq. (4) gives

$$\tau = \mu H \sin \theta \quad . \tag{6}$$

From this equation it can be seen that the turning effect is dependent on the magnitude of the field, the angle that the dipole makes with the field, and the magnitude of the dipole moment, which is a characteristic of the particular dipole in question.

In order to calculate the potential energy of the dipole in the field it is necessary to establish a reference geometry. The potential will be defined to be zero when the dipole or dipole moment is perpen-

dicular to the field ($\theta = 90°$). Thus, the potential is given by the
integral of the torque through the angle as

$$\varepsilon = \int_{90°}^{\theta} \tau \, d\theta \quad , \tag{7}$$

and substitution of Eq. (6) for τ yields

$$\varepsilon = \int_{90°}^{\theta} \mu H \sin \theta \, d\theta \quad . \tag{8}$$

Finally, integration gives

$$\varepsilon = -\mu H \cos \theta \quad , \tag{8a}$$

or expressed in vector notation

$$\varepsilon = -\underline{\mu} \cdot \underline{H} \quad . \tag{8b}$$

It can be seen that the potential energy has its minimum value when
the magnetic moment is pointing in the same direction as $\underline{H}$, then
$\varepsilon = -\mu H$. The maximum value occurs when μ and $\underline{H}$ are pointing in
opposite directions with a potential $\varepsilon = +\mu H$.

III. ORIGINS OF MAGNETISM

In the nineteenth century, Øersted discovered that a current
loop creates a magnetic field like that of a magnetic dipole (Fig. 2).
An electron in an orbit about a nucleus can be pictured as a small
current loop generating a magnetic field equivalent to a simple dipolar
bar magnet.

From Bohr's model of the hydrogen atom the following simple
qualitative treatment of the orbital origin of magnetism can be given.
Consider (Fig. 3) an electron in a circular orbit about a nucleus of
atomic number Z. Since the orbit is stable, the Coulombic attraction
between the electron and the proton must be equal and opposite in
direction to the centripetal force. In mathematical terms this is stated
simply as

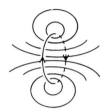

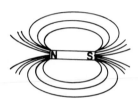

FIG. 2.   Representation of the similarity between the magnetic field of a simple dipolar bar magnet and that created by a current loop.

$$Ze^2/r^2 = m_e v^2/r \quad , \tag{9}$$

where e is the electron charge, r is the orbital radius, v is the tangential velocity of the electron, and m_e its rest mass.   Solving for the velocity yields

$$v = (Ze^2/m_e r)^{\frac{1}{2}} \quad . \tag{10}$$

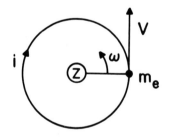

FIG. 3.   Bohr model for hydrogen-like atoms showing an electron of mass m_e and velocity v orbiting a nucleus, of charge Ze, with angular velocity ω and an orbital radius R. The electron current i is, by convention, opposite to the direction of the electron's rotation.

Noting that the angular velocity is $\omega = v/r$, then substitution for the tangential velocity gives

$$\omega = (Ze^2/m_e r^3)^{\frac{1}{2}} \quad . \tag{11}$$

If a current i is the rate at which charge e passes a given point in time, then

$$i = e\nu \tag{12}$$

for an electron with an orbital frequency ν. Recalling from elementary mechanics that the orbital frequency is given simply as

$$\nu = \omega/2\pi \quad , \tag{13}$$

then the orbital current can be written, in electromagnetic units, as

$$i = e\omega/2\pi c \quad , \tag{14}$$

where c is the speed of light. Substitution for ω by Eq. (11) gives

$$i = \frac{e(Ze^2/m_e r^3)^{\frac{1}{2}}}{2\pi c} \tag{15}$$

In analogy with a simple dipole magnet discussed in the previous section the dipole moment of a current loop is given as

$$\mu = iA \quad , \tag{16}$$

where $A = \pi r^2$ is the area of the electron orbit. Substitution for i and A yields

$$\mu_{\underset{\sim}{L}} = \frac{e^2(Zr/m_e)^{\frac{1}{2}}}{2c} \quad , \tag{17}$$

where the subscript $\underset{\sim}{L}$ has been introduced to designate that this moment is the orbital moment consistent with the orbital quantum number $\underset{\sim}{L}$. From classical physics it is known that the magnitude L of the orbital angular momentum is simply the product of the magnitude of the linear momentum, $p = mv$, times the radius of the orbit. Thus

$$L = m_e vr \quad , \tag{18}$$

and substitution of Eq. (10) for the velocity gives

$$L = (Ze^2 m_e r)^{\frac{1}{2}} \quad . \tag{19}$$

By eliminating r between Eqs. (17) and (19), the final expression is

$$\mu_{\underset{\sim}{L}} = eL/2m_e c \quad , \tag{20}$$

which relates the magnitude of the orbital angular momentum L to the orbital magnetic dipole moment $\mu_{\underset{\sim}{L}}$. In other words, the orbital angular momentum of an electron generates a magnetic dipole moment.

In the early 1920's Uhlenbeck and Goudsmit, in an effort to explain atomic spectra, proposed that the electron has associated with it an intrinsic angular momentum, independent of the orbital angular momentum. This intrinsic angular momentum qualitatively depicts an electron as a small negatively charged mass spinning about an axis. This notion of electron spin generates a magnetic field in the same qualitative sense that orbital motion produces a magnetic field: a charge in motion yields an electric current, which creates a magnetic field. Although this idea of a spinning electron is hardly in accord with quantum mechanics, Dirac in the late 1920's showed that when treated relativistically the quantum theory predicts an intrinsic electron angular momentum. From a qualitative point of view, the spin angular momentum, as well as the orbital angular momentum, of an electron produces a dipolar magnetic field. In analogy with the orbital moment given by Eq. (20), the spin moment is given quantitatively by

$$\mu_{\underset{\sim}{S}} = eS/m_e c \quad , \tag{21}$$

where S is the spin angular momentum magnitude.

A quantum mechanical treatment gives the following relations for the magnitude of an electron's orbital and spin angular momentum:

$$L = \sqrt{\underset{\sim}{L}(\underset{\sim}{L} + 1)}\,\hbar \tag{22}$$

and

$$S = \sqrt{\underset{\sim}{S}(\underset{\sim}{S} + 1)}\,\hbar \quad, \tag{23}$$

where $\underset{\sim}{L}$ and $\underset{\sim}{S}$ are the orbital and spin quantum numbers. Substitution into Eqs. (20) and (21) gives

$$\mu_{\underset{\sim}{L}} = \beta\sqrt{\underset{\sim}{L}(\underset{\sim}{L} + 1)} \tag{24}$$

and

$$\mu_{\underset{\sim}{S}} = 2\beta\sqrt{\underset{\sim}{S}(\underset{\sim}{S} + 1)} \quad, \tag{25}$$

where β, the Bohr magneton, is

$$\beta = e\hbar/2m_e c \quad. \tag{26}$$

IV. ANGULAR MOMENTUM AND THE DIPOLE MOMENT

For a classical system, as considered in the last section, the magnitude of the orbital and spin dipole moments μ_L and μ_S are proportional to the magnitude of the orbital and spin angular momenta L and S, respectively. The directions of their respective magnetic moments, however, would be opposite to those of the angular momentum vectors because of the intrinsic negative value of the electronic charge e, as expressed in Eqs. (20) and (21).

When an atom with a net nonzero orbital and/or spin angular momentum is placed in a uniform magnetic field, it interacts with the field like the simple magnetic dipole discussed in Section II, where it was shown that both the angle which the dipole makes with the field and the magnitude of the dipole moment are dependent variables determining the energy for a given field strength.

The magnetic energy of an arbitrary atomic dipole can then be written as

$$\varepsilon = -\underline{\mu}_{eff} \cdot \underline{H} \quad , \tag{27}$$

where $\underline{\mu}_{eff}$ is now the total effective magnetic dipole moment of the atom arising from the net orbital and/or spin angular momentum. The total angular momentum vector $\underline{J}$ is the vectorial sum of the orbital angular momentum vector $\underline{L}$ and the spin angular momentum vector $\underline{S}$. In a quantum mechanical description the magnitude of an electron's total angular momentum J is quantized and given as $\sqrt{J(J + 1)}\,\hbar$, where the quantum number $\underline{J}$ ranges, in integral steps, from the sum of the quantum numbers $\underline{L} + \underline{S}$ to the absolute value of their difference $|\underline{L} - \underline{S}|$. Thus the vector addition of $\underline{L}$ and $\underline{S}$ is specified quantum mechanically such that the possible values of $\underline{J}$ have only certain integral differences. In other words, the vector addition yields only certain discrete values of $\underline{J}$, indicating that $\underline{L}$ and $\underline{S}$ cannot assume arbitrary directions with respect to each other.

Just as two angular momentum vectors $\underline{L}$ and $\underline{S}$ in a quantum mechanical description cannot assume arbitrary directions with respect to each other, so too, the total angular momentum vector $\underline{J}$ can assume

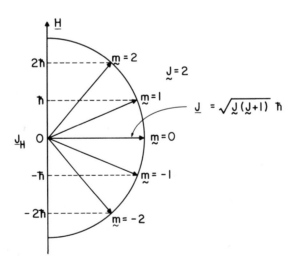

FIG. 4. The spatial quantization of the total angular momentum $\underline{J}$ with respect to a direction specified by an external field $\underline{H}$.

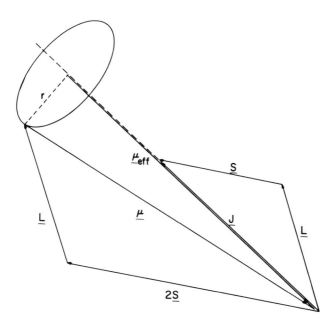

FIG. 5. The vector addition or coupling of the orbital $\underline{L}$ and spin $\underline{S}$ angular momenta yielding a resultant magnetic moment $\underline{\mu}$. The resultant magnetic moment $\underline{\mu}$ is not co-directional with the resultant angular momentum $\underline{J}$, because of the double contribution of the spin angular momentum to the magnetic moment. The effective component of the magnetic moment along $\underline{J}$ is specified by $\underline{\mu}_{eff}$.

only particular discrete directions with respect to an external magnetic field $\underline{H}$. $\underline{J}$ and its corresponding magnetic moment are spatially quantized such that J_H, the component of $\underline{J}$ along the magnetic field $\underline{H}$, is specified in integral steps. Thus

$$J_H = \underset{\sim}{m}\hbar \quad , \tag{28}$$

where $\underset{\sim}{m}$ ranges from $+J$ to $-J$ in integral steps. The spatial quantization of the total angular momentum is shown in Fig. 4 for the case when the quantum number $\underset{\sim}{J} = 2$.

At the time Goudsmit and Uhlenbeck proposed that the electron has an intrinsic spin, they also suggested that the ratio of the magnetic moment to the spin angular momentum is double that for the orbital angular momentum. Namely,

$$\frac{\mu_{\underset{\sim}{S}}}{\underline{S}} = \frac{2\mu_{\underset{\sim}{L}}}{\underline{L}} \qquad\qquad (29)$$

as indicated by Eqs. (20) and (21).   As shown in Fig. 5, the resultant
magnetic moment does not lie along the total angular momentum vector
$\underline{J}$ since the spin angular momentum contribution to the dipole moment
is twice that for the orbital angular momentum.   However, as a result
of a rapid precession of the dipole moment about the total angular
momentum vector $\underline{J}$, the long time average of the component of μ along
r, the precessional radius, is zero.   Consequently, it is only necessary to
consider the component μ_{eff} of μ in the $\underline{J}$ direction   when   dealing
with the magnetic moment of the atom.

The magnitude of the components of $\underline{L}$ and $\underline{S}$ along $\underline{J}$ are simply
L cos $(\underline{J}\underline{L})$ and   S cos $(\underline{J}\underline{S})$, respectively.   Thus,   the effective magnetic
moment is given as

$$\mu_{eff} = \frac{e}{2m_e c} [L \cos (\underline{J}\underline{L}) + 2S \cos (\underline{J}\underline{S})] \quad , \qquad\qquad (30)$$

where the factor of 2 in the spin term is a result of the double contribution of the spin angular momentum.

Application of the law of cosines to Fig. 5 shows that

$$L^2 = J^2 + S^2 - 2JS \cos (\underline{J}\underline{S}) \qquad\qquad (31)$$

and that

$$S^2 = J^2 + L^2 - 2JL \cos (\underline{J}\underline{L}) \quad . \qquad\qquad (32)$$

Rearrangement of the latter two equations yields

$$S \cos (\underline{J}\underline{S}) = \frac{J^2 + S^2 - L^2}{2J} \qquad\qquad (33)$$

and

$$L \cos (\underline{J}\underline{L}) = \frac{J^2 + L^2 - S^2}{2J} \quad , \qquad\qquad (34)$$

respectively.

Substitution of Eqs. (33) and (34) in Eq. (30) gives

$$\mu_{eff} = \frac{e}{2m_e c}\left(\frac{J^2 + L^2 - S^2}{2J} + 2\frac{J^2 + S^2 - L^2}{2J}\right) \quad . \tag{35}$$

Expressions for the magnitude of the angular momentum vectors in terms of the quantum numbers are given as follows

$$J = \sqrt{\underset{\sim}{J}(\underset{\sim}{J} + 1)}\,\hbar$$

$$L = \sqrt{\underset{\sim}{L}(\underset{\sim}{L} + 1)}\,\hbar \tag{36}$$

$$S = \sqrt{\underset{\sim}{S}(\underset{\sim}{S} + 1)}\,\hbar \quad ,$$

or

$$J^2 = \underset{\sim}{J}(\underset{\sim}{J} + 1)\,\hbar^2$$

$$L^2 = \underset{\sim}{L}(\underset{\sim}{L} + 1)\,\hbar^2 \tag{37}$$

$$S^2 = \underset{\sim}{S}(\underset{\sim}{S} + 1)\,\hbar^2 \quad .$$

Substitution of Eqs. (37) in Eqs. (35) yields

$$\mu_{eff} = g\frac{e}{2m_e c}J \quad , \tag{38}$$

where

$$g = 1 + \frac{\underset{\sim}{J}(\underset{\sim}{J} + 1) + \underset{\sim}{S}(\underset{\sim}{S} + 1) - \underset{\sim}{L}(\underset{\sim}{L} + 1)}{2\underset{\sim}{J}(\underset{\sim}{J} + 1)} \tag{39}$$

is referred to as the g-factor and is a measure of the relative contribution of the spin and orbital angular momenta to the effective magnetic moment.

Expression of Eq. (38) in vector notation as

$$\underline{\mu}_{eff} = g\frac{e}{2m_e c}\underline{J} \tag{40}$$

allows substitution in Eq. (27), giving

$$\varepsilon = -g\frac{e}{2m_e c}\underline{J}\cdot\underline{H}$$

$$= -g\frac{e}{2m_e c}J_H H \quad , \tag{41}$$

where J_H is again the component of $\underset{\sim}{J}$ in the direction of $\underline{H}$, as shown in Fig. 4.

Finally substitution from Eq. (28) yields

$$\varepsilon = -g \frac{e}{2m_e c} \underset{\sim}{m} \hbar H$$

$$= -g \beta \underset{\sim}{m} H \quad .$$

(42)

This equation expresses the energy of a quantum mechanical atom, with an effective magnetic moment arising from the relationship of angular momentum to the dipole moment, interacting with an external magnetic field.

V.   PARAMAGNETISM

Consider a solid consisting of N noninteracting atoms per unit volume, each with a net dipole moment. In the last section it was shown that the interaction energy for a single atomic dipole with a field is

$$\varepsilon_{\underset{\sim}{m}} = -g \beta \underset{\sim}{m} H \quad ,$$

(43)

where $\underset{\sim}{m}$ ranges from $+\underset{\sim}{J}$ to $-\underset{\sim}{J}$ in integral steps. When immersed in a uniform magnetic field at a temperature T, the probability that a particular atom of the system will be found in some energy state $\varepsilon_{\underset{\sim}{m}}$ is given by a Boltzmann distribution function,

$$P(\varepsilon_{\underset{\sim}{m}}) = A \exp(-\varepsilon_{\underset{\sim}{m}}/kT) \quad ,$$

(44)

which, when normalized, gives

$$\sum_{\underset{\sim}{m}=-\underset{\sim}{J}}^{\underset{\sim}{J}} P(\varepsilon_{\underset{\sim}{m}}) = 1 \quad .$$

(45)

This implies that

$$A = \frac{1}{\displaystyle\sum_{\underset{\sim}{m}=-\underset{\sim}{J}}^{\underset{\sim}{J}} \exp(-\varepsilon_{\underset{\sim}{m}}/kT)} \qquad . \qquad (46)$$

Then

$$P(\varepsilon_{\underset{\sim}{m}}) = \frac{\exp(-\varepsilon_{\underset{\sim}{m}}/kT)}{\displaystyle\sum_{\underset{\sim}{m}=-\underset{\sim}{J}}^{\underset{\sim}{J}} \exp(-\varepsilon_{\underset{\sim}{m}}/kT)} \qquad , \qquad (47)$$

and substitution for $\varepsilon_{\underset{\sim}{m}}$ yields

$$P(\varepsilon_{\underset{\sim}{m}}) = \frac{\exp(g\beta \underset{\sim}{m} H/kT)}{\displaystyle\sum_{\underset{\sim}{m}=-\underset{\sim}{J}}^{\underset{\sim}{J}} \exp(g\beta \underset{\sim}{m} H/kT)} \qquad . \qquad (48)$$

This probability function arises as a result of the tendency of the field to orient the atomic dipoles, as opposed to the tendency of the thermal agitations to randomize them. Consequently, a dimensionless quantity can be defined as the ratio of the magnetic ordering energy to the thermal randomizing energy as follows:

$$Y = g\beta H/kT \quad . \qquad (49)$$

The probability of a particular energy state is then given as

$$P(\underset{\sim}{m}) = \frac{\exp(\underset{\sim}{m}Y)}{\displaystyle\sum_{\underset{\sim}{m}=-\underset{\sim}{J}}^{\underset{\sim}{J}} \exp(\underset{\sim}{m}Y)} \qquad . \qquad (50)$$

Now note that the average value of a quantity x is given in terms of its probability P(x) as

$$x_{av} = \frac{\displaystyle\sum_{all\ x} P(x)x}{\displaystyle\sum_{all\ x} P(x)} = \sum_{all\ x} P(x)x \quad , \qquad (51)$$

where the sum of the probabilities has been normalized to unity, namely,

$$\sum_{all\ x} P(x) = 1 \quad . \tag{52}$$

Then, recalling from Fig. 4 that the component of $\underset{\sim}{J}$ along the field direction is specified by $\underset{\sim}{m}$, the magnetic moment along the field is given simply as

$$\mu_H = g\beta\underset{\sim}{m} \quad . \tag{53}$$

The average magnetic moment along the field direction μ_{Hav} can then be written as follows:

$$\mu_{Hav} = \sum_{\underset{\sim}{m} = -\underset{\sim}{J}}^{\underset{\sim}{J}} P(\epsilon_{\underset{\sim}{m}}) g\beta\underset{\sim}{m} = \frac{\sum_{\underset{\sim}{m} = -\underset{\sim}{J}}^{\underset{\sim}{J}} \exp(\underset{\sim}{m}Y) g\beta\underset{\sim}{m}}{\sum_{\underset{\sim}{m} = -\underset{\sim}{J}}^{\underset{\sim}{J}} \exp(\underset{\sim}{m}Y)} \quad . \tag{54}$$

Next a magnetic partition function Q is defined as the summation over all the magnetic energy levels for a single atom as

$$Q = \sum_{\underset{\sim}{m} = -\underset{\sim}{J}}^{\underset{\sim}{J}} \exp(\underset{\sim}{m}Y) = \sum_{\underset{\sim}{m} = -\underset{\sim}{J}}^{\underset{\sim}{J}} \exp(g\beta\underset{\sim}{m}H/kT) \quad . \tag{55}$$

The numerator in Eq. (54) can be expressed in terms of Q as follows:

$$\sum_{\underset{\sim}{m} = -\underset{\sim}{J}}^{\underset{\sim}{J}} \exp(\underset{\sim}{m}Y) g\beta\underset{\sim}{m} = kT \frac{\partial Q}{\partial H} \quad . \tag{56}$$

Then the average magnetic moment along the field is given simply as

$$\underline{\mu}_{Hav} = \frac{kT}{Q} \frac{\partial Q}{\partial H} = kT \frac{\partial \ln Q}{\partial H} = kT \frac{\partial \ln Q}{\partial Y} \frac{\partial Y}{\partial H} = g\beta \frac{\partial \ln Q}{\partial Y} \quad . \tag{57}$$

In order to obtain a value for the average magnetic moment, it is necessary to evaluate the partition function Q. To evaluate Q, let

$$x = \exp(Y) \quad . \tag{58}$$

Then

$$Q = \sum_{\underset{\sim}{m} = -\underset{\sim}{J}}^{\underset{\sim}{J}} x^{\underset{\sim}{m}} = x^{-\underset{\sim}{J}} + x^{(-\underset{\sim}{J}+1)} + \ldots + x^{\underset{\sim}{J}} \quad . \tag{59}$$

Since the ratio of any term to the preceding term is x, this can be recognized as a simple geometric series. Multiplying both sides by x gives

$$Qx = x^{(-J+1)} + x^{(-J+2)} + \ldots + x^{(J+1)} \quad . \tag{60}$$

The subtraction of Eq. (60) from Eq. (59) gives

$$Q(1-x) = x^{-J} - x^{(J+1)} \quad , \tag{61}$$

implying that

$$Q = \frac{x^{-J} - x^{(J+1)}}{1-x} \quad . \tag{62}$$

Multiplication by $x^{-\frac{1}{2}}/x^{-\frac{1}{2}}$ yields the symmetric form

$$Q = \frac{x^{-(J+\frac{1}{2})} - x^{(J+\frac{1}{2})}}{x^{-\frac{1}{2}} - x^{\frac{1}{2}}} \quad , \tag{63}$$

and substitution for x gives

$$Q = \frac{\exp[-(\underset{\sim}{J} + \frac{1}{2})Y] - \exp[(\underset{\sim}{J} + \frac{1}{2})Y]}{\exp(-Y/2) - \exp(Y/2)} \quad . \tag{64}$$

The latter expression can be written in terms of a hyperbolic function, recalling that

$$\sinh x = [\exp(x) - \exp(-x)]/2 \quad . \tag{65}$$

Then

$$Q = \frac{\sinh[(\underset{\sim}{J} + \frac{1}{2})Y]}{\sinh(Y/2)} \quad . \tag{66}$$

Taking the natural logarithm yields

$$\ln Q = \ln \sinh[(\underset{\sim}{J} + \frac{1}{2})Y] - \ln \sinh(Y/2) \quad , \tag{67}$$

and now differentiating with respect to Y gives

$$\frac{\partial \ln Q}{\partial Y} = \frac{(\underset{\sim}{J} + \frac{1}{2}) \cosh[(\underset{\sim}{J} + \frac{1}{2})Y]}{\sinh[(\underset{\sim}{J} + \frac{1}{2})Y]} - \frac{\frac{1}{2}\cosh(Y/2)}{\sinh(Y/2)}$$

(68)

$$= (\underset{\sim}{J} + \frac{1}{2}) \coth[(\underset{\sim}{J} + \frac{1}{2})Y] - \frac{1}{2}\coth(Y/2) \quad .$$

Defining a function $B_{\underset{\sim}{J}}(Y)$ as

$$B_{\underset{\sim}{J}}(Y) = \frac{2\underset{\sim}{J} + 1}{2\underset{\sim}{J}} \coth[(\underset{\sim}{J} + \frac{1}{2})Y] - \frac{1}{2\underset{\sim}{J}} \coth(Y/2) \quad ,$$

(69)

then

$$\frac{\partial \ln Q}{\partial Y} = \underset{\sim}{J} B_{\underset{\sim}{J}}(Y) \quad ,$$

(70)

where $B_{\underset{\sim}{J}}(Y)$ is known as a Brillouin function. Substitution of Eq. (70) into Eq. (57) gives the average magnetic moment for a single atom as a function of the temperature, field strength, and quantum number:

$$\mu_{Hav} = g\beta \underset{\sim}{J} B_{\underset{\sim}{J}}(Y) \quad .$$

(71)

The magnetization or magnetic moment per unit volume is then simply

$$M_{Hav} = N\mu_{Hav} = g\beta \underset{\sim}{J} B_{\underset{\sim}{J}}(Y)N \quad ,$$

(72)

where N is the number of magnetic atoms per unit volume.

In the derivation of Eq. (72) the assumption was made that every atom has the same $\underset{\sim}{J}$ quantum state. This is to say that $\underline{L}$ and $\underline{S}$ are coupled strongly and that a large amount of energy Δ would be required in order to excite the atom to a higher energy state $\underline{J}'$. It is convenient to distinguish between two cases: (1) when the transition energy Δ to a higher $\underset{\sim}{J}$ state is much greater than the thermal energy kT and (2) when the transition energy is much less than the thermal energy (to be treated in Section XI).

When the transition energy is much greater than the thermal energy, the probability that the atoms will be in their ground $\underset{\sim}{J}$ quantum state is large, and Eq. (72) for the magnetic moment will be accurate, since almost all the atoms will have the same J quantum number. Within this ground state there will be a further splitting into a magnetic multiplet, as was shown in Fig. 4, where the angular momentum was spatially quantized such that there existed $2J + 1$ distinct orientations of J with respect to the external field H. This again can be separated into two limiting cases; first, where Y, the ratio of the magnetic ordering energy to the thermal energy, is much greater than unity, and second, where Y is much less than unity. For a system with all the atoms in a given ground J state, if $Y \gg 1$, then

$$B_J(Y) = 1 \quad , \tag{73}$$

since

$$\coth X \cong 1 \quad \text{for} \quad X \gg 1 \, . \tag{74}$$

Therefore, the magnetization is given as

$$M_{sat} = M_{Hav} = g\beta J N \quad , \tag{75}$$

which is a constant for a given N and J. This limiting case is an example of saturation magnetization where each atom is contributing its maximum possible magnetic moment to the magnetization. This saturation behavior can be shown graphically by a plot of the Brillouin function as a function of the ratio of the magnetic ordering energy to the thermal randomizing energy. As shown in Fig. 6, for large values of Y the Brillouin function asymptotically approaches a nearly constant value of unity, resulting in magnetic saturation.

On the other hand, for a ground state J with $Y \ll 1$,

$$B_J(Y) = (J + 1)Y/3 \quad , \tag{76}$$

since

$$\coth X \cong (3 + X^2)/3X \quad \text{for} \quad X \ll 1 \, . \tag{77}$$

The magnetization is then given as

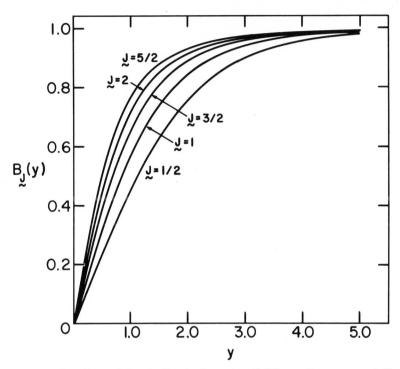

FIG. 6. Plot of the Brillouin function $B_J(Y)$ vs its argument Y for various different total angular moment states $\underset{\sim}{J}$. Note in particular the asymptotic approach of all curves to a saturation value of unity.

$$M_{Hav} = g\beta \underset{\sim}{J}(\underset{\sim}{J} + 1)YN/3 = g^2\beta^2\underset{\sim}{J}(\underset{\sim}{J} + 1)HN/3kT \quad . \tag{78}$$

A paramagnetic solid consists of N atoms, each with a magnetic dipole moment. When placed in an external magnetic field, the elementary atomic dipoles tend to align – hampered by thermal vibrations – producing a characteristic magnetic moment for the solid. Examination of Eq. 78 shows a direct proportionality between the magnetization and the field:

$$M_{Hav} = \kappa H \quad , \tag{79}$$

where the proportionality constant κ is defined as the magnetic susceptibility per unit volume, a measure of the extent to which a material is susceptible to induced magnetization:

$$\kappa = g^2 \beta^2 J(J + 1)N/3kT \quad . \tag{80}$$

If the density of the material is given by ρ then the susceptibility per gram or mass susceptibility is given as

$$\chi = \kappa/\rho \tag{81}$$

and the susceptibility per gram atom and per gram mole are, respectively,

$$\chi_A = \chi \cdot AW \quad , $$
$$\chi_M = \chi \cdot MW \quad , \tag{82}$$

where AW and MW are the atomic and molecular weights. Thus, for a paramagnetic material

$$\chi_M = C/T \tag{83}$$

is known as the Curie law where the Curie constant C is given by

$$C = g^2 \beta^2 J(J + 1)N\,MW/3k\rho \quad . \tag{84}$$

Equation (83) has been derived by assuming that each atom is independent. If the atomic dipoles were allowed to interact with each other, the magnitude of the field for a single atom would be the sum of the applied external field and the internal field produced by all the other magnetic atoms in the substance. If the additional dipole-dipole interaction is considered, the system would be described as follows:

$$\chi_M = \frac{C}{(T - \theta)} \quad . \tag{85}$$

Equation (85) is known as the Curie-Weiss law and θ is the Weiss constant.

VI.   FERROMAGNETISM

In contrast to the behavior of paramagnets, where a net magnetization results only when the material is immersed in a field, ferromagnets exhibit a net magnetization in the absence of an applied field.
The interaction responsible for this is not the weak magnetic dipole-
dipole interaction mentioned in the last paragraph of Section V, but
results from the Pauli exclusion principle (no two electrons are allowed
to occupy the same quantum state).

When atoms with a net spin magnetic moment are condensed to
form a regular solid, the valence electrons on neighboring atoms interact with each other, the core electrons being negligibly affected. If
atomic orbitals on neighboring atoms overlap, valence electrons in the
same orbital states will be allowed to approach each other if their
spins are antiparallel, since antiparallel spins will put each in a different total quantum state. However, if electrons are allowed to
approach each other closely, a large coulombic repulsion results in an
unstable configuration. On the other hand, if the electrons are in
parallel spin states, then they will not be allowed to approach each
other since the Pauli exclusion principle will dictate that they have
different orbital states. In order to reduce the coulombic repulsion
between electrons, their spins assume a parallel configuration producing ferromagnetism. This basic electrostatic spin orientation interaction between neighboring atoms is referred to as the exchange
interaction. It is dependent on the amount of overlap between the
orbitals on adjacent atoms (a function of the distance between the
atoms in the solid).

Early in the twentieth century Weiss proposed that, in a ferromagnet, the atomic dipoles can be qualitatively viewed as experiencing
a large magnetic field produced by the exchange interaction. The
extent of spontaneous magnetization is then found by application of the
statistical treatment of paramagnetism given in Section V, where the

field H is now the external applied field H_a plus a large molecular field due to the exchange interaction between valence electrons of neighboring atoms in the solid. Equation (72) for the paramagnetic magnetization applies to ferromagnetic materials and is restated as follows:

$$M = g\beta J B_J(Y)N \quad , \tag{86}$$

where

$$Y = g\beta H/kT \tag{87}$$

with

$$H = H_a + \gamma M \quad . \tag{88}$$

The exchange interaction is introduced as a magnetic field γM proportional to the magnetization with a molecular field constant γ. Substitution of Eqs. (87) and (88) into Eq. (86) gives

$$M = g\beta J B_J\left[\frac{g\beta}{kT}(H_a + \gamma M)\right] N \quad . \tag{89}$$

In the absence of an applied field ($H_a = 0$), Eq. (89) simplifies to

$$M_{spon} = g\beta J B_J\left[\frac{g\beta}{kT} \gamma M_{spon}\right] N \quad , \tag{90}$$

where the magnetization is now the spontaneous magnetization of the ferromagnetic material. Now recall Eq. (75), which gives the saturation magnetization M_{sat} when all the N dipole moments are aligned:

$$M_{sat} = g\beta J N \quad . \tag{91}$$

This allows a ratio between the intensity of spontaneous magnetization at a given temperature and the maximum possible magnetization to be given as

$$\frac{M_{spon}}{M_{sat}} = B_J\left[\frac{g\beta}{kT} \gamma M_{spon}\right] \quad . \tag{92}$$

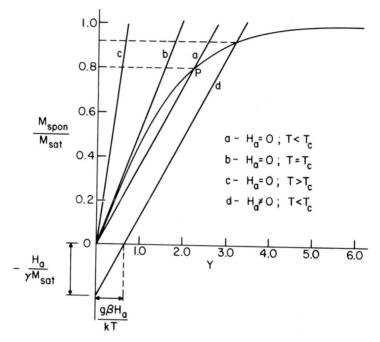

FIG. 7.   Graphical solution of the implicit function for the
reduced magnetization.

The latter equation is an implicit function of the dependence of the
spontaneous magnetization on temperature.   The solutions of Eq. (92)
can be found by a simultaneous plot of

$$\frac{M_{spon}}{M_{sat}} = \frac{kT}{g\beta\gamma M_{sat}} Y \tag{93}$$

and Eq. (92) itself, as functions of Y, where now

$$Y = g\beta\gamma M_{spon}/kT \quad . \tag{94}$$

In Fig. 7, Eqs. (92) and (93) are plotted as functions of Y.   The
straight lines with slope proportional to T represent Eq. (93) for several
different temperatures.   The points of intersection between the straight
lines and the Brillouin function are clearly the simultaneous solutions
of the equations.   For all lines, where $H_a = 0$, there exist the trivial

solution $M_{spon}/M_{sat} = 0$. For line a, the point P indicates the solution with a corresponding value $M_{spon}/M_{sat} = 0.8$. The line b, for $T = T_c$, is tangent to the Brillouin function at the origin where $M_{spon}/M_{sat} = 0$. This defines the critical temperature T_c, such that, for $T < T_c$, a spontaneous magnetization exists.

The transition temperature T_c, between a state of zero spontaneous magnetization to a net spontaneous magnetization can be found by equating Eqs. (92) and (93) as

$$B_{\underset{\sim}{J}}(Y) = \frac{kT}{g\beta\gamma M_{sat}} Y \quad , \tag{95}$$

which on rearrangement defines a temperature necessary for a particular simultaneous solution as

$$T = \frac{g\beta\gamma M_{sat}}{kY} B_{\underset{\sim}{J}}(Y) \quad . \tag{96}$$

Near the critical temperature the argument Y of the Brillouin function $B_{\underset{\sim}{J}}(Y)$ is much less than unity, and Eq. (76), which approximates the Brillouin function for small values of its argument Y as

$$B_{\underset{\sim}{J}}(Y) = (\underset{\sim}{J} + 1)Y/3 \quad , \tag{97}$$

can be applied. By substituting Eq. (97) into Eq. (96), the critical temperature is found to be

$$T_c = g\beta\gamma M_{sat}(\underset{\sim}{J} + 1)/3k \quad . \tag{98}$$

This critical, or Curie, temperature T_c is the temperature at which the spontaneous magnetization appears and disappears as the ferromagnetic material passes through it.[*]

Rearranging Eq. (98) as

$$g\beta\gamma/k = 3T_c/(\underset{\sim}{J} + 1)M_{sat} \tag{99}$$

---

[*]Magnetic hysteresis usually occurs when transcending the Curie temperature.

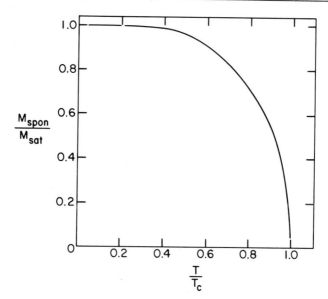

FIG. 8. Theoretical plot of the reduced magnetization M_{spon}/M_{sat} as a function of the reduced temperature T/T_C for a ferromagnetic material with $J = \frac{1}{2}$.

allows Eq. (92) to be rewritten as

$$\frac{M_{spon}}{M_{sat}} = B_J \left[\frac{3}{(J + 1)} \frac{M_{spon}}{M_{sat}} \frac{T_c}{T} \right] . \tag{100}$$

The implication of this equation is interesting; it shows that according to the molecular field theory the reduced magnetization M_{spon}/M_{sat} depends only on the form of the Brillouin function. In other words, a plot of the reduced magnetization vs the reduced temperature (Fig. 8) for several ferromagnetic materials with the same J value will be independent of the parameters which might vary for different materials, such as γ, N, and g. Figure 8 also shows that, for any temperature T below the Curie temperature T_c, a ferromagnetic material is always spontaneously magnetized. It is possible for a ferromagnetic material to exhibit zero total magnetic moment below its Curie temperature. To interpret this phenomenon we must assume that the material is composed of

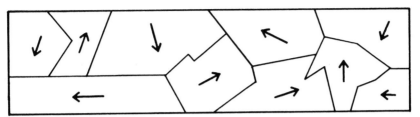

FIG. 9. Schematic representation of domain structure for a ferromagnetic material. The arrows represent the alignment of the dipoles
within the bounded domains.

separate magnetic domains, each with a net magnetic moment, oriented
with respect to each other so as to produce no external moment. Figure 9 schematically represents a two-dimensional ferromagnetic material
exhibiting domain structure; the independent magnetic moments in each
domain cancel.

In the ordered state ($T < T_c$), the presence of a small applied
field H_a changes Eq. (93) for the straight line necessary to solve the
implicit function to

$$\frac{M_{spon}}{M_{sat}} = \frac{kT}{g\beta\gamma M_{sat}}\left[Y - \frac{g\beta H_a}{kT}\right] = \frac{kT}{g\beta\gamma M_{sat}}Y - \frac{H_a}{\gamma M_{sat}} \quad . \quad (101)$$

The additional term $-H_a/\gamma M_{sat}$ corresponds to an ordinal intercept, as
shown in Fig. 7 for line d, from which an estimate of the molecular
field constant γ can be made. At constant temperature (constant slope)
application of an external field H_a shifts line a to give line d (Fig. 7).
As would be expected, this shift produces an increase in the ratio
M_{spon}/M_{sat} since the field experienced by the dipoles is now larger,
tending to align a greater number of them.

For $T > T_c$, the ferromagnetic material will be in a paramagnetic
state with zero spontaneous magnetization (line c of Fig. 7). The
application of an external field H_a plus the field produced by the
exchange interaction will induce a paramagnetic magnetization. In
this temperature range the ratio of the magnetic ordering energy to the
thermal randomizing energy will be small, and only the linear portion

of the Brillouin function will be required. The average magnetization along the field is given by Eq. (78) as follows:

$$M_{Hav} = g\beta \underline{J}(\underline{J} + 1)YN/3 \quad , \tag{102}$$

where

$$Y = g\beta H/kT \ll 1 \quad . \tag{103}$$

Substituting for Y and rewriting in terms of the saturation magnetization gives

$$M_{Hav} = M_{sat}(\underline{J} + 1)g\beta H/3kT \quad , \tag{104}$$

where

$$M_{sat} = g\beta \underline{J}N \quad . \tag{105}$$

Since the molecular field is no longer much greater than the applied field both must be retained, thus

$$H = H_a + \gamma M_{Hav} \quad , \tag{106}$$

and substitution in Eq. (104) for H gives on rearrangement

$$\kappa = M_{Hav}/H_a = C/(T - T_c) \quad , \tag{107}$$

where T_c is defined by Eq. (98) and

$$C = T_c/\gamma \quad . \tag{108}$$

As suggested in Section V, an interaction between the dipoles, in the latter case the exchange interaction, changes the form of the susceptibility from the Curie law, $\kappa = C/T$, to the Curie-Weiss law behavior, $\kappa = C/(T - \theta)$.

VII. ANTIFERROMAGNETISM

In the last section, it was shown that the role of the exchange interaction was to align elementary dipoles on neighboring atoms in a parallel configuration. This parallel alignment produces a spontaneous magnetization below the Curie temperature; a material exhibiting this behavior was termed a ferromagnet. In the discussion of ferromagnets it was noted that the exchange interaction was dependent on the distance between the atoms. The atoms had to be close enough for atomic orbital overlap, allowing application of the Pauli exclusion principle. However, if a material is so structured that all the elementary dipoles are separated by an intermediate ion (too far apart for direct overlap), a process termed superexchange can occur. Superexchange is a mechanism by which the spins on nearest neighbor dipoles are coupled through an intermediate.

The mechanism of superexchange can be qualitatively viewed as follows: suppose that two cations with partially filled d-orbitals are each sigma bonded to a p-orbital on an intermediate anion. If the p-shell of the intermediate anion is completely filled, then each p-orbital will contain two electrons of antiparallel spin. The exclusion principle now dictates that, if an electron is to be transferred between p and d orbitals forming a bond, it must have spin opposite to the electron in the already partially occupied d-orbitals. Thus, as shown in Fig. 10, a simultaneous partial covalent bond on opposite sides of

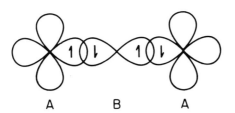

FIG. 10. Representation of a superexchange mechanism, showing antiparallel orientation of spins on nearest neighbor cations A via an intermediate anion B.

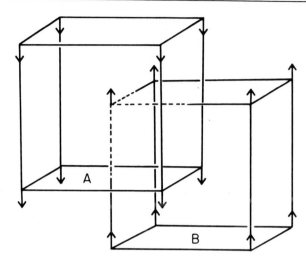

FIG. 11. Schematic diagram of two identical interpenetrating magnetic sublattices with opposing magnetic moments. This type of interpenetrating sublattices is characteristic of antiferromagnetic materials.

the intermediate anion will correlate the d-electron spins of the cations. The correlation results in the d-electrons of the cations being oriented antiparallel. This implies that the spin moment on the nearest neighbor cations will be in opposition, resulting in zero net spontaneous magnetization. Materials exhibiting this type of internal arrangement are called antiferromagnets.

A typical antiferromagnetic material may be pictured (Fig. 11) as two identical interpenetrating magnetic sublattices with spin magnetic moments in direct opposition, yielding no net spontaneous magnetization. The intermediate anions responsible for the superexchange are then equidistant between the apices of the interpenetrating magnetic sublattices.

Qualitatively, the behavior of antiferromagnetic materials can be treated in terms of the molecular field approximation. A first-order approximation would be to assume that the superexchange or antiferromagnetic coupling between nearest neighbor magnetic ions is so large as to render all other interactions negligible. A more refined model is

one that allows for next nearest neighbor interactions, namely, interactions within the sublattices. It is this more refined theory that will
be presented here.

Consider an ion of the A sublattice in Fig. 11. It will experience
an effective magnetic field H_A given as

$$H_A = H_a - \Gamma M_B - \gamma M_A \quad , \tag{109}$$

where the second term takes into account the molecular fields proportional to the magnetization of the B sublattice. The third term expresses the proportionality to the magnetization of the A sublattice
itself in the same sense that it was applied in the treatment of ferromagnetism. The first term is, of course, the external applied field.
Similarly, for an ion in the B sublattice the effective field is given as

$$H_B = H_a - \Gamma M_A - \gamma M_B \quad . \tag{110}$$

The proportionality constants Γ and γ are measures of the interactions
between nearest neighbor A-B ions and next nearest neighbor A-A or
B-B ions, respectively. The same proportionality constants appear in
both Eqs. (109) and (110) because the A and B sublattices are identical.
The signs of the proportionality constants Γ and γ can be determined
by noting that $\Gamma M_{A \, or \, B}$ corresponds to a molecular field generated by
superexchange and therefore favors antiparallel alignment. This antiparallel alignment implies that the effective field H_A or H_B should be
diminished. Thus, as expressed in Eqs. (109) and (110), Γ must be a
positive quantity. On the other hand, γ is the molecular field constant
resulting from a ferromagnetic coupling within the sublattice. It should
correspond to an increase in the effective field, implying a negative
sign as written in Eqs. (109) and (110).

Following the development leading to Eq. (89), expressions for
the magnetization M_A and M_B for each sublattice as a function of
temperature are

$$M_A = g\beta \underline{S} B_{\underline{S}} \left[\frac{g\beta}{kT} (H_a - \Gamma M_B - \gamma M_A) \right] N_A \tag{111}$$

and

$$M_B = g\beta \underline{S} B_{\underline{S}} \left[\frac{g\beta}{kT} (H_a - \Gamma M_A - \gamma M_B) \right] N_B \quad . \tag{112}$$

These equations differ from Eq. (89) in that the total angular momentum $\underline{J}$ has been reduced to its spin only component $\underline{S}$. This results from the fact that in most antiferromagnetic materials the field created by the crystal lattice reduces or quenches the orbital angular momentum to the extent of being negligible. This means that only the spin term contributes to the total angular momentum. Thus

$$\underline{J} = \underline{L} + \underline{S}$$

reduces to

$$\underline{J} = \underline{S} \quad , \tag{113}$$

since

$$\underline{L} = 0 \quad . \tag{114}$$

The g-factor as given in Eq. (33) also reduces to its spin only value of $g = 2$.

Dividing Eqs. (111) and (112) by the saturation magnetization for each lattice gives

$$\frac{M_A}{M_{A_{sat}}} = B_{\underline{S}} \left[\frac{g\beta}{kT} (H_a - \Gamma M_B - \gamma M_A) \right] \tag{115}$$

and

$$\frac{M_B}{M_{B_{sat}}} = B_{\underline{S}} \left[\frac{g\beta}{kT} (H_a - \Gamma M_A - \gamma M_B) \right] \quad , \tag{116}$$

where

$$M_{A_{sat}} = g\beta \underline{S} N /2 = g\beta \underline{S} N_A \quad , \tag{117}$$

and similarly

$$M_{B_{sat}} = g\beta\underset{\sim}{S}N/2 = g\beta\underset{\sim}{S}N_B \quad . \tag{118}$$

In the absence of an applied field the antiferromagnetic coupling tends to align the A and B sublattices in an antiparallel arrangement, which, together with the equivalence of the A and B sublattices, results in no net spontaneous magnetization. Thus, in the absence of an external field

$$M = M_{A_{spon}} + M_{B_{spon}} = 0 \quad , \tag{119}$$

which implies

$$M_{A_{spon}} = -M_{B_{spon}} \quad . \tag{120}$$

Rewriting Eqs. (115) and (116) for $H_a = 0$ and $M_A = -M_B$ and noting that $\Gamma > 0 > \gamma$, as discussed previously, gives

$$\frac{M_{A_{spon}}}{M_{sat}} = B_{\underset{\sim}{S}}\left[\frac{g\beta}{kT}(\Gamma - \gamma)M_{A_{spon}}\right] \tag{121}$$

and

$$\frac{M_{B_{spon}}}{M_{sat}} = B_{\underset{\sim}{S}}\left[\frac{g\beta}{kT}(\Gamma - \gamma)M_{B_{spon}}\right] \quad , \tag{122}$$

where

$$M_{sat} = M_{A_{sat}} = M_{B_{sat}} \quad . \tag{123}$$

Equations (121) and (122) are of the same form as Eq. (92) with a net molecular field constant $(\Gamma - \gamma)$ replacing γ. Solutions for these equations can be found by application of the graphical method discussed in Section VI as it applies to Fig. 7.

A development analogous to that leading to Eq. (98) defines a temperature called the Néel point as

$$T_n = g\beta(\Gamma - \gamma)M_{sat}(\underset{\sim}{S} + 1)/3k \quad . \tag{124}$$

This equation gives the temperature at which a transition from an anti-ferromagnetic state to a paramagnetic state takes place.

At temperatures below the Néel point, an applied field can assume any arbitrary angle relative to the preferred magnetization axes along which the magnetizations of the A and B sublattices are directed. Two distinct cases of special interest are: (1) when the applied field is parallel to this preferred axis, and (2) when it is perpendicular to it.

If the applied field is parallel to the preferred axis in the direction of the A sublattice magnetization, then the effective field experienced by A atoms will be increased by an amount H_a.

The antiparallel coupling of the B lattice magnetization relative to the A magnetization direction necessitates a decrease of H_a in the effective field experienced by B atoms. The change in the effective field produces a change in the magnetization of the sublattices. If the field is applied parallel to the A magnetization direction, then $M_A > M_B$. In Fig. 7, line d shows that an applied field shifts the simultaneous solution, changing the ratio of M_{spon}/M_{sat}. For small shifts, which displace line d in opposite directions for the A and B sublattices, the increase in magnetization for the sublattice aligned parallel to the field will be nearly equal to the decrease of magnetization in the other sublattice. Thus

$$M_A = M_{A_{spon}} + \partial M_A \tag{125}$$

and

$$M_B = M_{B_{spon}} + \partial M_B , \tag{126}$$

where

$$\partial M_A \approx \partial M_B , \tag{127}$$

and, as noted before for the antiferromagnetic state,

$$M_{A_{spon}} = - M_{B_{spon}} . \tag{128}$$

Thus an external field applied parallel to the preferred axis results in a
net magnetization

$$M = M_A + M_B \approx 2 \partial M_A \quad . \tag{129}$$

Near the absolute zero of temperature the Brillouin function approaches
a nearly constant value of unity. ∂M_A will then be essentially zero
for a small field-induced shift. Thus at very low temperatures the
parallel susceptibility for the antiferromagnetic state is given as

$$\kappa_\parallel = M/H_a = 2 \partial M_A / H_a = 0 \quad . \tag{130}$$

For temperatures approaching the Néel temperature the Brillouin
function can be approximated (as noted in Section V) as follows:

$$B_{\underset{\sim}{S}}(Y) = \left(\frac{S+1}{3}\right) Y \quad . \tag{131}$$

Equations (111) and (112) can then be approximated as

$$M_A = M_{sat} \left(\frac{S+1}{3}\right) Y_A \tag{132}$$

and

$$M_B = M_{sat} \left(\frac{S+1}{3}\right) Y_B \quad , \tag{133}$$

where

$$Y_A = \frac{g\beta}{kT} [H_a - \Gamma M_B - \gamma M_A] \tag{134}$$

and

$$Y_B = \frac{g\beta}{kT} [H_a - \Gamma M_A - \gamma M_B] \quad . \tag{135}$$

Differentiating with respect to H_a yields

$$\frac{\partial M_A}{\partial H_a} = M_{sat} \left(\frac{S+1}{3}\right) \frac{\partial Y_a}{\partial H} = \frac{c}{T} \left[1 - \Gamma \frac{\partial M_B}{\partial H_a} - \gamma \frac{\partial M_A}{\partial H_a}\right] , \tag{136}$$

where

$$c = M_{sat} \left(\frac{S+1}{3}\right) \frac{g\beta}{k} \quad . \tag{137}$$

Similarly

$$\frac{\partial M_B}{\partial H_a} = \frac{c}{T}\left[1 - \Gamma\frac{\partial M_A}{\partial H_a} - \gamma\frac{\partial M_B}{\partial H_a}\right] \quad . \tag{138}$$

The simultaneous solution of Eqs. (136) and (138) yields

$$\frac{\partial M_A}{\partial H_a} = \frac{\partial M_B}{\partial H_a} = \frac{c}{T + (\gamma + \Gamma)c} \quad . \tag{139}$$

Therefore the susceptibility for an antiferromagnetic state with the field parallel to the preferred axis for temperatures approaching the Néel temperature is given as

$$\kappa_\| = \frac{\partial M}{\partial H_a} = \frac{\partial M_A}{\partial H_a} + \frac{\partial M_B}{\partial H_a} = \frac{2c}{T + (\gamma + \Gamma)c} \quad . \tag{140}$$

Substitution gives

$$\kappa_\| = C/(T - \theta) \quad , \tag{141}$$

where

$$\theta = -T_n(\Gamma + \gamma)/(\Gamma - \gamma) \quad . \tag{142}$$

The Néel temperature T_n was given previously by Eq. (124).

If the field is applied perpendicular to the preferred axis of magnetization, the sublattice magnetizations will each be equally rotated through some small angle φ, resulting in a small component of induced magnetization in the direction of the field. Neglecting any changes in the magnitude of the sublattice magnetizations, the magnitude of the induced moment parallel to the applied field is

$$M_H = M_A \sin \varphi + M_B \sin \varphi \quad . \tag{143}$$

The perpendicular susceptibility is given by

$$\kappa_\perp = M_H/H_a = (M_A + M_B) \sin \varphi/H_a \quad . \tag{144}$$

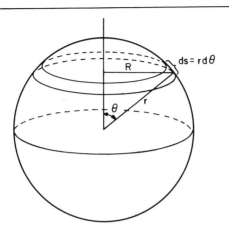

FIG. 12. A sphere of radius vector r representing all possible random orientations with respect to a specified direction.

The angle φ through which the sublattice magnetizations are rotated will be determined by an equilibrium between the applied field and the intermolecular fields which will oppose it. At equilibrium

$$H_a = \Gamma M_A \sin \varphi + \Gamma M_B \sin \varphi \quad , \tag{145}$$

or

$$\frac{1}{\Gamma} = \frac{M_A + M_B}{H_a} \sin \varphi \quad . \tag{146}$$

Substitution in Eq. (144) yields

$$\kappa_{\perp} = 1/\Gamma \quad . \tag{147}$$

A polycrystalline sample has a random orientation of crystallites relative to the applied field direction. A sphere with radius vector r represents all the possible random orientations. The number of orientations in a range $d\theta$ of θ with respect to a fixed direction will be proportional to the surface area of the annular ring, shown in Fig. 12, relative to the surface area of the entire sphere. The infinitesimal width of the annulus is

$$ds = r \, d\theta \quad . \tag{148}$$

Multiplication by its length gives an area

$$dS = 2\pi R \, ds = 2\pi R r \, d\theta = 2\pi r^2 \sin\theta \, d\theta \quad , \tag{149}$$

since

$$R = r \sin\theta \quad . \tag{150}$$

The ratio of Eq. (149) to the total surface of the sphere gives, for the relative number of crystallites lying within the range $d\theta$ of θ,

$$\frac{2\pi r^2 \sin\theta \, d\theta}{4\pi r^2} = \tfrac{1}{2} \sin\theta \, d\theta \quad . \tag{151}$$

The susceptibility in the direction of the field for a single crystal oriented at some arbitrary angle θ with respect to it is determined by resolving the field into its parallel and perpendicular components. These components are then used to calculate the parallel and perpendicular susceptibilities which in turn are resolved into a resultant susceptibility in the direction of the field as

$$\kappa_\theta = \kappa_{\parallel} \cos^2\theta + \kappa_{\perp} \sin^2\theta \quad . \tag{152}$$

For a polycrystalline sample an average susceptibility κ_{av} can be calculated by integrating the angular susceptibility times the relative number of orientations at that angle over all possible angles. Hence

$$\kappa_{av} = \int_0^\pi \tfrac{1}{2} \sin\theta \, \kappa_\theta \, d\theta = \int_0^{\pi/2} \sin\theta \, \kappa_\theta \, d\theta \quad . \tag{153}$$

Substitution of Eq. (152) for κ_θ yields

$$\kappa_{av} = \int_0^{\pi/2} \kappa_{\parallel} \cos^2\theta \sin\theta \, d\theta + \int_0^{\pi/2} \kappa_{\perp} \sin^3\theta \, d\theta = (\kappa_{\parallel} + 2\kappa_{\perp})/3 \tag{154}$$

VIII.   FERRIMAGNETISM

The last section dealt with antiferromagnetic materials, which were described in terms of two identical interpenetrating magnetic sublattices such that, in the absence of an applied field, their magnetic moments canceled. Another class of materials, termed ferrimagnetic, results from an interpenetration of sublattices with unequal magnetic moments. Thus ferrimagnetic materials, in contrast to antiferromagnetic materials, will exhibit a net spontaneous magnetization below some critical temperature. This net spontaneous magnetization will be the resultant of the magnetization of the sublattices as follows:

$$\underline{M} = \underline{M}_A + \underline{M}_B \ .$$ (155)

Equations (115) and (116) can be applied to ferrimagnetic materials and are rewritten as

$$M_A = M_{Asat} B_{\underset{A}{S}} \left[\frac{g\beta}{kT} (H_a - \Gamma M_B - \gamma_A M_A) \right]$$ (156)

and

$$M_B = M_{Bsat} B_{\underset{B}{S}} \left[\frac{g\beta}{kT} (H_a - \Gamma M_A - \gamma_B M_B) \right] \ .$$ (157)

The subscripts on the intramolecular field constants and spin quantum numbers have been added to indicate the possible dissimilarity between these quantities in the respective sublattices. Dissimilarities in these quantities could lead to ferrimagnetic behavior. Another possibility would be a difference in the number of magnetic atoms of type A and those of type B. This number difference, $N_A \neq N_B$, appears in the theory via the saturation magnetization of each sublattice, where

$$M_{Asat} = g\beta \underset{\sim}{S}_A N_A$$ (158)

and

$$M_{Bsat} = g\beta \underset{\sim}{S}_B N_B \ .$$ (159)

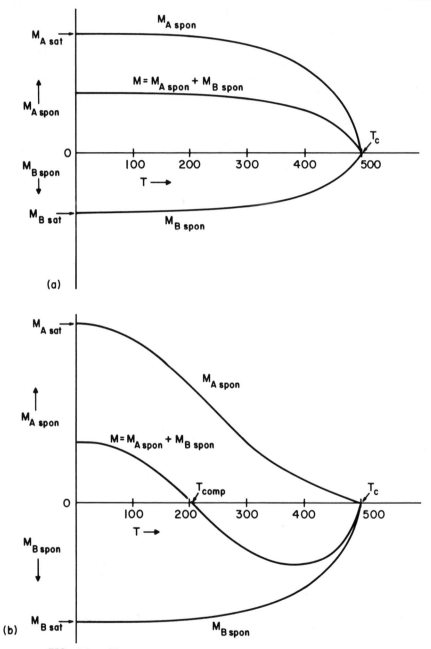

FIG. 13.   Characteristic plots of sublattice magnetizations and resultant magnetizations for ferrimagnetic materials.

In general any combination of these differences could lead to a ferrimagnetic state provided that their effects do not precisely cancel.

Following the method developed in Section VI, graphical solutions
for the magnetizations as functions of temperature can be determined for
Eqs. (156) and (157). Noting that the directions of magnetization for
the A and B sublattices in ferrimagnetic materials are antiparallel
allows Eq. (155) to be written as

$$M = M_A - M_B \ .$$

(160)

This equation together with the solutions of Eqs. (156) and (157) will
determine the form of the resultant magnetization as a function of temperature. A typical plot of the combination of the sublattice magnetizations to give a resultant magnetization is shown in Fig. 13. The
magnetizations shown are the spontaneous magnetizations by virtue of
an assumed zero magnitude applied field. For the case b, shown in
Fig. 13, the intermediate temperature where the resultant spontaneous
magnetization assumes a zero value is termed the compensation temperature. At this temperature the magnetizations of the A and B sublattices
just cancel, while the zero resultant spontaneous magnetization at
higher temperatures corresponds to the Curie point of the ferrimagnetic
material.

For temperature $T > T_c$, the material will be in a paramagnetic
state and only the initial linear portion of the Brillouin function will be
required in the theory. Consequently, Eqs. (156) and (157) for an
applied field parallel to the preferred magnetization axis may be rewritten as

$$M_A = M_{Asat} \left(\frac{S_A + 1}{3} \right) \frac{g\beta}{kT} (H_a - \Gamma M_B - \gamma_A M_A)$$

(161)

and

$$M_B = M_{Bsat} \left(\frac{S_B + 1}{3} \right) \frac{g\beta}{kT} (H_a - \Gamma M_A - \gamma_B M_B) \ .$$

(162)

Rearranging the latter equations gives

$$M_A = \frac{C_A H_a - C_A \Gamma M_B}{T + \theta_A} \tag{163}$$

and

$$M_B = \frac{C_B H_a - C_B \Gamma M_A}{T + \theta_B} \tag{164}$$

where

$$C_A = M_{Asat} \left(\frac{\underset{\sim}{S}_A + 1}{3} \right) \frac{g\beta}{k} \tag{165}$$

$$\theta_A = \gamma_A C_A \quad ,$$

and

$$C_B = M_{Bsat} \left(\frac{\underset{\sim}{S}_B + 1}{3} \right) \frac{g\beta}{k} \tag{166}$$

$$\theta_B = \gamma_B C_B \quad .$$

If

$$\Gamma = \alpha \gamma_A \tag{167}$$

and

$$\Gamma = \beta \gamma_B \quad , \tag{168}$$

then

$$M_A = \frac{C_A H_a - \alpha \theta_A M_B}{T + \theta_A} \tag{169}$$

and

$$M_B = \frac{C_B H_a - \beta \theta_B M_A}{T + \theta_B} \quad . \tag{170}$$

Simultaneous solution of these equations gives

$$M_A = \frac{H_a (C_A T + C_A \theta_B - \alpha \theta_A C_B)}{T^2 + T\theta_A + T\theta_B + \theta_A \theta_B (1 - \alpha\beta)} \tag{171}$$

and

$$M_B = \frac{H_a(C_B T + C_B \theta_A - \beta \theta_B C_A)}{T^2 + T\theta_A + T\theta_B + \theta_A \theta_B (1 - \alpha\beta)} \quad . \tag{172}$$

Finally, the expression for the volume susceptibility for a ferrimagnetic material immersed in an applied field which is parallel to the preferred axis of magnetization is as follows:

$$\kappa = \frac{M_A - M_B}{H_a} = \frac{C_A[T + \theta_B(1 + \beta)] - C_B[T + \theta_A(1 + \alpha)]}{T^2 + T(\theta_A + \theta_B) + \theta_A \theta_B (1 - \alpha\beta)} \quad . \tag{173}$$

An important distinguishing point arises as shown by Eq. (173); namely, the magnetic susceptibility for a ferrimagnetic material is not a linear function of the temperature. The Curie point for a ferrimagnetic material is defined as that temperature at which a resultant spontaneous magnetization appears in the absence of an applied field. Thus at $T = T_c$ the denominator in Eq. (173) must go to zero so that the susceptibility approaches an infinite value:

$$T_c^2 + T_c(\theta_A + \theta_B) + \theta_A \theta_B (1 - \alpha\beta) = 0 \quad . \tag{174}$$

This equation has two roots, the larger real positive root being the Curie temperature.

IX. DIAMAGNETISM

Consider an atom in which the net magnetic moment is zero in the absence of an external field, that is, the spin and orbital moments add vectorially to give a zero resultant moment. By placing such an atom in a magnetic field $\underline{H}$, the electrons, assumed to be in circular orbits, will experience a force given by

$$\underline{F}_H = e(\underline{v} \times \underline{H}) \quad , \tag{175}$$

which, depending on the direction of rotation of the electron, will either

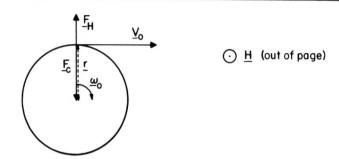

FIG. 14.   Vector diagram of charged particle orbiting in a plane perpendicular to a uniform magnetic field.

add to or subtract from the centripetal force that the nucleus exerts on the electron.   The situation can be visualized as in Fig. 14, where the centripetal force $\underline{F}_c$ is given as

$$\underline{F}_c = m_e \underline{v}^2/\underline{r} = m_e \omega_0^2 \underline{r} \quad . \tag{176}$$

The additional force, $\underline{F}_H$, due to the magnetic field acting on the electron, is shown to be opposite in direction to $\underline{F}_c$. The direction is specified by the right-hand rule, remembering the intrinsic negative charge on the electron.   Had the orbital velocity been in the opposite direction, a similar analysis would show the additional force $\underline{F}_H$ to be in the same direction as the centripetal force $\underline{F}_c$.

In order that the electron remain in an orbit of the same radius, which can be shown to remain constant, the nucleus must now furnish a force

$$\underline{F}_c \pm \underline{F}_H = m_e \omega^2 \underline{r} \quad , \tag{177}$$

where the choice of sign depends on the direction of rotation. ω is the new angular velocity necessarily assumed by the electron to maintain the orbit of constant radius.

The maximum magnitude of the additional force $\underline{F}_H$ is given by Eq. (175) as

$$F_H = evH \quad . \tag{178}$$

Now

$$\omega = v/r \quad , \tag{179}$$

and substitution for v in Eq. (178) gives

$$F_H = e\omega rH \quad . \tag{180}$$

Replacing F_C and F_H in Eq. (177) yields

$$m_e\omega_0^2 r \pm e\omega rH = m_e\omega^2 r \quad . \tag{181}$$

Rearrangement gives

$$\pm e\omega rH = m_e r(\omega_0^2 - \omega^2) = m_e r(\omega_0 + \omega)(\omega_0 - \omega) \quad , \tag{182}$$

and now letting

$$\Delta\omega = \omega_0 - \omega \tag{183}$$

implies

$$\Delta\omega = e\omega H/m_e(\omega_0 + \omega) \quad . \tag{184}$$

Finally, realizing that even the strongest magnetic field experimentally obtainable will not to any appreciable extent vary the orbital angular velocity of the electron, then to an excellent approximation

$$\omega_0 \cong \omega \quad , \tag{185}$$

which implies that

$$\Delta\omega = eH/2m_e \quad . \tag{186}$$

The latter equation is independent of both ω and r and can be viewed as a change in the angular velocity of an electron in an atom when immersed in a magnetic field. Within the approximation, $\Delta\omega$ corresponds to a change in the orbital angular momentum of the electron and thus an induced magnetic moment.

According to Lenz's law, the induced magnetic moment will be directed in opposition to the field that produced it. Thus in direct contrast to paramagnetic materials, diamagnetic materials produce a magnetic moment in opposition to the external field. Although diamagnetism is a property of all matter, its effect is very small and easily measurable only when no paramagnetic moment is present.

X. PARAMAGNETISM OF CONDUCTION ELECTRONS

Another type of paramagnetism of interest in solid state theory is that which arises from the conduction electrons in a metal. Each electron has spin $m = \pm\frac{1}{2}$ and $g = 2$. According to the quantum mechanical description, the magnetic moment of an electron can point either parallel or antiparallel to an external field. When pointing parallel to the field $(m = +\frac{1}{2})$, the energy is

$$\varepsilon_\uparrow = -g\beta \underset{\sim}{m} H = -\beta H \quad , \tag{187}$$

whereas for the antiparallel configuration $(m = -\frac{1}{2})$ the energy is

$$\varepsilon_\downarrow = \beta H \quad . \tag{188}$$

Equations (187) and (188) imply that the magnetic moment of a free electron is one Bohr magneton, consistent with its definition as the fundamental unit of magnetic moment. Thus

$$\mu_e = \beta \quad , \tag{189}$$

and then the average magnetic moment or magnetization for a system of N electrons is

$$M_{Hav} = N\mu_e(P_\uparrow - P_\downarrow) \quad . \tag{190}$$

The probabilities of finding an electron in either the parallel or antiparallel state (arrows up or down) are, respectively,

$$P_\uparrow = N_\uparrow / N \tag{191}$$

and

$$P_\downarrow = N_\downarrow / N \quad , \tag{192}$$

where

$$N = N_\uparrow + N_\downarrow \quad . \tag{193}$$

Substitution of Eqs. (191) and (192) into Eq. (190) gives the magnetization as

$$M_{Hav} = \mu_e (N_\uparrow - N_\downarrow) = \beta(N_\uparrow - N_\downarrow) \quad . \tag{194}$$

Equation (194) states the expected fact that the average magnetic moment per unit volume will be simply the difference in populations of the electrons with spin up and spin down times the magnetic moment per electron spin.

Since electrons are quantum mechanical particles which obey the Pauli exclusion principle, using the classical Boltzmann distribution gives an unsatisfactory answer for the magnetization of conduction electrons. However, if the Fermi-Dirac distribution, which is based on the Pauli principle of one allowed particle per quantum state, is used, the theory is corrected as required. Using the Fermi-Dirac distribution, it is found that the probability that a conduction electron will change spin state when immersed in a field is very small. This small probability results from the fact that the opposite spin state is already occupied and thus, if the electron were to turn, the Pauli exclusion principle of no two electrons in the same state would be violated. Consequently, only those electrons within the order of the thermal energy from the top of the conduction band are allowed to orient parallel to the applied field. Thus, there exists only a small difference in the number of electrons with spin $N_\uparrow$ and those with spin $N_\downarrow$, resulting in a small Pauli spin magnetization.

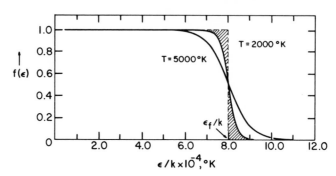

FIG. 15.   Typical Fermi-Dirac distribution for free electrons in a metal.

If the energy of an electron in the absence of the field is denoted as ε, then the energies for a spin up and spin down electron when immersed in the field are, respectively,

$$\varepsilon_\uparrow = \varepsilon - \beta H \tag{195}$$

and

$$\varepsilon_\downarrow = \varepsilon + \beta H \ . \tag{196}$$

The average number of electrons in each quantum state of energy ε is given by the Fermi-Dirac distribution as

$$f(\varepsilon) = \frac{1}{\exp[(\varepsilon - \varepsilon_f)/kT] + 1} \tag{197}$$

where the Fermi energy ε_f is reached if all the electrons are placed into the lowest possible energy levels.   Thus, the Fermi energy is shown in Fig. 15 as the energy at which the two cross-hatched areas are equal.   It is interesting to note that, for energies much larger than the Fermi energy, the Fermi distribution reduces to a Boltzmann distribution, since

$$\exp[(\varepsilon - \varepsilon_f)/kT] \gg 1 \tag{198}$$

for

$$(\varepsilon - \varepsilon_f) \gg 0 \ , \tag{199}$$

and the additional plus one term in the denominator of Eq. (197) is insignificant, giving

$$f(\varepsilon) = \exp\left[-(\varepsilon - \varepsilon_f)/kT\right] \quad . \tag{200}$$

On the other hand, if

$$(\varepsilon - \varepsilon_f) \ll 0 \quad , \tag{201}$$

then

$$f(\varepsilon) \cong 1 \quad , \tag{202}$$

as shown in Fig. 15.

Now defining a function $g(\varepsilon)\,d\varepsilon$ as the number of quantum states in the interval ε to $\varepsilon + d\varepsilon$, the number of electrons in the interval is

$$N(\varepsilon)\,d\varepsilon = f(\varepsilon)\,g(\varepsilon)\,d\varepsilon \quad . \tag{203}$$

In the absence of a field, the density of states at some energy ε is equally divided between electrons with spin up and those with spin down. The conduction electrons can be separated into two distinct Fermi-Dirac distributions, one with spin up and one with spin down. The numbers of spin up and spin down electrons in the presence of a field are, respectively,

$$N_\uparrow = \int_0^\infty \tfrac{1}{2} f(\varepsilon - \beta H)\,g(\varepsilon)\,d\varepsilon \tag{204}$$

and

$$N_\downarrow = \int_0^\infty \tfrac{1}{2} f(\varepsilon + \beta H)\,g(\varepsilon)\,d\varepsilon \quad , \tag{205}$$

where the factor of one-half appears because the total number of electrons has been split into spin up and spin down distributions. Substitution of Eqs. (204) and (205) into Eq. (194) gives

$$\begin{aligned}
M_{Hav} &= \beta \int_0^\infty \tfrac{1}{2}[f(\varepsilon - \beta H) - f(\varepsilon + \beta H)]g(\varepsilon)\,d\varepsilon \\
&\cong \beta^2 H \int_0^\infty -\frac{\partial f}{\partial \varepsilon}\,g(\varepsilon)\,d\varepsilon \quad ,
\end{aligned} \tag{206}$$

provided that βH is small with respect to ε_f, that is, that the magnetic energy is much less than the Fermi energy. Expansion of $g(\varepsilon)$ in a Taylor series about ε_f to first order followed by integration gives

$$M_{Hav} = \beta^2 Hg(\varepsilon) \quad . \tag{207}$$

Finally, substitution of the proper density of state function for an electron gas gives

$$M_{Hav} = 3\beta^2 N/2kT_f \tag{208}$$

where $T_f = \varepsilon_f/k$.

It can be shown that, if the electrons are immersed in a field, an induced moment of $-\frac{1}{3} M_{Hav}$ results from a distortion of their wave functions. The total magnetic moment that would be found experimentally, neglecting the diamagnetism of the core electrons, is given as

$$M_{Hav} = \beta^2 N/kT_f \quad . \tag{209}$$

A qualitative picture for the paramagnetism of conduction electrons is shown in Fig. 16. The energy of the spin up distribution is lowered by an amount βH, while the energy of the spin down distribution is raised by this same amount. The electrons then flow from the spin down distribution to the spin up distribution, maintaining the Fermi level. Thus a net magnetization arises, as given by Eq. (209). It should be noted that this Pauli magnetization is independent of temperature, as indicated in Eq. 209 and as shown in Fig. 15 for two different temperature Fermi distributions.

XI.   GENERAL THEORY OF PARAMAGNETISM

In the foregoing treatment of paramagnetism it was noted that the spin and orbital angular momenta could couple to give several distinct states with differing total angular momentum $\underline{J}$. If the energy between

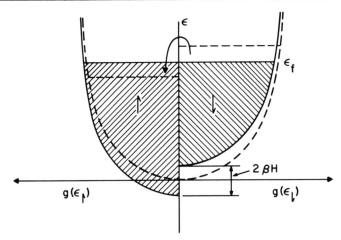

FIG. 16.   Variation of the density of states as a function of
energy for free electrons at 0 °K.

the $\underset{\sim}{J}$ states is in great excess of the thermal energy only the ground
state will be significantly populated, as is the case for normal paramagnetism treated in Section V.   However, if the energy splitting
between the various $\underset{\sim}{J}$ states is less than the thermal energy, then the
distribution of the $\underset{\sim}{J}$ states must be considered when deducing the magnetic moment.   When the transition energy between $\underset{\sim}{J}$ states is much
less than the thermal energy, the ratio of the magnetic ordering energy
to the thermal energy will be much less than unity.   This can be understood by noting that the transition energies between $\underset{\sim}{J}$ states are much
greater than the magnetic transition energies within a given $\underset{\sim}{J}$ state.
Thus the reasoning leading to Eq. (78) will be valid, giving a magnetization as

$$M_{Hav} = \sum_i g_i^2 \beta^2 \underset{\sim}{J}_i (\underset{\sim}{J}_i + 1) HN_i / 3kT \quad .\tag{210}$$

The summation i over all the $\underset{\sim}{J}$ states is now necessary to take into
account the distribution of the atoms over the set of accessible $\underset{\sim}{J}$
states.   When in thermal equilibrium, each $\underset{\sim}{J}$ state will be occupied
by a number of atoms N_i given by a Boltzmann distribution function.

A rigorous quantum mechanical treatment shows that, due to a
distortion of the electronic wave functions by an applied magnetic field,
an extra term appears in the general formulation for the mean magnetic
moment, yielding

$$M_{Hav} = \sum_{i} N_i H[g_i^2 \beta^2 J_i(J_i + 1)/3kT + \alpha_i] \quad . \tag{211}$$

The additional term α_i can be viewed as a magnetic polarization resulting from the field-induced distortion of the electronic wave functions.

Applying this more general equation for the magnetization, it can
be seen that, if each atom is in its ground J state, the summation will
be over a single term, yielding

$$M_{Hav} = NH(g^2 \beta^2 J(J + 1)/3kT + \alpha) \quad . \tag{212}$$

This equation is equivalent to Eq. (78) with the addition of a correction
term for the field-induced magnetic moment.

The polarization term for a given J state is independent of temperature, and therefore its contribution to the magnetization will also
exhibit temperature independence. It is generally negligibly small and
commonly referred to as Van Vleck temperature-independent paramagnetism.

If the transition energy between J states is much less than the
thermal energy, then, according to a Boltzmann distribution, all the J
states will be nearly equally populated. This nearly equal population
results in an almost complete cancellation of the polarization term. A
negligible residue will remain as a result of small population differences in the J states. The quantum mechanical relation for the polarization as derived from second-order perturbation theory is

$$\alpha_i = \sum_{j \neq i} \frac{|\mu_{ij}|^2}{E_i - E_j} \quad . \tag{213}$$

Qualitatively, it can be seen that summation over i and j will
contribute both negative and positive terms depending on the magnitude

of the J state energies in the denominator. For essentially equally populated J states the negative terms will nearly equal the positive terms, the net contribution to the magnetization summing to zero. When the thermal energy is such that it populates all the J levels equally, effectively there is no coupling between $\underline{L}$ and $\underline{S}$ and they can be considered to interact separately with the applied field. Then by an analytical procedure analogous to that employed in the derivation of Eq. (78), ignoring the spin-orbit coupling would give an equation for the magnetization as follows:

$$M_{Hav} = g_S^2 \beta^2 \underline{S}(\underline{S} + 1)HN/3kT + g_L^2 \beta \underline{L}(\underline{L} + 1)HN/3kT$$
$$= \frac{N\beta^2 H}{3kT}[4\underline{S}(\underline{S} + 1) + \underline{L}(\underline{L} + 1)] \quad , \tag{214}$$

since $g_L = 1$ for the orbital only contribution whereas $g_S = 2$ for a spin only contribution, as found from Eq. (39).

Finally, for the case when the transition energies between J states are comparable to the thermal energy, the number of atoms in a single energy level E_i above the ground state follows a Boltzmann distribution:

$$N_i \propto \exp(-E_i/kT) \quad , \tag{215}$$

where i ranges through all the accessible J states. Now for each J state there are $2\underline{J} + 1$ $\underline{m}$ states, since m ranges in integral steps from $+\underline{J}$ to $-\underline{J}$ as noted previously. Each of these $2\underline{J} + 1$ components, or orientations of J with respect to the field, are separate states and must be counted, so that

$$N_i \propto (2\underline{J} + 1) \exp(-E_i/kT) = C(2\underline{J}_i + 1)\exp(-E_i/kT) \quad , \tag{216}$$

where C, the proportionality constant, is given as

$$C = \frac{N}{\sum_i (2\underline{J}_i + 1) \exp(-E_i/kT)} \quad , \tag{217}$$

since

$$\sum_i N_i = N \ . \tag{218}$$

Thus, substitution for C in Eq. (216) yields

$$N_i = \frac{N(2\underset{\sim}{J}_i + 1) \ \exp(-E_i/kT)}{\sum_i (2\underset{\sim}{J}_i + 1) \ \exp(-E_i/kT)} \ . \tag{219}$$

If the thermal energy is of the order of the $\underset{\sim}{J}$-state transition energy, then the ratio of the magnetic energy to the thermal energy will be much, much less than unity, and Eq. (211) will be valid, giving, on substitution of Eq. (219) for the number of atoms in a particular $\underset{\sim}{J}$ state, a magnetization as follows:

$$M_{Hav} = \sum_i \left(\frac{N(2\underset{\sim}{J}_i + 1) \ \exp(-E_i/kT)H[g_i^2 \beta^2 \underset{\sim}{J}_i(\underset{\sim}{J}_i + 1)/3kT + \alpha_i]}{\sum_i (2\underset{\sim}{J}_i + 1) \ \exp(-E_i/kT)} \right) \tag{220}$$

It is evident that the temperature dependence of this magnetization is complicated, and the magnetic susceptibility will not follow the Curie law as found for normal paramagnetism.

For all the magnetization equations that have been considered, a small diamagnetic correction term would have to be subtracted, since, as was noted in Section IX, diamagnetism is a property of all materials.

XII. MAGNETIC MEASUREMENTS

Many different methods are available to investigate the magnetic properties of materials. These include magnetic susceptibility, neutron diffraction, and Mössbauer effect measurements. Of primary importance are, of course, the susceptibility measurements, since these measurements alone establish the basic magnetic character of the material in question.

The magnetic susceptibility of a material is defined as a proportionality constant between the magnitude of the induced magnetization, I, and the magnitude of the applied field, H_a. Thus,

$$I = \kappa H_a \quad , \tag{221}$$

where the magnetic susceptibility per unit volume, κ, is a measure of the extent to which a material is susceptible to induced magnetization. The magnitude of the induced magnetization, I, is simply the induced dipole moment μ per unit volume V; that is,

$$I = \mu/V = m\ell/V = m/A \quad , \tag{222}$$

where m is the induced pole strength and A is the area over which it is induced.

The flux density, or number of lines of force per unit area, is defined to be numerically equal to the magnitude of the field at the point in question. For a unit pole the flux density at 1 cm is 4π Maxwells, since the surface area of a 1-cm radius sphere is 4π cm^2. Thus if a material is placed in a magnetic field H_a the magnitude of the field within the material, B, is given as

$$B = H_a + 4\pi I \quad . \tag{223}$$

Division by H_a yields

$$B/H_a = 1 + 4\pi I/H_a \quad . \tag{224}$$

Defining a permeability P as the ratio of the field within the substance to the applied field gives

$$P = B/H_a = 1 + 4\pi I/H_a \quad . \tag{225}$$

Rearrangement of Eq. (221) and substitution in Eq. (225) gives

$$P = 1 + 4\pi\kappa \quad . \tag{226}$$

This implies that

$$\kappa = (P - 1)/4\pi \quad . \tag{227}$$

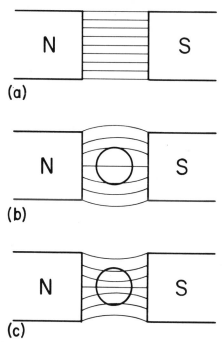

FIG. 17. Schematic diagram showing the resulting flux densities
for: (a) a vacuum, (b) a diamagnetic material, and (c) a paramagnetic
material immersed in a uniform field.

Hence the volume susceptibility for a vacuum is zero, since we have
$P = B/H_a = 1$. The permeability of a diamagnetic material, for which
$P < 1$, implies a negative volume susceptibility, whereas the permeability of a paramagnetic material, for which $P > 1$, implies a positive
volume susceptibility. The corresponding flux densities for a vacuum,
a diamagnetic material, and a paramagnetic substance, all immersed
in a uniform field H_a, are shown in Fig. 17.

One of the principal methods for measuring the magnetic susceptibility is the uniform field or Gouy technique. This method consists
of suspending from a balance a sample of length Z such that one end
extends into a region of strong uniform magnetic field, while the other
end is in a region where the field is negligible. The experimental
apparatus is schematically depicted in Fig. 18. The actual measure-

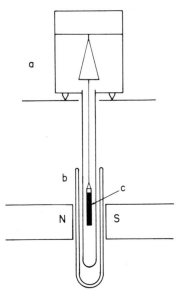

FIG. 18. Diagram of a Gouy magnetic susceptibility apparatus consisting of a balance (a) for measuring the difference in force on the sample (c) in the presence or absence of an external field. An insulating jacket (b) surrounds the sample such that measurements at various temperatures can be obtained.

ments consist of measuring the force on the sample with varying temperature and applied field strength. The linear displacing force experienced by the sample will be directly proportional to the product of its volume V, the applied field H_a, and the material's volume susceptibility κ_1. Thus if the field gradient generated along the sample is given as dH_a/dZ then the incremental force dF_z for a sample suspended in a vacuum is

$$dF_z = \kappa_1 H_a \, dV \, dH_a/dZ \quad . \tag{228}$$

If the sample is suspended in an atmosphere with permeability other than unity and corresponding susceptibility κ_2, then it displaces a volume of that medium such that Eq. (228) for the incremental linear displacing force becomes

$$dF_z = (\kappa_1 - \kappa_2)H_a \, dV \, dH_a/dZ = (\kappa_1 - \kappa_2)A H_a \, dH_a \tag{229}$$

where $A = dV/dZ$ is the cross-sectional area of the sample. Therefore the total displacing force on the entire sample can be found by integrating the field over the entire length of the sample:

$$F_z = (\kappa_1 - \kappa_2)A \int_{H_1}^{H_2} H_a \, dH_a = (\kappa_1 - \kappa_2)A(H_2^2 - H_1^2)/2 \quad . \qquad (230)$$

Substitution of wg for the force F_z, where w is the apparent change in mass when the field is applied and g is the gravitational constant, gives after rearrangement

$$\kappa_1 = \frac{2wg}{A(H_2^2 - H_1^2)} + \kappa_2 = \frac{2Zwg}{V(H_2^2 - H_1^2)} + \kappa_2 \quad . \qquad (231)$$

In Eq. (81) it was shown that

$$\chi = \kappa/\rho = \kappa V/W \quad , \qquad (232)$$

where W is the mass of the sample. This implies that the gram susceptibility χ can be written as

$$\chi = \frac{2Zg}{(H_2^2 - H_1^2)} \frac{w}{W} + \frac{\kappa_2 V}{W} \quad . \qquad (233)$$

Now if the surrounding medium is selected such that its permeability is unity, then $\kappa_2 = 0$ and

$$\chi = Cw/W \quad , \qquad (234)$$

where the constant C is given by

$$C = \frac{2Zg}{(H_2^2 - H_1^2)} \quad . \qquad (235)$$

The value of the constant can be obtained from a calibration measurement of a sample of known susceptibility, eliminating the necessity of directly measuring H_1 and H_2. It is important in making Gouy measurements that the samples have uniform density, area, and length.

Another common method for measuring the susceptibility of small solid samples is the Faraday or nonuniform field method. In essence

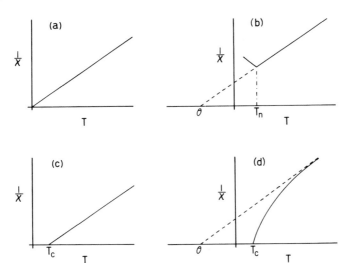

FIG. 19.   Plots of the reciprocal of the gram susceptibility as a function of temperature for: a, a paramagnetic material; b, an antiferromagnet; c, a ferromagnetic material; d, a ferrimagnet. Van Vleck paramagnetism, Pauli paramagnetism, and diamagnetism would yield straight lines with zero slope (temperature independent) with diamagnetism having a characteristic negative $1/\chi$ intercept.

it is the same as the Gouy method, except that the entire sample resides in a constant maximum region $H_a dH_a/dZ$. The generation of the field necessary for this method is accomplished by use of a specially designed pole face.   The constancy of $H_a dH_a/dZ$ over the volume of the sample makes integration of Eq. (229), or the field over the sample, unnecessary. Paralleling the development for the Gouy method, the gram susceptibility of the Faraday method is given by

$$\chi = Cw/W \quad , \tag{236}$$

where now

$$C = \frac{g}{H_a dH_a/dZ} \quad . \tag{237}$$

Other methods for measurement of magnetic susceptibility are available; however, they enjoy only limited popularity.

Susceptibility data can be most easily interpreted by plotting the

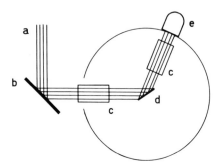

FIG. 20. Schematic diagram of a neutron diffractometer, where a, a polychromatic neutron flux from a reactor, is made monochromatic by a single crystal monochromator b. The monochromatic flux is then collimated by a collimator c and diffracted off the sample d. The diffracted flux is again collimated and the flux density, or intensity of the diffracted beam as a function of angle, is received by a detector e.

inverse susceptibility $1/\chi$ vs temperature T. Figure 19 shows the general form of the plots for the various types of magnetism discussed. The type of magnetism can be inferred qualitatively from the form of the plot, giving information about the internal structure of the sample under investigation.

One of the most valuable methods for the investigation of magnetic structure is neutron diffraction. This method consists of scattering low energy neutrons off a sample of the material and noting the intensity of the neutron flux as a function of the scattering angle θ. A schematic diagram of a neutron diffraction apparatus is shown in Fig. 20. The neutrons' energies are such that their de Broglie wavelengths are of the order of the interatomic spacings in the material being investigated. Thus the neutrons' magnetic moments interact with the atomic magnetic moments according to a Bragg relationship in a sense similar to that of x-ray diffraction. In addition to the magnetic interaction the neutrons also have a nuclear scattering pattern which must be separated from the total pattern to leave only a magnetic pattern. This can be accomplished by first obtaining a scattering pattern for a paramagnetic state of the material and then another pattern for the material below its ordering temperature. The extra lines that appear below the ordering temperature are a result of the magnetic scattering interaction. The magnetically scattered lines can be indexed and intensity analyzed, in a manner similar to an x-ray structure determination, and the final result gives the magnetic structure of the material.

The application of Mössbauer spectroscopy has yielded valuable information for certain magnetic materials. The method consists of measuring the nuclear resonant absorption of gamma rays emitted from an appropriate radioactive source. The energy of the monochromatic source is Doppler varied by use of a velocity transducer. The absorption of the sample is then detected and plotted as a function of the Doppler energy shift. The interactions of the nuclear electric and magnetic moments of the absorbing nucleus with its electrons, with the lattice of neighboring atoms, or with externally applied fields can lift the degeneracy of its nuclear energy levels. An analysis of this energy level splitting gives insight into the magnetic and electronic structures of the material being investigated.

BIBLIOGRAPHY

J. C. Anderson, Magnetism and Magnetic Materials, Chapman and Hall, London, 1968.

G. E. Bacon, Neutron Diffraction, Oxford Univ. Press, London and New York, 1962.

L. F. Bates, Modern Magnetism, Cambridge Univ. Press, London and New York, 1963.

R. M. Bozorth, Ferromagnetism, Van Nostrand, Princeton, New Jersey, 1951.

A. Earnshaw, Introduction to Magnetochemistry, Academic Press, New York, 1968.

B. N. Figgis, Introduction to Ligand Fields, Wiley-Interscience, New
York, 1966.

J. B. Goodenough, Magnetism and the Chemical Bond, Wiley-Interscience,
New York, 1963.

T. L. Hill, Introduction to Statistical Thermodynamics, Addison-Wesley,
Reading, Massachusetts, 1962.

C. Kittel, Introduction to Solid State Physics, Wiley, New York, 1967.

D. H. Martin, Magnetism in Solids, Iliffe Books, London, 1968.

J. P. McKelvey, Solid State and Semiconductor Physics, Harper and Row,
New York, 1966.

L. N. Mulay, Magnetic Susceptibility, Wiley-Interscience, New York,
1963.

F. Reif, Fundamentals of Statistical and Thermal Physics, McGraw-Hill,
New York, 1965.

M. Sachs, Solid State Theory, McGraw-Hill, New York, 1963.

P. W. Selwood, Magnetochemistry, Wiley-Interscience, New York, 1956.

J. M. Ziman, Principles of the Theory of Solids, Cambridge Univ. Press,
London and New York, 1965.

Chapter 6

INTRODUCTION TO MAGNETIC RESONANCE IN SOLIDS

Paul H. Kasai

Union Carbide Research Institute
Tarrytown, New York

357

I.   INTRODUCTION

Magnetic susceptibility, discussed in the preceding section, is
a bulk, static interaction of a system with a static magnetic field.   In
this chapter we shall discuss the phenomenon called "magnetic resonance," which may be viewed as an atomistic and dynamic interaction
of the magnetic dipole moment of electrons or nuclei within a system
with an oscillating magnetic field.   Its application in solid state
physics has proven to be extremely useful.   It provides, for instance,
one of the very few techniques capable of making direct exploration of
the environment of an impurity or defect within a solid.

A.   Magnetic Moments of Electrons and Nuclei

We shall first review the nature of the magnetic moments of electrons and nuclei.   In classical mechanics, when a charged mass rotates
in a circular orbit, it creates a magnetic moment, and there is a definite

ratio between this magnetic moment and the angular momentum associated with the rotation. Let m be the mass of the rotating particle, r the
radius of the orbit, and v its velocity. The angular momentum J is
given by

$$J = mvr \quad . \tag{1}$$

The magnetic moment μ, on the other hand, is given by the product of
the electric current and the area of the circular current loop. Therefore,
if the charge of the particle is q, we obtain

$$\mu = \text{current} \times \text{area}$$

$$= \left(\frac{qv}{2\pi r}\right)(\pi r^2) = \frac{qvr}{2} \quad . \tag{2}$$

Inspection of Eqs. (1) and (2) leads us to write

$$\mu = \left(\frac{q}{2m}\right)J \quad . \tag{3}$$

We can thus see that the magnetic moment and the angular momentum
are parallel to each other, and their ratio, q/2m, depends neither on
the velocity nor on the radius of the motion.

For an electron with mass m_e and charge -e, there are two
sources of angular momentum. One source is its orbital motion whose
angular momentum is denoted by L, and the other is the intrinsic spinning motion whose angular momentum is denoted by S. Equation (3) is
rigorously valid for the magnetic moment associated with the orbital
motion of an electron. However, it has been shown, both by experiments and from a quantum mechanical theory, that the magnetic moment
associated with the spin of an electron is roughly twice as large (2.0023)
as that given by Eq. (3). We also know from quantum mechanics that
angular momentum is a property quantized in units of $\hbar$ (= $h/2\pi$) where
h is Planck's constant. It has, therefore, become customary to write

$$\mu_e = \gamma\hbar J = -g\left(\frac{e\hbar}{2m_e}\right)J \quad , \tag{4}$$

where J is now the quantum number characterizing the "total" angular momentum of the electron. The constant γ is called the gyromagnetic ratio, and the constant quantity $e\hbar/2m_e$ is known as the Bohr magneton. The factor g, the Landé g-factor, is a dimensionless quantity; it is 1 if an electron has only an orbital moment, and 2.0023 if it has only a spin moment.

A similar expression is used to represent the magnetic moment of nuclei:

$$\mu_n = \gamma\hbar J = g_n\left(\frac{e\hbar}{2m_p}\right)J \ . \tag{5}$$

Here m_p is the mass of a proton, and the quantity $e\hbar/2m_p$ is called the nuclear magneton. Expressed in this fashion, owing to the composite nature of a nucleus, the interpretation of the nuclear g-factor, g_n, is not as straightforward as that for an electron. Its value for a proton, for instance, is 5.5854. Even a neutron, though electrically neutral, has been found to possess a magnetic moment with $g_n = -3.826$.

B. Magnetic Resonance

1. Simple Classical Description

We shall next consider what happens when these "atomic magnetic dipoles" are placed in a static magnetic field. It is immediately clear that there will be a torque, which tries to align the dipoles with the field, as shown in Fig. 1. Both the direction and the magnitude of this torque can be most conveniently expressed in terms of vector notations.

$$\underline{T} = \underline{\mu} \times \underline{H} \ . \tag{6}$$

We must recall, at this point, that our magnetic dipole moment μ has its origin in the angular momentum J and, as we know from classical mechanics, the effect of a torque upon a system is to change its angular momentum in the following way:

$$\frac{d(\text{angular momentum})}{dt} = \text{torque} \ .$$

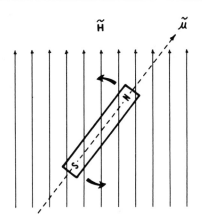

FIG. 1. When subjected to a static magnetic field $\underline{H}$, a magnetic dipole $\underline{\mu}$ experiences a torque given by $\underline{\mu} \times \underline{H}$.

Therefore the equation of motion for our magnetic dipole $\underline{\mu}$ in a magnetic field $\underline{H}$ can be written as

$$\frac{d(J\hbar)}{dt} = \underline{\mu} \times \underline{H}$$

or, using Eq. (4) or (5),

$$\frac{d\underline{\mu}}{dt} = \gamma\underline{\mu} \times \underline{H} \quad . \tag{7}$$

Since this differential equation is a vector equation, it can be written in terms of the vector components along the x, y, and z directions, respectively. Therefore,

$$\frac{d\mu_x}{dt} = \gamma(\mu_y H_z - \mu_z H_y) \tag{8a}$$

$$\frac{d\mu_y}{dt} = \gamma(\mu_z H_x - \mu_x H_z) \tag{8b}$$

$$\frac{d\mu_z}{dt} = \gamma(\mu_x H_y - \mu_y H_x) \quad . \tag{8c}$$

Let us identify the direction of the static field H_0 with the z axis. We then have $H_x = H_y = 0$ and $H_z = H_0$. Equations (8) now take the forms

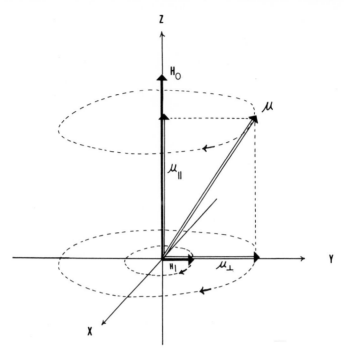

FIG. 2.   Under the influence of a magnetic field H_0, a gyrating magnetic dipole μ precesses about the direction of the field.

$$\frac{d\mu_x}{dt} = \gamma\mu_y H_0 \tag{9a}$$

$$\frac{d\mu_y}{dt} = -\gamma\mu_x H_0 \tag{9b}$$

$$\frac{d\mu_z}{dt} = 0 \quad . \tag{9c}$$

As straightforward differentiation proves, a set of solutions to this set of differential equations can be written as

$$\mu_x = \mu_\perp \sin(\omega_0 t + \phi) \tag{10a}$$

$$\mu_y = \mu_\perp \cos(\omega_0 t + \phi) \tag{10b}$$

$$\mu_z = \mu_\parallel \quad , \tag{10c}$$

where $\omega_0 = \gamma H_0$ and $\mu_{\parallel}$ and $\mu_{\perp}$ represent, respectively, the parallel and perpendicular components of the magnetic moment μ relative to the direction of the magnetic field H_0. The phase angle ϕ is determined by the initial condition. What these equations describe is a constant magnetic moment μ precessing about the direction of the magnetic field with the angular velocity $\omega_0 = \gamma H_0$ keeping a constant azimuthal angle from the field. See Fig. 2. This is the general result of subjecting a gyrating magnetic moment to a magnetic field, and the frequency of the precession, $\omega_0/2\pi$, is known as the Larmor frequency.

Suppose we now apply to this system an additional small magnetic field H_1 which is rotating in the x-y plane with the angular velocity ω. If the direction of the rotation of the second field is opposite to that of the precession of the magnetic moment, or even if it is rotating in the same sense, if its velocity ω is different from that of the precession ω_0, the position of H_1 relative to μ will change in time such that there will be no net effect of H_1 upon the precessing magnetic moment μ. However, if H_1 is rotating in the same direction and with exactly the same angular velocity as that of μ, the position of H_1 as viewed from the rotating μ would appear fixed in space. Hence there will be a torque effected by H_1, which would lead to a change in the orientation of μ and therefore in a change of $\mu_{\parallel}$. This means an absorption (or an emission) of energy by the magnetic moment, since its potential energy E in the static magnetic field H_0 is given by $E = -H \cdot \mu = H\mu_{\parallel}$. This phenomenon of resonant absorption (or emission) of energy by a magnetic moment in a static field H_0 is called "magnetic resonance." It occurs when the frequency $(\omega/2\pi)$ of a weak field rotating in the x-y plane satisfies the condition

$$\omega = \gamma H_0 \quad . \tag{11}$$

In actual magnetic resonance experiments, instead of a rotating field, a field oscillating linearly in the x-y plane is employed. This

accomplishes the same result since the linearly oscillating field can be
resolved into two circularly polarized lights rotating in opposite senses.

Substitution of Eqs. (4) and (5) into Eq. (11) gives the following
convenient expressions for the resonance frequency of an electron and
a nucleus, respectively:

$$f_e = \frac{\omega}{2\pi} = 1.40 \left(\frac{MHz}{G}\right) g_e H_0 \tag{12a}$$

$$f_n = \frac{\omega}{2\pi} = 0.76 \left(\frac{KHz}{G}\right) g_n H_0 \quad . \tag{12b}$$

The magnetic field H_0 commonly used in actual experiments ranges from
3000 to 15,000 G. Thus the frequency of the oscillating field is in the
microwave region $(10 \sim 35$ GHz) for ESR (electron spin resonance) experiments, and is in the vhf region $(10 \sim 100$ MHz) for NMR (nuclear magnetic resonance) experiments.

2. Simple Quantum Mechanical Description

The energy of a magnetic moment μ placed in a magnetic field H_0
is $-\mu \cdot H_0$. Therefore, for an electronic or nuclear magnetic dipole, the
Hamiltonian representing this interaction takes the following form, iden-
tifying the direction of the magnetic field with the z axis,

$$\begin{aligned}
\mathcal{H} &= -\mu \cdot H_0 \\
&= -(\gamma \hbar) J \cdot H_0 \\
&= -\gamma \hbar H_0 J_z \quad .
\end{aligned} \tag{13}$$

Hence the energy levels or the eigenvalues of this Hamiltonian are
given by a simple expression

$$E = -\gamma \hbar H_0 m_z \quad , \tag{14}$$

where m_z is the azimuthal quantum number of the angular momentum J
and can take any one of the $2J + 1$ values, $+J, +J - 1, \ldots, -J$. Thus
there will be $2J + 1$ equally spaced energy levels allowed for our magnetic dipoles with the separation between successive levels being

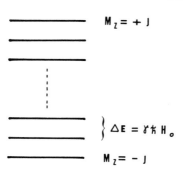

FIG. 3. Energy levels of a magnetic dipole associated with angular momentum J in a static magnetic field H_0.

$\gamma \hbar H_0$, as shown in Fig. 3. We next consider the effect of applying an electromagnetic radiation that is linearly polarized in the plane perpendicular to the magnetic field H_0. We must now add to our Hamiltonian a perturbing term which represents the interaction between the magnetic moment μ and the magnetic part of the oscillating electromagnetic radiation, say $H_x \cos \omega t$. Thus we have

$$\mathcal{H}_{pert} = -\mu \cdot H_x \cos \omega t$$

$$= -\gamma \hbar J \cdot H_x \cos \omega t \qquad (15)$$

$$= -\gamma \hbar H_x \cos \omega t \, J_x \ .$$

The operator J_x has nonvanishing elements between states m_z and m'_z only if $m'_z = m_z \pm 1$. Thus the selection rule governing the transition between these states is $\Delta m_z = \pm 1$.

The following condition must, therefore, be satisfied by the electromagnetic radiation inducing the transition:

$$h\nu = \gamma \hbar H_0$$

or

$$\omega = \gamma H_0 \ . \qquad (16)$$

This, of course, is identical to Eq. (11).

II. MAGNETIC RESONANCE: MACROSCOPIC DESCRIPTION

So far we have discussed the phenomenon of magnetic resonance in terms of the behavior of an individual electron or nuclear magnetic dipole under the influence of a magnetic field. In this section we shall consider the phenomenon in terms of macroscopic properties of an actual sample which contains a large number of such atomic dipoles. Let us first define and discuss some of the macroscopic properties relevant to the resonance phenomenon.

A. Magnetization

Magnetization $\underline{M}$ of a sample is defined as the vector sum of the magnetic moments of all the dipoles of the same kind (same γ and J) within a unit volume:

$$\underline{M} = \sum_i \underline{\mu}_i \quad . \tag{17}$$

If a sample is placed in a static magnetic field $\underline{H}_0 = H_0 \underline{z}$ and brought to thermal equilibrium at temperature T ($^\circ K$), it is clear from Eqs. (10), which describe the behavior of an individual dipole, that

$$M_x = M_y = 0 \quad \text{but} \quad M_z \neq 0 \quad .$$

In a quantum mechanical description, each dipole in the state m_J has the energy given by (14) and contributes a magnetization of $\gamma \hbar m_z$ to M_z. The number of dipoles in each m_J level, say N_{m_J}, is determined by the Boltzmann equation:

$$\frac{N_{m_J}}{N_{total}} = \frac{\exp(-\gamma \hbar H_0 m_J / kT)}{\sum_{m_J} \exp(-\gamma \hbar H_0 m_J / kT)} \quad . \tag{18}$$

Hence

$$M_z = \sum_{m_J} \gamma \hbar m_J N_{m_J} \quad . \tag{19}$$

When $J = \frac{1}{2}$ and $\gamma \hbar H_0 \ll kT$, it can be shown that

$$\Delta N = N_{+\frac{1}{2}} - N_{-\frac{1}{2}}$$

$$= N_{total} \frac{\mu_0 H_0}{kT} \tag{20}$$

and

$$M_z = N_{total} \frac{\mu_0^2 H_0}{kT} \tag{21}$$

where

$$\mu_0 = \frac{\gamma \hbar}{2} .$$

B. Relaxation Processes

In the absence of the magnetic field, all the m_J levels are equally populated. Therefore, the attainment of the Boltzmann distribution (18) and hence the magnetization given by (19) or (21) under the influence of the static magnetic field requires the presence of some mechanism through which an excess energy can be transferred from the magnetic dipole system to the surrounding lattice system. This process is called a spin-lattice or longitudinal relaxation process. Under the resonance condition, the transition probability induced by the oscillating magnetic field is the same for an upward transition (e.g., $m_J = -\frac{1}{2} \rightarrow m_J = +\frac{1}{2}$) as for a downward transition (e.g., $m_J = +\frac{1}{2} \rightarrow m_J = -\frac{1}{2}$). Therefore a net absorption of energy and hence the detection of the magnetic resonance phenomenon are possible only so long as there are excess numbers of dipoles in the lower level over that in the upper level. Thus we can well appreciate the importance of the role played by the spin-lattice relaxation process during the magnetic resonance experiment also. Were it not for this relaxation mechanism, under the resonance condition, all the m_J levels would rapidly become equally populated, and the absorption would cease.

Before we go on with the exploration of the behavior of the macroscopic magnetization under the resonance condition, we must consider another type of relaxation process. Suppose after the Boltzmann distribution and hence the equilibrium value of M_z, say M_0, have been reached, the direction of the magnetic field H_0 is abruptly changed by 90°. At this instance there will be a macroscopic magnetization $M_\perp$ equal to M_0 in the direction perpendicular to the newly oriented magnetic field. The $M_\perp$ would persist and precess about the direction of the new field at the rate given by γH_0 so long as all the dipoles within the system experience exactly the same field and hence maintain the coherence in their precessing motions. The actual field felt by an individual dipole is not H_0, however, but is $H_0 \pm \delta$ where δ represents a fluctuation in the local field resulting from the presence of neighboring dipoles. Therefore some dipoles precess at a rate faster than that given by γH_0, while others precess at a slower rate. Thus the magnetization $M_\perp$ within the plane would eventually vanish due to the loss of coherence in the precessing motion. This type of relaxation is called spin-spin relaxation or transverse relaxation. Note that this type of relaxation does not involve energy transfer with the lattice. Of course, while this process is going on, the spin-lattice relaxation mechanism is also in action building up a new magnetization M_z in the direction of the new field.

The rates of both the spin-lattice relaxation and the spin-spin relaxation processes are supposed to be first order with respect to their respective deviations from the equilibrium values. Thus,

$$\left(\frac{dM_z}{dt}\right)_{\text{spin-lattice}} = -\frac{1}{T_1}(M_z - M_0) \tag{22a}$$

$$\left(\frac{dM_x}{dt}\right)_{\text{spin-spin}} = -\frac{M_x}{T_2} \tag{22b}$$

$$\left(\frac{dM_y}{dt}\right)_{\text{spin-spin}} = -\frac{M_y}{T_2} \quad . \tag{22c}$$

T_1 and T_2, the reciprocals of the rate constants, are called, respectively, the spin-lattice and spin-spin relaxation times. For a nuclear system in a solid, T_1 is of the order of seconds and T_2 is $10^{-4} \sim 10^{-5}$ sec. For a system of electrons the corresponding values are usually much shorter than these, as expected.

C. Bloch Equation and Its "Slow Passage" Solution

From the definition of the magnetization (17) and Eq. (7), which describes the interaction between an individual magnetic dipole and a magnetic field, the interaction between the magnetization M and a magnetic field H can be directly put forward as

$$\frac{dM}{dt}\bigg|_{\text{magnetic field}} = \gamma \underline{M} \times \underline{H} \tag{23}$$

We will now combine Eqs. (22) and (23) and examine the behavior of the magnetization M under the combined influence of the static magnetic field H_0 and a small magnetic field H_1 rotating in the plane perpendicular to the static field H_0. We shall write, therefore,

$$\underline{H} = H_0 \underline{z} + (H_1 \cos \omega t)\underline{x} - (H_1 \sin \omega t)\underline{y} \quad . \tag{24}$$

The direction of the rotation of the field H_1 as given by this equation is shown in Fig. 4. Substituting Eq. (24) into (23), combining the result with Eq. (22), and writing in terms of the components along the x, y, and z directions, we obtain

$$\dot{M}_x = \gamma(M_y H_0 + M_z H_1 \sin \omega t) - \frac{M_x}{T_2} \tag{25a}$$

$$\dot{M}_y = \gamma(M_z H_1 \cos \omega t - M_x H_0) - \frac{M_y}{T_2} \tag{25b}$$

$$\dot{M}_z = \gamma(-M_x H_1 \sin \omega t - M_y H_1 \cos \omega t) - \frac{M_z - M_0}{T_1} \quad . \tag{25c}$$

The set of equations (25) were first proposed by F. Bloch and are commonly referred to as the "Bloch equations." Although the equations

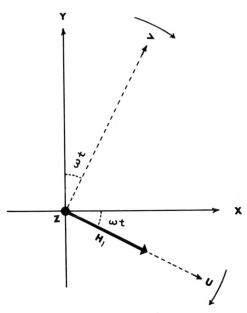

FIG. 4. Direction of the rotating field H_1, and the coordinate system (u, v, z) rotating with it.

were developed in connection with his theory of nuclear induction, they are equally applicable to the study of electron spin resonance phenomena.

In solving the Bloch equations (25), and discussing the result, it is most convenient to consider the coordinate system u, v, z which is affixed to the rotating magnetic field H_1 and rotates with it. See Fig. 4. The components of the transverse magnetization M_u and M_v in this system are related to M_x and M_y in the laboratory fixed system by the following equations:

$$M_u = M_x \cos \omega t - M_y \sin \omega t \tag{26a}$$

$$M_v = M_x \sin \omega t + M_y \cos \omega t \ . \tag{26b}$$

Differentiation of these equations with respect to time gives

$$\dot{M}_u = \dot{M}_x \cos \omega t - \dot{M}_y \sin \omega t - \omega M_v \tag{27a}$$

$$\dot{M}_v = \dot{M}_x \sin \omega t + \dot{M}_y \cos \omega t + \omega M_u \ . \tag{27b}$$

From the Bloch equations (25a) and (25b) it can be shown easily that

$$\dot{M}_x \cos \omega t - \dot{M}_y \sin \omega t = \gamma H_0 M_v - M_u/T_2 \tag{28a}$$

$$\dot{M}_x \sin \omega t + \dot{M}_y \cos \omega t = -\gamma H_0 M_u + \gamma H_1 M_z - M_v/T_2 \quad . \tag{28b}$$

Now substituting Eqs. (28) into (27), we obtain the Bloch equations in the rotating coordinate system:

$$\dot{M}_u = (\omega_0 - \omega)M_v - M_u/T_2 \tag{29a}$$

$$\dot{M}_v = -(\omega_0 - \omega)M_u - M_v/T_2 + \omega_1 M_z \tag{29b}$$

$$\dot{M}_z = -\omega_1 M_v - (M_z - M_0)/T_1 \tag{29c}$$

where

$$\omega_0 = \gamma H_0 \qquad \text{and} \qquad \omega_1 = \gamma H_1 \quad .$$

Experimentally the resonance phenomenon is observed by varying in time either the frequency $\omega/2\pi$ of the rotating field H_1 or the magnitude of the static field H_0. Note that the right side of Eqs. (29) is independent of time. Therefore, if the rate of sweep in ω or H_0 is slow in comparison with the relaxation times T_1 and T_2, these components of the magnetization may be assumed to maintain their respective steady state values, i. e., $\dot{M}_u = \dot{M}_v = \dot{M}_z = 0$. This condition is called the "slow passage condition" and is usually met in actual experiments. Thus setting $\dot{M}_u = \dot{M}_v = \dot{M}_z = 0$ in Eq. (29), we obtain the following "steady state" solution:

$$M_u = M_0 \frac{T_2^2 \omega_1 (\omega_0 - \omega)}{1 + \eta + T_2^2 (\omega_0 - \omega)^2} \tag{30a}$$

$$M_v = M_0 \frac{T_2 \omega_1}{1 + \eta + T_2^2 (\omega_0 - \omega)^2} \tag{30b}$$

$$M_z = M_0 \frac{1 + T_2^2 (\omega_0 - \omega)^2}{1 + \eta + T_2^2 (\omega_0 - \omega)^2} \tag{30c}$$

where

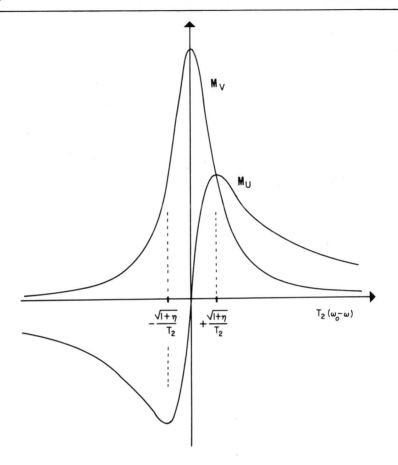

FIG. 5. M_u and M_v shown as functions of $T_2(\omega_0 - \omega)$.

$$\eta = T_1 T_2 \omega_1^2 \quad .$$

Equations (30) show that, under the combined influence of the static field H_0 and the rotating field H_1, there will be a transverse magnetization $\underline{M}_\perp = \underline{M}_u + \underline{M}_v$ rotating in the laboratory space with the same frequency as that of $\underline{H}_1$. $\underline{M}_u$ represents the part rotating in phase with $\underline{H}_1$ and $\underline{M}_v$ the part rotating 90° out of phase with $\underline{H}_1$. Figure 5 shows $\underline{M}_u$ and $\underline{M}_v$ as functions of $T_2(\omega_0 - \omega)$. We see that M_u has the shape of a dispersion curve while M_v has the shape of an absorption curve.

From the transformation equations (26), we obtain the equations for inverse transformation:

$$M_x = M_u \cos \omega t + M_v \sin \omega t \tag{31a}$$

$$M_y = M_v \cos \omega t - M_u \sin \omega t \; . \tag{31b}$$

The rate of energy consumption due to the magnetization is given in the laboratory coordinate system as

$$\frac{dE}{dt} = H_{lab} \cdot \frac{dM_{lab}}{dt} \; . \tag{32}$$

Using Eqs. (24) and (31), and recalling that $\dot{M}_z = \dot{M}_u = \dot{M}_v = 0$, we obtain

$$\frac{dE}{dt} = (H_1 \cos \omega t)(-\omega M_u \sin \omega t + \omega M_v \cos \omega t) +$$

$$(-H_1 \sin \omega t)(-\omega M_v \sin \omega t - \omega M_u \cos \omega t) \tag{33}$$

$$= H_1 \omega M_v \; .$$

We thus see that $\underline{M_v}$ is indeed related to the absorption of energy by our dipole system from the oscillating field H_1.

The absorption curve M_v has its maximum value when

$$\omega_0 - \omega = 0 \qquad \text{or} \qquad \omega = \gamma H_0 \tag{34}$$

and

$$M_{v,max} = M_0 \frac{T_2 \omega_1}{1 + \eta} \; . \tag{35}$$

Also it can be shown that the line width $\Delta \omega_{\frac{1}{2}}$ of this absorption curve, the width at half-maximum, is given by

$$\Delta \omega_{\frac{1}{2}} = \frac{2(1 + \eta)^{\frac{1}{2}}}{T_2} \; . \tag{36}$$

As for the dispersion curve M_u, one can show from Eq. (30a) that its extrema occur at $\omega_0 - \omega = \pm(1 + \eta)^{\frac{1}{2}}/T_2$ and are given by

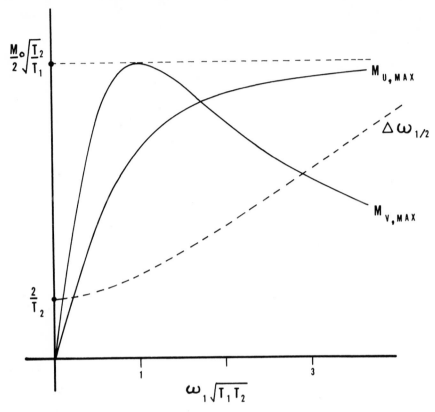

FIG. 6. $M_{u, max}$, $M_{v, max}$, and $\Delta\omega_{\frac{1}{2}}$ shown as functions of the power of the oscillating field $\omega_1(T_1 T_2)^{\frac{1}{2}}$.

$$M_{u,max} = \pm M_0 \frac{T_2 \omega_1}{2(1 + \eta)^{\frac{1}{2}}} \qquad . \tag{37}$$

Note that the points of extrema of M_u coincide with the half-maximum points of M_v (Fig. 5).

Equation (34) is identical to the resonance condition developed in the preceding sections. We thus see that, in this macroscopic description, the magnetic resonance is a detection of the transverse magnetization $M_\perp$ which develops as one sweeps in time either ω_0 or ω.

In usual experiments, the amplitude of the oscillating field H_1 is very small. Hence $\eta = T_1 T_2 \omega_1^2 \ll 1$, and one obtains a very useful relation

$$\Delta\omega_{\frac{1}{2}}\text{,weak } H_1 = \frac{2}{T_2} \quad . \tag{38}$$

The values of $M_{v,max}$, $M_{u,max}$, and $\Delta\omega_{\frac{1}{2}}$ are shown in Fig. 6 as a function of $\eta^{\frac{1}{2}} = \omega_1(T_1 T_2)^{\frac{1}{2}}$. Some of the most important aspects of magnetic resonance experiments are clearly revealed in this figure. For the observation of M_v, the optimum condition is

$$H_1 = \frac{1}{\gamma(T_1 T_2)^{\frac{1}{2}}} \quad .$$

Beyond this point the signal strength decreases with increasing power level of the oscillating field. This phenomenon is called the saturation effect and is directly related to a decrease in the excess number of dipoles in the lower level over that in the upper level. Note that the value of $M_{u,max}$ does not suffer from this effect. However, the magnitude of $\Delta\omega_{\frac{1}{2}}$, which represents the width of the absorption curve for M_v, and the separation of the extremum points for M_u, increases monotonically with increasing power level of H_1. Therefore, an increase in the power level of H_1 leads to broadening of both the absorption and the dispersion curves. This phenomenon is known as saturation broadening. Relaxation times T_1 and T_2 are indeed often measured through study of such aspects as the line width and the saturation effects of magnetic resonance signals.

D. Experimental Arrangements

The first magnetic resonance experiment in a solid was performed in 1945 by Zavoisky, who observed strong absorption due to the electron spin resonance phenomenon in transition metal salts. The first successful observation of a nuclear magnetic resonance signal was reported simultaneously in 1946 by Bloch and Purcell. Schematic arrangements of commonly used nuclear magnetic resonance and electron spin resonance spectrometers are shown, respectively, in Figs. 7 and 8.

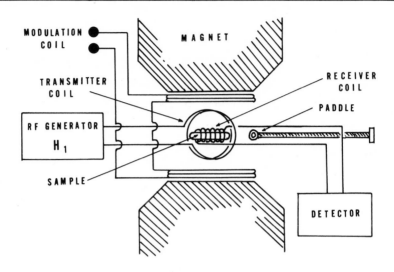

FIG. 7.   Schematic arrangement of an NMR spectrometer.

The most essential parts of the arrangements in Fig. 7 are the two coils, the transmitter coil and the receiver coil, placed perpendicular to each other and to the static field H_0. The transmitter coil provides a linearly oscillating field which can be thought of as two fields rotating in opposite directions. Hence the linearly oscillating field as such could not induce any emf at the receiving coil. However, if one of the rotating fields could meet the resonance condition (34), there would be a net transverse magnetization $M_\perp$ rotating in the plane perpendicular to the receiving coil. Thus in this arrangement the magnetic resonance phenomenon is studied by detecting the rf current induced on the receiver coil by the rotating field $M_\perp$.

As stated earlier, the magnetic resonance condition in a usual ESR experiment requires an oscillating field of microwave frequency. Therefore, a waveguide system is used for its propagation, and the absorption cell is usually a resonant cavity of a cylindrical or rectangular form. The resonance phenomenon is studied by monitoring the microwave power reflected back from the cavity. Note the directions of the flow of the microwave power indicated in Fig. 8. The resonant mode of

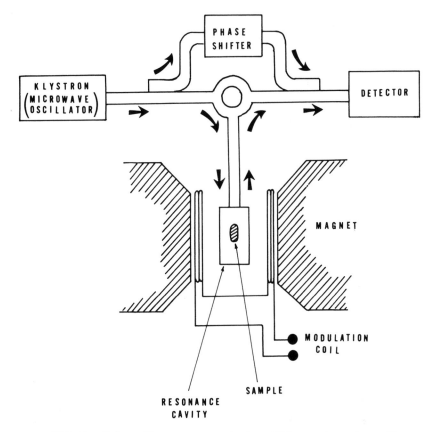

FIG. 8. Schematic arrangement of an ESR spectrometer. The arrows indicate the directions of the propagation of the microwave.

the cavity and the position of the sample inside the cavity are chosen such that the sample experiences an optimum amount of the magnetic part of the microwave field oscillating perpendicular to the static field H_0.

In either case the detection system can be adjusted so that it responds to either the absorption part of the transverse magnetization or the dispersion part. This is done, in the arrangements of Fig. 7, by allowing a controlled amount of the oscillating field H_1 to leak directly from the transmitter coil to the receiver coil, and by adjusting the phase of this leakage current through the use of the "coupling paddle." The

same situation is accomplished in the case of the ESR arrangements
shown in Fig. 8 by means of the leakage path, which involves a phase-
shifter. In both of these arrangements, the magnitude of the oscillating
field detected by the detector is given by the vector sum M_T of the
leakage current or radiation M_ℓ, and the current induced by the transverse magnetization or the radiation reflected back from the cavity M_i.
Since the frequencies of M_ℓ and M_i are the same, the following relation would hold.

$$M_T^2 = M_\ell^2 + M_i^2 + 2M_\ell M_i \cos \phi \quad . \tag{39}$$

Here ϕ represents the relative phase angle between M_ℓ and M_i. For
an ordinary experiment $M_i \ll M_\ell$. Hence Eq. (39) can be rearranged
to yield

$$M_T \cong M_\ell + M_i \cos \phi \quad . \tag{40}$$

We thus see that M_T is most sensitive to that part of M_i which is in
phase with M_ℓ and that the adjustment of the phase of M_ℓ can lead to
an observation of either M_u or M_v selectively.

Another practical feature often found in these spectrometers is the
modulation technique employed in order to increase the sensitivity. The
signals are modulated by a weak, low frequency current going through
the modulation coil (Figs. 7 and 8) and are detected by a "lock-in"
detector which singles out only the signals appropriately modulated.
A spectrum obtained in this way has the shape of the first derivative of
the absorption or the dispersion curve if the amplitude of the modulating
field is small compared to the line width.

III.   ELECTRON SPIN RESONANCE

Although each electron has a magnetic moment given by Eq. (4), a
prerequisite for the observation of electron spin resonance is that the

system under investigation contain electrons that are not paired spinwise. In spite of this limitation, the area of research now covered by this technique is so vast that we can discuss here only some of the more important and general topics relevant to the problems in solid state chemistry.

A. Quenching of the Orbital Moment and the Spin-Orbit Interaction

As mentioned earlier, the g-value of Eq. (4) is defined in such a way as to represent the total magnetic moment arising from both the orbital motion and the spin of the unpaired electron. However, in molecules and crystals, one encounters more often than not a situation where the contribution from the orbital motion is very small even though the orbital occupied by the unpaired electron is not an s-orbital. The orbital moment is then said to be "quenched." To illustrate the phenomenon, let us consider a tetravalent vanadium ion, V^{4+}. It has one 3d electron outside the argon core. Hence, if it were a free ion, the electronic configuration of the ground state would be $^2D_{\frac{3}{2}}$ and the g-value would be given by the well-known Landé formula:

$$g = 1 + \frac{J(J+1) + S(S+1) - L(L+1)}{2J(J+1)}$$

$$= 0.80$$

where J, L, and S are the quantum numbers for the total, orbital, and spin angular momenta, respectively. Now suppose that we place this ion in an environment (a molecule or a crystal) where the arrangement of the surrounding atoms or ions is such that the fivefold degeneracy of the 3d orbitals becomes split in a manner depicted in Fig. 9. We shall further suppose that the splittings indicated are much larger than the so-called spin-orbit coupling interaction, which will be discussed later, or the Zeeman interaction $\mu \cdot H$. Our 3d electron would then occupy the $3d_{x^2-y^2}$ orbital, and the direction of the quantization of

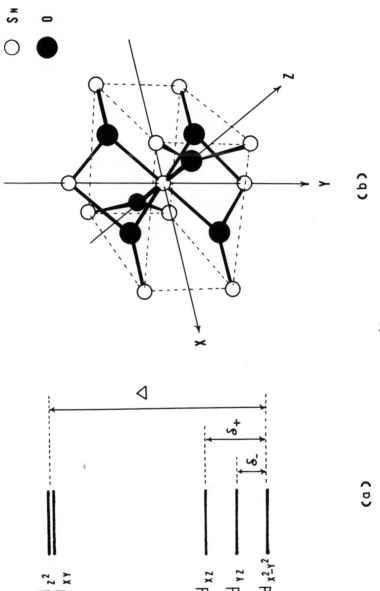

FIG. 9.   (a) Energy level splitting of V^{4+} ion ($3d^1$) placed substitutionally in an SnO_2 crystal. (b) Crystal structure of SnO_2.

the orbital moment would stay unaltered even in the presence of the

external field. We must recall now that the five originally degenerate

3d "space directed" orbitals are related to the orbitals characterized

by the azimuthal quantum number by the following relations:

$$d_{z^2} \quad = |0\rangle$$

$$d_{x^2-y^2} = \frac{1}{\sqrt{2}} (|+2\rangle + |-2\rangle)$$

$$d_{xy} \quad = \frac{1}{\sqrt{2}} (|+2\rangle - |-2\rangle) \tag{41}$$

$$d_{xz} \quad = \frac{1}{\sqrt{2}} (|+1\rangle + |-1\rangle)$$

$$d_{yz} \quad = \frac{1}{\sqrt{2}} (|+1\rangle - |-1\rangle)$$

Here $|M_1\rangle$ represents the 3d orbital with an azimuthal quantum num-

ber M_1. We see immediately that, for any one of these "space directed"

orbitals, the expectation value of the component of the orbital angular

momentum $\langle \phi / L_i / \phi \rangle$ along any direction (i = x, y, or z) is identically

zero. Thus for an electron occupying one of these orbitals, the orbital

contribution to the magnetic moment should also vanish. Indeed one

may conclude that the orbital moment becomes quenched when the

arrangement of the neighboring atoms or ions is such that one of the

"space directed" orbitals represents uniquely the most stable energy

level. Thus if our V^{4+} ion is coupled to an oxide ion to form a diatomic

species $(VO)^{2+}$, the orbital moment of the 3d electron would not be

quenched. In this particular arrangement, if we identify the molecular

axis with the z direction, there are two degenerate orbitals, $3d_{x^2-y^2}$

and $3d_{xy}$, with the same lowest energy. On the other hand, if V^{4+}

is surrounded by four oxide ions in the x-y plane, defining a square,

the orbital moment would be quenched, since the $3d_{z^2}$ orbital repre-

sents uniquely the lowest energy level. In the following section we

shall be concerned only with cases where the orbital ground state is
nondegenerate.

We shall next consider the spin-orbit interaction. It is the interaction between the spin and the orbital motions of an electron and can
be represented by a Hamiltonian of the form $\lambda L \cdot S$. Here λ has the
unit of energy and its magnitude depends upon the state and size of the
atom or ion being considered. The origin of this mechanism is a rela-
tivistic one. Its net effect relevant to magnetic resonance phenomena
is that an orbital motion in the direction opposite to that of the spin
becomes desirable. We thus see the existence of two opposing mechanisms, the quenching mechanism due to a crystal field, and the spin-
orbit interaction which tries to create an orbital motion in the direction
opposite to that of the spin. The latter mechanism is, of course,
responsible for the separation between the $^2D_{\frac{3}{2}}$ level and the $^2D_{\frac{5}{2}}$ level
of a free V^{4+} ion in its ground-state configuration. In a crystalline
environment, however, the magnitude of the splittings among the space-
directed orbitals of same L caused by a crystal field is 1,000-10,000
cm^{-1} and is usually much larger than that caused by the spin-orbit interaction, the latter being 100-1,000 cm^{-1}. Hence, within the zeroth
order of approximation, the orbital moment of an ion in a crystal can be
considered quenched if its orbital ground state is nondegenerate. We
should also note that, because of this large separation among the space-
directed orbitals, the ground state is the only state that is significantly
populated at an ordinary condition.

B. Spin Hamiltonian

Thus for an electron spin resonance of a paramagnetic species
with a nondegenerate orbital ground state one needs to consider only
the effect of the perturbation term $\mathcal{H}'$ upon the ground-state orbital ϕ_0:

$$\mathcal{H}' = \lambda L \cdot S + \mu \cdot H$$

$$= \lambda L \cdot S + (L + g_e S)\beta \cdot H$$

$$(42)$$

where

$$\beta = \frac{e\hbar}{2m_e} \qquad \text{and} \qquad g_e = 2.0023 \quad .$$

The equality $\mu = (L + g_e S)\beta$ follows directly from the discussion leading to Eq. (4). Through a straightforward, though rather tedious, evaluation of the perturbation term $\mathscr{H}'$ in terms of the ground state and the excited orbital states of same L and S, it has been shown that the resonance phenomena can be adequately described by an effective "spin Hamiltonian" of the form

$$\mathscr{H}_{spin} = \beta H \cdot \underline{g} \cdot S + S \cdot \underline{D} \cdot S \quad . \tag{43}$$

Here $\underline{g}$ and $\underline{D}$ are tensor quantities, and, as such, could assume a diagonal form in the principal axes system. The spin Hamiltonian can then be written as

$$\mathscr{H}_{spin} = \beta(g_x S_x H_x + g_y S_y H_y + g_z S_z H_z) +$$

$$D[S_z^2 - \tfrac{1}{3}S(S + 1)] + E(S_x^2 - S_y^2) \quad , \tag{44a}$$

where

$$g_i = g_e + 2\lambda \Lambda_{ii} \tag{44b}$$

$$D = \frac{\lambda^2}{2}(\Lambda_{xx} + \Lambda_{yy} - 2\Lambda_{zz}) \tag{44c}$$

$$E = \frac{\lambda^2}{2}(\Lambda_{xx} - \Lambda_{yy}) \tag{44d}$$

and

$$\Lambda_{ii} = \sum_{n \neq 0} \frac{\langle \phi_0/L_i/\phi_n \rangle \langle \phi_n/L_i/\phi_0 \rangle}{E_0 - E_n} \quad . \tag{44e}$$

Equation (44a) describes the interaction of the paramagnetic species in a crystal with the external magnetic field having components of H_x, H_y, and H_z along the principal axes. The advantage of the spin Hamiltonian is that it reduces the appearance of the problem to that of a spin-only case. The effects of the spin-orbit coupling interaction are manifested in anisotropic deviations of the g-value from that of a true spin-only case, as shown in Eq. (44b), and the appearance of the field independent terms, those involving D and E in Eq. (44a). These field-independent terms are important only when $S > \frac{1}{2}$ and are responsible for the zero field splitting and fine structure of the resonance signal. We shall not show the derivation of Eqs. (44) but will indicate the origins of the parameters defined there. As stated earlier, to the zeroth order of approximation, the ground state orbital is ϕ_0. It would be the $d_{x^2-y^2}$ orbital in our example of V^{4+} shown in Fig. 9. When the $\lambda L \cdot S$ term is treated by the first-order perturbation theory, the perturbed ground-state function Φ_0 takes the form of

$$\Phi_0 \cong \phi_0 + \sum_n \frac{\langle \phi_0 / \lambda L \cdot S / \phi_n \rangle}{E_0 - E_n} \phi_n \quad . \tag{45}$$

Evaluating the Zeeman term $(L + g_e S)\beta \cdot H$ in terms of this new wave function and retaining the terms through the first power in $\lambda / \Delta E$ lead to the definition of g_i given in (44b). Next note that, because of the quenching of the orbital moment, there is no first-order correction to the energy values from the perturbation term $\lambda L \cdot S$. The second-order corrections to the energy levels due to this term, however, do not vanish. They have the form given by

$$\lambda^2 \sum_n \frac{\langle \phi_0 / L \cdot S / \phi_n \rangle \langle \phi_n / L \cdot S / \phi_0 \rangle}{E_0 - E_n}$$

and result in the expressions for D and E in Eq. (44).

Once the spin Hamiltonian (44a) is established, the Hamiltonian matrix can be obtained by evaluating the Hamiltonian in terms of the $(2S + 1)$ spin function, i.e., $M_s = +S, \ldots, M_s = -S$. The eigenvalues of the $(2S + 1)$ Zeeman levels are then obtained through diagonalization of this matrix. The procedure is straightforward, but again somewhat tedious. Hence we shall not show the derivation but will show the result and discuss some of its features most often encountered in actual experiments.

C. Anisotropic g-Tensor

As stated earlier, when $S = \frac{1}{2}$ the fine-structure terms do not contribute to the resonance phenomenon. The energy values obtained from the spin Hamiltonian (44a) for the two Zeeman levels are

$$W(M_s) = g\beta H M_s \quad , \tag{46}$$

where

$$M_s = \pm\frac{1}{2}$$

and

$$g^2 = g_z^2 \cos^2 \theta + (g_x^2 \cos^2 \phi + g_y^2 \sin^2 \phi)\sin^2 \theta \quad ;$$

g_z, g_x, and g_y are defined in Eqs. (44b), and the angles θ and ϕ define the direction of the magnetic field relative to the principal axes. The resonance transition ($M_s = +\frac{1}{2} \longleftrightarrow M_s = -\frac{1}{2}$) occurs when the energy of the electromagnetic radiation is equal to the energy difference:

$$h\nu = \Delta W = g\beta H_{res}$$

or

$$H_{res} = \frac{h\nu}{g\beta} \quad . \tag{47}$$

Thus, in the usual experimental procedure where the frequency of the radiation is held constant and the magnetic field is swept at a constant rate, the resonance field changes depending upon the orientation of the sample crystal (Fig. 13). Figure 9(b) illustrates the structural arrangement of atoms within an SnO_2 crystal. Suppose that our tetravalent vanadium ion V^{4+} is incorporated in this crystal substitutionally at a cation site. It is not difficult to see that the electrostatic potential imposed by the negative charge of the surrounding oxide ions should split the orbital energy of the 3d electron exactly in the manner depicted in Fig. 9(a). Evaluation of the expressions (44e) in terms of the ground-state wave function $d_{x^2-y^2}$ and the excited states shown in the figure give the following results:

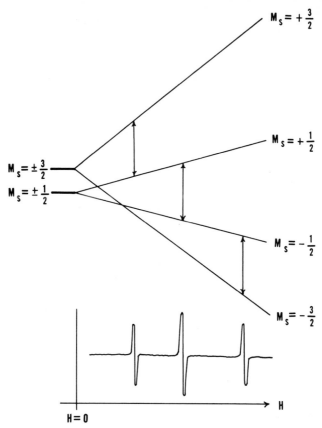

FIG. 10. Energy level splitting of Cr^{3+} (S = 3/2) in chromium alum at zero and increasingly stronger magnetic fields applied parallel to the symmetry axis of the crystal field. The axial field removes a part of the spin degeneracy at zero field and causes the fine structure of the ESR spectrum.

$$g_x = g_e - \frac{2\lambda}{\delta+}$$

$$g_y = g_e - \frac{2\lambda}{\delta-} \tag{48}$$

$$g_z = g_e - \frac{8\lambda}{\Delta} .$$

The values experimentally measured for V^{4+} in SnO_2 are 1.94, 1.91, and 1.94 for g_x, g_y, and g_z, respectively. We thus see that, if the value of the spin-orbit coupling constant λ is known, the magnitude of the crystal field splitting can be assessed from the observed g-tensor.

D. Zero-Field Splitting and Fine Structure

When $S > \frac{1}{2}$, the effect of the field-independent terms becomes important. When the environment of the paramagnetic center possesses an axial symmetry, hence $E = 0$, and if $|g\beta H \cdot S| \gg |D|$, the spin Hamiltonian (44a) can be shown to yield the following eigenvalues:

$$W(M_s) = g\beta H M_s + \frac{D}{2}[M_s^2 - \frac{1}{3}S(S+1)]\left[\frac{3g_\parallel^2}{g^2}\cos^2\theta - 1\right] \tag{49}$$

where

$$g^2 = g_\parallel^2 \cos^2\theta + g_\perp^2 \sin^2\theta$$

$$g_\parallel = g_z$$

$$g_\perp = g_x = g_y \quad .$$

The symmetry axis is identified with the z axis, and θ is the angle between this axis and the magnetic field. Let us apply this equation to a case of $S = 3/2$ with an isotropic g-tensor, that is, $g_\parallel = g_\perp$. Shown in Fig. 10 are the energy levels obtained for each M_s state at zero field and as an increasingly strong field is applied parallel to the z axis. We see that, because of the splitting at zero field, the spacings between the adjacent Zeeman levels are successively different. The fine structure of the resonance spectrum results from this difference, as indicated in the figure. In general, with the selection rule $\Delta M_s = \pm 1$, there would be 2S fine-structure components, and the resonance condition for each transition $M_s \longleftrightarrow M_s - 1$ is given by

$$h\nu = g\beta H + \frac{D}{2}(2M_s - 1)(3\cos^2\theta - 1) \quad . \tag{50}$$

The intensity of each fine-structure component is proportional to $[S(S+1) - M(M-1)]$. As shown in Eq. (50), the spacing between

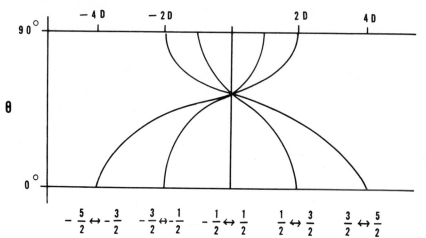

$$-\frac{5}{2}\leftrightarrow-\frac{3}{2} \quad -\frac{3}{2}\leftrightarrow-\frac{1}{2} \quad -\frac{1}{2}\leftrightarrow\frac{1}{2} \quad \frac{1}{2}\leftrightarrow\frac{3}{2} \quad \frac{3}{2}\leftrightarrow\frac{5}{2}$$

FIG. 11. Angular dependency of the fine-structure components of Mn^{2+} ion (S = 5/2) in an axially symmetric crystal field.

fine-structure components depends upon the direction of the magnetic field. Figure 11 shows the fine-structure pattern as a function of θ for S = 5/2. Note the coalescence of the pattern at $\theta = \cos^{-1}(1/\sqrt{3})$.

E.   Hyperfine Structure of the Electron Spin
Resonance Spectrum

In discussing ESR phenomena we have thus far assumed that the electronic system in question does not involve a magnetic nucleus. When the system possesses a magnetic nucleus, one must consider the hyperfine interaction, the interaction between the magnetic moment of the electron and that of the nucleus. In a typical ESR experiment the magnitude of the electronic Zeeman splitting $g\beta H \cdot S$ is much larger than the hyperfine interaction. Therefore the situation should be analogous to the Pashen-Bach effect of the hyperfine structure of free atoms where the electronic angular momentum J and the nuclear angular momentum I precess independently of each other, and the interaction energy can be given by $AM_J M_I$. This analogy and the spin Hamiltonian (43) or (44) suggest that the hyperfine interaction in a molecular or crystalline environment might be given in the following forms:

$$\mathcal{H}_{hfs} = S \cdot \underline{A} \cdot I$$

$$= AS_z I_z + A_x S_x I_x + A_y S_y I_y \quad . \tag{51}$$

Furthermore, if the distribution of the unpaired electron about the magnetic nucleus is axially symmetric, it may be written as

$$\mathcal{H}_{hfs} = A_\parallel I_z S_z + A_\perp (I_x S_x + I_y S_y) \quad . \tag{52}$$

Let us investigate the nature and magnitude of these hyperfine coupling constants, $A_\parallel$ and $A_\perp$. As we know from classical mechanics, the magnetic field $\underline{H}_{dip}$ due to a magnetic dipole μ at a point r away from the dipole is given by

$$\underline{H}_{dip} = -\frac{1}{r^3} \left[\mu - \frac{3(r \cdot \mu) r}{r^2} \right] \quad . \tag{53}$$

Since the magnetic moment of a nucleus is given by $\mu = g_n \beta_n I$, we obtain the Hamiltonian $\mathcal{H}_{dip}$ representing the interaction of the electron's magnetic moment $g_e \beta_e S$ and the magnetic field due to the nucleus as follows:

$$\mathcal{H}_{dip} = g_e \beta_e S \cdot \underline{H}_{dip}$$

$$= g_e \beta_e g_n \beta_n \left[\frac{3(r \cdot S)(r \cdot I)}{r^5} - \frac{S \cdot I}{r^3} \right] \quad . \tag{54}$$

Here r represents the separation between the electron and the nucleus and, hence, can be eliminated from the expression by evaluating $\mathcal{H}_{dip}$ over the space coordinates of the ground-state wave function Φ_0. Let us suppose that the wave function Φ_0 possesses an axial symmetry about the nucleus. We then have the relations $\langle x^2/r^5 \rangle = \langle y^2/r^5 \rangle$, $\langle xy/r^5 \rangle = 0$, etc. Equation (54) can then be written as

$$\mathcal{H}_{dip} = 2A_{dip} I_z S_z - A_{dip}(I_x S_x + I_y S_y) \quad , \tag{55}$$

where

$$A_{dip} = g_e \beta_e g_n \beta_n \left[\langle \frac{3z^2}{2r^5} \rangle - \langle \frac{1}{2r^3} \rangle \right] = g_e \beta_e g_n \beta_n \langle \frac{3\cos^2 \theta - 1}{2r^3} \rangle \quad .$$

Here θ is the angle between $\underline{r}$ and the symmetry axis. Note that, if the electron is localized at a point $\underline{R}$ away from the nucleus, $\theta = 0$, $\langle 1/r^3 \rangle = 1/R^3$, and we obtain $A_{dip} = g_e \beta_e g_n \beta_n / R^3$, a well-known result for a dipole-dipole interaction separated by a distance R. We should next note that the expressions (53) and (54) are clearly inadequate when the electron is "inside" the nucleus. The magnetic interaction in this particular situation was first treated by Fermi and is known as the Fermi contact term. It is given by

$$\mathcal{H}_{cont} = \frac{8\pi}{3} g_e \beta_e g_n \beta_n |\Psi(0)|^2 I \cdot S \tag{56}$$

or

$$\mathcal{H}_{cont} = A_{iso}(I_z S_z + I_x S_x + I_y S_y) \quad,$$

where

$$A_{iso} = \frac{8\pi}{3} g_e \beta_e g_n \beta_n |\Psi(0)|^2 \quad.$$

$|\Psi(0)|^2$ is the probability of the electron's being at the nucleus and is nonzero only if the unpaired electron occupies an s-orbital of the magnetic nucleus. The Fermi contact interaction represents the interaction between the magnetic moment of the electron at the center of the nucleus and the magnetic field generated by the positive charge circulating within the volume of the nucleus. The origin of the current loop is, of course, the angular momentum I of the nucleus. As one can recognize from Eq. (56), the hyperfine interaction arising from the Fermi contact term is not dependent upon the direction of the external field and is often called an isotropic part of the hyperfine interaction. Thus a Hamiltonian for the hyperfine interaction, in general, should be given as the sum of Eqs. (54) and (56):

$$\mathcal{H}_{hfs} = \mathcal{H}_{cont} + \mathcal{H}_{dip} \quad,$$

and for an axially symmetric case, it can be written in the form originally anticipated:

$$\mathcal{H}_{hfs} = A_\parallel I_z S_z + A_\perp (I_x S_x + I_y S_y) \quad ,$$

where we now have

$$A_\parallel = A_{iso} + 2A_{dip}$$

$$A_\perp = A_{iso} - A_{dip} \quad . \tag{57}$$

Let us consider an axially symmetric case involving one unpaired electron ($S = \frac{1}{2}$) and one magnetic nucleus with spin I. The total Hamiltonian takes the form

$$\mathcal{H}_{spin} = g_\parallel S_z H_z + g_\perp (S_x H_x + S_y H_y) +$$

$$A_\parallel S_z I_z + A_\perp (S_x I_x + S_y I_y) + a\, I \cdot H \quad . \tag{58}$$

The last term in this expression represents the Zeeman interaction between the nuclear magnetic moment and the magnetic field. The magnitude of this interaction is very small, and as long as the ESR transition does not involve a change in the magnetic quantum number of the nucleus, it can be omitted from the discussion. Evaluation of the remaining expressions in terms of the $(2S + 1)(2I + 1)$ spin functions and diagonalization of the resulting matrix give the following eigenvalue expression for each state defined by the electronic and nuclear magnetic quantum numbers M_S and M_I:

$$W(M_S, M_I) = g\beta H M_S + A M_S M_I \quad , \tag{59}$$

where

$$M_S = \pm\tfrac{1}{2} , \qquad M_I = +I, \ldots, -I$$

$$g^2 = g_\parallel^2 \cos^2 \theta + g_\perp^2 \sin^2 \theta$$

$$A^2 = \frac{A_\parallel^2 g_\parallel^2}{g^2} \cos^2 \theta + \frac{A_\perp^2 g_\perp^2}{g^2} \sin^2 \theta \quad .$$

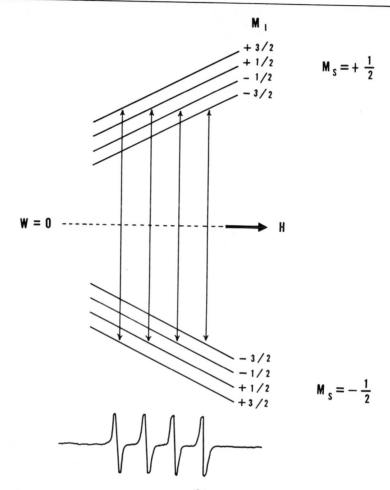

FIG. 12.   Energy levels of Cu^{2+} ion (S = 1/2, I = 3/2) as a function of the external magnetic field.

The directions of the magnetic field and the symmetry axis define the angle θ.   The energy levels of the six states for S = 1/2, I = 3/2 together with the allowed ESR transitions are depicted in Fig. 12.   The ESR spectrum in this case would be a quartet of equal intensity separated by A.   Figure 13 shows a spectrum obtained when ZnO powder doped with Li is irradiated with uv light at 77 °K.   The spectrum is attributed to ionized acceptor centers due to substitutionally incorpo-

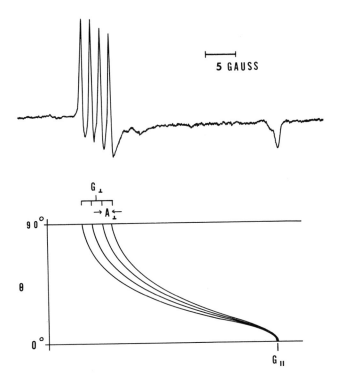

FIG. 13. ESR spectrum due to Li ($I = 3/2$) acceptor centers in ZnO powder (microwave frequency = 9.2 GHz). The g and A tensors of the center are $g_\parallel = 2.0026$, $g_\perp = 2.0254$ and $A_\parallel \cong 0$, $A_\perp = 1.85$ G, respectively. The angular dependence of the four hyperfine components is depicted in the lower figure.

rated Li^+ ions ($I = 3/2$). The angular dependence of the four hyperfine components due to the anisotropy of both the g and A tensors of the center is shown schematically in the same figure.

IV. NUCLEAR MAGNETIC RESONANCE

As is well known to chemists, high resolution nuclear magnetic resonance spectroscopy has become one of the most powerful techniques for studying the structures of molecules, particularly those of organic compounds. High resolution nuclear magnetic resonance spectra, however, are usually obtained only from liquid samples in which the effect of the local magnetic field due to neighboring magnetic nuclei is reduced to a very small average value through the large translational and rotational freedoms that the molecules possess in a liquid. The most important, and perhaps the most familiar, terms which characterize the high resolution nuclear magnetic resonance spectrum of a molecule are the chemical shift, a shift of a resonance position from that expected for an isolated nucleus due to the effect of the surrounding electrons, and the splitting of a resonance line into various multiplets as the result of an indirect, electron-coupled, nuclear spin-spin interaction. Both of these effects are extremely small in magnitude compared to the broadening which exists in a solid arising from the direct dipole-dipole interaction between nearby magnetic nuclei.

Other important differences between nuclear magnetic resonance in liquids as compared to solid materials are the much longer spin-lattice relaxation times and much shorter spin-spin relaxation times in the solids. These differences also arise from the lack of translational freedom in a solid. Thus a crystal may have a spin-lattice relaxation time of $10 \sim 100$ sec and a spin-spin relaxation time of $10^{-4} \sim 10^{-5}$ sec, while in a liquid form, the two relaxation times are of the same order of magnitude and in the neighborhood of a few seconds.

Another phenomenon which often complicates the nuclear magnetic resonance spectrum of a solid is the effect of an electric quadrupole moment. The electric quadrupole moment is a measure of a deviation from spherical symmetry of the positive charge within the nucleus under consideration, and needs to be considered only with nuclei with spin larger than $\frac{1}{2}$.

In this section we focus on three aspects, line broadening, the relaxation process, and the effect of the nuclear electric quadrupole moment, in order to better understand the characteristics of nuclear magnetic resonance of solids.

A. Line Broadening

The broadening of the magnetic resonance absorption of one magnetic nucleus by a second magnetic nucleus can be caused by two different mechanisms. Let us consider a solid in which the magnetic nuclei
with $I = \frac{1}{2}$ are distributed in pairs such that the broadening of the
resonance line of nucleus A is caused primarily by its partner nucleus B.
We recall from Eqs. (10a) ~ (10c) that, in the presence of a large external
magnetic field H_0, the magnetic dipole of a nucleus can be divided into
a static component $\mu_{\parallel}$ which is parallel to H_0, and a rotating component
$\mu_{\perp}$ which rotates in the plane perpendicular to H_0. The first mechanism
arises from the static component. The magnetic field at the point of the
nucleus A due to the static component of the nucleus B is given by

$$\underline{H}_s = \pm \frac{1}{r^3} \left[\frac{3(\underline{r} \cdot \underline{\mu}_{\parallel})\underline{r}}{r^2} - \underline{\mu}_{\parallel} \right] \quad . \tag{60}$$

We have already encountered an expression of this form in discussing
the hyperfine interaction in electron spin resonance. The $\pm$ sign corresponds to the two allowed orientations of the nucleus B, and r is the
separation between the nuclei. The total field H_T at the nucleus A is
given by the sum of the externally applied field H_0 and the local field
$\underline{H}_s$. When $H_0 \gg \underline{H}_s$, one can approximate $H_T = H_0 + H_{s,\parallel}$ where
$H_{s,\parallel}$ is the component of $\underline{H}_s$ parallel to H_0. Thus, noting that $\mu_{\parallel} \equiv \mu$
for $I = \frac{1}{2}$, the total field at the nucleus A can be given as

$$H_T = H_0 \pm \frac{\mu}{r^3}(3 \cos^2 \theta - 1) \quad , \tag{61}$$

where θ is the angle between r and H_0. The nuclear magnetic resonance line of A, therefore, should appear as a doublet with a separation
of $2\mu(3 \cos^2 \theta - 1)/r^3$. The second mechanism arises from the rotating
component $\mu_{\perp}$. This mechanism is of importance only when the two
nuclei A and B possess the same resonance frequency. Under such a
circumstance, the oscillating field due to one nucleus can induce the
resonance transition of the other provided that the first nucleus undergoes

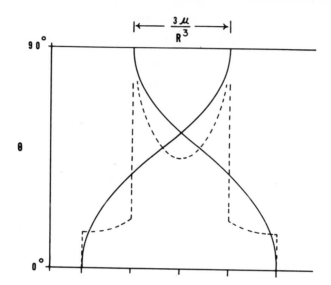

FIG. 14. Dependence on θ of the resonance positions of the doublet arising from a pair of identical nuclei separated by R (solid lines) and the absorption pattern expected from a powder sample containing such pairs (dotted line).

the transition in the reverse direction. This concerted process of exchanging the spin states between a pair of nuclei is called "spin exchange." The magnitude of the rotating component is similar to that of the static component, and its effect at the other nucleus should have the same functional dependence as that given by Eq. (61). A rigorous quantum mechanical treatment of this spin-exchange term shows that its contribution to the doublet separation is exactly half of that caused by the static component. Thus, in our example, if the nuclei A and B have the same resonance frequency a doublet would be observed with the separation given by

$$\frac{3\mu}{r^3}(3\cos^2\theta - 1) \quad . \tag{62}$$

Note that if a given crystal contains pairs of such nuclei, say two protons in each pair, both the direction and the magnitude of the H-H vector can be assessed by studying the dependence of the doublet separation upon the crystal orientation with respect to the applied field H_0. For two protons at a distance of about 1Å, this doublet separation is of the order of $10\,G$.

Figure 14 shows the dependence upon θ of the resonance positions of the doublet as given by (62). If a sample is in a polycrystalline form so that the orientation of an individual pair relative to the external field is completely random, the observed spectrum would represent the envelope of many individual lines spread over this range, as indicated in the same figure. The height of the envelope is proportional to $\left|\dfrac{d\theta}{dH}\right| \sin\theta$ where $\left|\dfrac{d\theta}{dH}\right|$ can be obtained from (62).

It is clear that, when the number of interacting nuclei increases, the number of components increases and eventually the line structure becomes too complex to be resolved. Particularly with polycrystalline powder what one observes most often is a broad featureless resonance signal. Even in this situation, however, useful information can be obtained by measuring the second moment of the signal, the mean square derivation $\langle \Delta H^2 \rangle$, defined as

$$\langle \Delta H^2 \rangle = \int_{-\infty}^{\infty} h^2 S(h)\, dh \quad . \tag{63}$$

Here $h = H_0 - H_0^*$, H_0^* is the applied field at the center of the resonance, and $S(h)$ is the normalized shape function of the observed signal. Let us consider a polycrystalline sample containing only one kind of magnetic nuclei having $I = \frac{1}{2}$. From Eqs. (61) and (62), the resonance position, $h = H_0 - H_0^*$, of a given component can be written as

$$h = \sum_{j \neq i} \frac{3\mu_j}{2r_{ij}^3} (3\cos^2\theta_{ij} - 1) \quad , \tag{64}$$

where μ_j is a component of the jth magnetic nucleus parallel to H_0. Since each μ_j can be either $\pm \frac{1}{2}\gamma h$, it follows that

$$\langle \Delta H^2 \rangle = \frac{9}{4}\mu^2 \sum_j \frac{(3\cos^2\theta_{ij} - 1)^2}{r_{ij}^6} \tag{65}$$

Substituting the value of $(3 \cos^2 \theta - 1)^2$ averaged over all the directions, we obtain for a polycrystalline sample

$$
\begin{aligned}
\langle \Delta H^2 \rangle &= \frac{18}{5} \mu^2 \sum_j \frac{1}{r_{ij}^6} \\
&= \frac{9}{10} \gamma^2 h^2 \sum_j \frac{1}{r_{ij}^6} \quad .
\end{aligned}
\tag{66}
$$

The exact quantum mechanical value for the second moment of a powder sample containing one kind of nuclei with arbitrary spin I has been shown to be

$$
\langle \Delta H^2 \rangle = \frac{6}{5} \frac{I(I+1)}{N} \gamma^2 h^2 \sum_j \frac{1}{r_{ij}^6} \quad ,
\tag{67}
$$

where N is the number of such nuclei in the unit cell. For a simple cubic crystal with lattice constant a, for example, N = 1 and

$$
\sum_j \frac{1}{r_{ij}^6} = \frac{8.5}{a^6} \quad .
$$

If the positions of nuclei are not as rigidly held as assumed in the foregoing discussion, but are in motion, the internuclear distance as well as the direction will vary, resulting in a partial averaging of the dipolar broadening. This effect is called motional narrowing.

B. Relaxation Mechanism

As stated earlier, the spin-lattice relaxation is the mechanism by which the longitudinal or the parallel component of the magnetization can change such that the "temperature" of the spin system defined by the distribution of the nuclear magnets among the available energy levels approaches the temperature of the lattice. In order to understand the relaxation mechanism, one must realize that the z component of a precessing nuclear magnet can be changed only by subjecting it to a magnetic field alternating at a frequency near its Larmor precession. For the spin-lattice relaxation the necessary oscillating field is provided

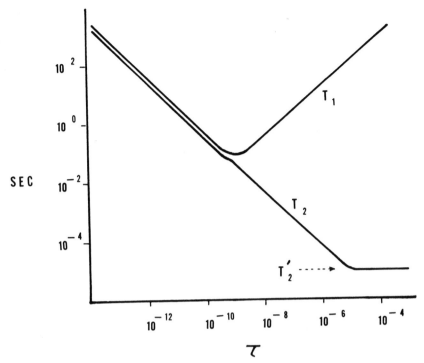

FIG. 15. Variation of relaxation times T_1 and T_2 as a function of correlation time according to the theory of N. Bloembergen, E. M. Purcell, and R. V. Pound [Phys. Rev., 73, 679 (1948)]. The scales are logarithmic and the times are in seconds.

by the thermal fluctuation of the positions of magnetic centers. At extremely low temperature, the frequency of the lattice motion is low, and there is very little fluctuation near the Larmor frequency. The spin-lattice relaxation is ineffective and T_1 is long. On the other hand, when the temperature is high, most of the fluctuations are far above the Larmor frequency, and again the spin-lattice relaxation time T_1 is long. The frequency spectrum of the thermal motion is given by the Debye spectrum density function

$$f(\omega) \propto \frac{\tau}{1 + \omega^2 \tau^2} \quad ,$$

where $f(\omega)$ is the intensity of the lattice motion occurring at frequency ω.

τ is the correlation time, the average time required for the motion to occur, and is proportional to η/T where η is the viscosity of the material and T the absolute temperature. One can see immediately that $f(\omega)$ is small for the two extreme cases cited above, $\omega\tau \gg 1$ (low T) and $\omega\tau \ll 1$ (high T). In between, there is an optimum η/T for each Larmor frequency at which the "magnetic noise" of the lattice is most effective in promoting the spin-lattice relaxation. This relation is shown in Fig. 15 where $\log T_1$ is plotted against $\log \tau$.

As stated earlier, the transverse relaxation time, or the spin-spin relaxation time, refers to the rate at which the coherence of the transverse magnetization $M_\perp$ is lost. Two mechanisms that promote such a relaxation process have already been discussed. They are the dipole-dipole interaction, and the spin exchange mechanism discussed in connection with line broadening. The dipolar interaction between every pair of nuclear magnets within an ensemble of nuclei produces a scatter in the exact magnetic field that each individual nuclear magnet experiences. Individual magnets, therefore, would not precess at exactly the same Larmor frequency. This results in a decay in the total transverse magnetization $M_\perp$. The spin-exchange mechanism also contributes to the transverse relaxation process, since the exchange of the longitudinal components between two nuclear magnets results in a mixing of their transverse components. The spin-lattice relaxation process also contributes to the transverse relaxation process. It stems from the fact that the longitudinal relaxation occurs at random with respect to the transverse components of the individual nuclear magnets. It has been shown that

$$\frac{1}{T_2} = \frac{1}{T_2'} + \frac{1}{2T_1} \quad .$$

Here $1/T_2'$ represents the portion of the transverse relaxation enacted by the dipolar and the exchange interactions between the nuclear magnets. In a liquid sample where the movements of molecules (hence the nuclei) are large and rapid, the fluctuation of the local field is such

that its "average" value during each Larmor cycle can become vanish-
ingly small. In a liquid, therefore, the contribution to the transverse
relaxation process by the dipolar interaction is small. The exchange
process in a liquid is also inefficient since the time during which two
interacting nuclei are close to each other is short. On the other hand,
in a rigid solid, both processes are efficient and T_2 approaches T_2',
the inverse of the line width given by the line broadening discussed for
a rigid lattice. The dependence of T_2 upon τ is also shown in Fig. 15.

When a sample contains paramagnetic ions, the relaxation times
are considerably shortened. This is because of a similar but much
larger interaction between the magnetic moments of the paramagnetic
electrons and the nuclei.

For a nucleus with an electric quadrupole moment ($I > \frac{1}{2}$), an
additional broadening effect and spin-lattice relaxation mechanism
must be considered. The latter comes about from the interaction between the nuclear quadrupole moments and the fluctuating electric field
caused by the lattice motion.

Throughout the foregoing discussions, it was assumed that the
externally applied field H_0 is homogeneous over the entire volume of
the sample. In practice one must consider an additional transverse
relaxation mechanism provided by the inhomogeneity of the laboratory
magnetic field.

C. Nuclear Magnetic Resonance in Metals:
Knight Shift and Overhauser Effect

Magnetic resonance signals of nuclei in metals have been found
to occur at a slightly different frequency for a given magnetic field than
for the same nuclei in a dielectric environment. The shift observed is
always positive, much larger than the normal chemical shift, and is
proportional to the applied field. It is called the Knight shift, after
the discoverer, and is attributed to the hyperfine interaction between
the nuclei and the conduction electrons when the latter are in the

immediate neighborhood of the nuclei. In the absence of an applied
magnetic field, there are just as many conduction electrons with the
$+\frac{1}{2}$ spin state as there are those with the $-\frac{1}{2}$ spin state. In the presence of a magnetic field H, however, there are slightly more conduction
electrons in the $-\frac{1}{2}$ spin state than in the $+\frac{1}{2}$ spin state, since the former
state is more stable than the latter by $g\beta H$. The Knight shift results
from this unbalance of the spin states among the conduction electrons.
As one may surmise, the shift is large when the conduction electrons
originate from an s-orbital of the atoms and small, and may even be
anisotropic, if they originate from other types of orbitals.

Suppose one now applies a microwave frequency that induces the
electron spin resonance transitions of these conduction electrons and
at the same time observes the nuclear magnetic resonance signal
described. One of the relaxation mechanisms for the electron spin
system is provided by the hyperfine interaction with the nuclei, and it
results in the exchange of the spins, i. e., $\Delta M_S = -1$ while $\Delta M_I = +1$.
Note that, as more and more electrons are pumped to the upper $+\frac{1}{2}$ spin
state by the microwave field, approaching the saturation condition, this
relaxation mechanism leads to an increase in the longitudinal magnetization of the nuclear spin system, hence an enhancement of the NMR
signal. This phenomenon, known as the Overhauser effect, has been
observed experimentally. In order for the effect to be observable, the
spin-exchange process must prevail over other relaxation mechanisms
for the nuclear spin system. One should also note that, as the electron
spin system approaches the saturation condition, the Knight shift discussed earlier should become smaller, and vanish when the complete
saturation is achieved.

D. Nuclear Magnetic Resonance by
Pulsed Radio Frequency Waves

As shown earlier, the relaxation times T_1 and T_2 can be measured
from the saturation effect and the line width, respectively. A much more

precise measurement of these quantities is afforded by means of pulsed NMR spectroscopy. The novelty of this technique is in the application of the oscillating magnetic field H_1 in a strong pulsed form as opposed to a weak continuous form used in ordinary spectrometers.

Consider a sample placed in a steady magnetic field H_0 with its nuclear magnetization brought to a state of thermal equilibrium. A strong magnetic field H_1 oscillating exactly at the Larmor frequency of the nuclei is then suddenly applied in a pulsed form lasting for a period much shorter than T_1 or T_2 of the sample. The effect of the pulse can be followed by considering the Bloch equations (29) simplified to

$$\dot{M}_u = 0$$
$$\dot{M}_v = \omega_1 M_z \qquad\qquad (68)$$
$$\dot{M}_z = -\omega_1 M_v \quad .$$

The simplifications are possible because $\omega = \omega_0$, the duration of the pulse is much shorter than T_1 or T_2 , and the initial magnetization is such that $M_u = M_v = 0$, and $M_z = M_0$. Solutions of the differential equations (68) with the initial condition cited above give

$$M_u = 0$$
$$M_v = M_0 \sin \omega_1 t \qquad\qquad (69)$$
$$M_z = M_0 \cos \omega_1 t \quad .$$

Equations (69) state that, with the application of the pulse, the longitudinal magnetization M_0 will begin to nutate about the u axis with an angular velocity proportional to H_1 (since $\omega_1 = \gamma H_1$). Hence by adjusting either the strength of H_1 or the length of the pulse, the magnetization M_0 can be rotated about the u axis by any desired angle.

Let us consider a situation where a "90° pulse" has just been applied. At the end of the pulse the magnetization is such that

$$M_u = 0$$

$$M_v = M_0 \tag{70}$$

$$M_z = 0 \quad .$$

Note that the resonance signal is now induced at the receiver coil even though the oscillating field H_1 is absent. This signal is called the free induction signal. Immediately after the pulse, the signal is that induced by the transverse magnetization M_1 equal to M_0. After the cessation of the pulse, however, the magnetization will continue to change according to the Bloch equations (29). Because of the conditions $\omega = \omega_0$, $H_1 = 0$, and the initial magnetization given by (70), however, the equations simplify to the following forms:

$$\dot{M}_u = 0$$

$$\dot{M}_v = - \frac{M_v}{T_2} \tag{71}$$

$$\dot{M}_z = - \frac{M_z - M_0}{T_1} \quad .$$

Solution of these equations, assuming the magnetization given by (70) for $t = 0$, yields

$$M_u(t) = 0$$

$$M_v(t) = M_0 \exp(-t/T_2) \tag{72}$$

$$M_z(t) = M_0 [1 - \exp(-t/T_1)] \quad .$$

We thus see that T_2 can be measured by observing the decay of the free induction signal. The T_1 can be obtained by measuring M_z, which has built up during a period t following the application of the initial 90° pulse. This can be done easily by applying a second 90° pulse and observing the intensity of the free induction signal that follows.

E. Nuclear Quadrupole Resonance

We have thus far discussed the interaction of nuclear magnetic
moments with a magnetic field. Nuclei do not possess electric dipole
moments, so the energy of any nucleus is independent of its orientation in a uniform electric field, the first derivative of the electric
potential. For nuclei with spin $I \geq 1$, however, one needs to consider
the interaction of the nuclear quadrupole moment Q with the electric
field gradient q. The quadrupole moment Q of a nucleus is defined as

$$Q = \int \rho r^2 (3 \cos^2 \theta - 1) d\tau \quad , \tag{73}$$

where $\rho(r, \theta)$ is the charge density per unit volume, r is the distance
of the volume element $d\tau$ from the origin, and θ is the angle between
this radius vector and the spin axis. The distribution of nuclear charge
is axially symmetric, the axis of symmetry being the axis of spin. Thus
Q is a measure of the deviation from a spherical distribution of a nuclear
charge, and is positive for a nucleus with a prolate (elongated) charge
distribution, zero for a spherical distribution, and negative for an
oblate (flattened) distribution. When a nucleus with a quadrupole
moment is located at a place where there is a gradient q in the electric
field, a net torque is exerted on the nucleus in a form similar to that
given by a magnetic dipole field. A classical expression for this interaction can be shown to be

$$\Delta W_{quad} = \frac{eqQ}{4} (\tfrac{3}{2} \cos^2 \theta - \tfrac{1}{2}) \quad , \tag{74}$$

where

$$q = q_{zz} = \frac{\partial E_z}{\partial z} = \frac{\partial^2 V}{\partial z^2} \quad .$$

Here the z axis is the direction along which the electric field changes,
and θ is the angle between this axis and the nuclear spin axis. The
energy of the nucleus thus depends upon its orientation with respect
to the direction of the field gradient, hence the component of the spin

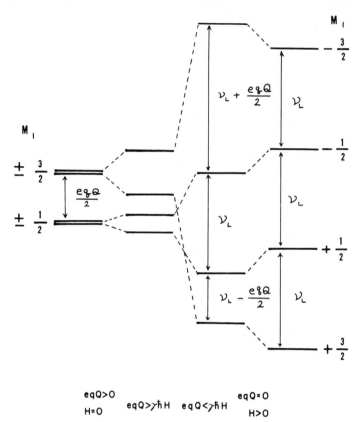

FIG. 16. Energy level splitting of a nucleus ($I = 3/2$) with a quadrupole moment interaction eqQ in the presence of zero, weak, and strong magnetic fields applied parallel to the z axis of the quadrupole interaction.

along this direction. A quantum mechanical treatment gives the energy of the interaction in the following form:

$$\Delta W_{quad} = \frac{eqQ}{4I(2I - 1)} [3M_I^2 - I(I + 1)] \quad . \tag{75}$$

Therefore, the energy levels of a nucleus with a quadrupole moment can be split even in the absence of magnetic field. The energy, however, is independent of the sign of M_I, so that a twofold degeneracy remains unless $M_I = 0$. The twofold degeneracy may be removed by the application of a magnetic field. The situation is illustrated in Fig. 16 for a nucleus with $I = 3/2$.

In pure quadrupole resonance spectroscopy, one observes the transition between the levels split by a quadrupole moment interaction. The transitions are induced by precisely the same mechanism responsible for nuclear magnetic resonance. Hence the selection rules are $\Delta M_I = \pm 1$, and the transition probability depends upon the magnetic moment of the nucleus and not upon its quadrupole moment. Only the resonance frequency is determined by the quadrupole moment Q of the nucleus and the electric field gradient q of the environment. Representative transitions have been observed in the range $10 \sim 1000$ MHz. Once assignments are made on a pure quadrupole resonance spectrum, the evaluation of the coupling constant $|eqQ|$ is straightforward, and, if Q is known, the measured value of q provides information about the electronic structure of the ligands.

LaPlace's equation states

$$\Delta V = \frac{\partial^2 V}{\partial x^2} + \frac{\partial^2 V}{\partial y^2} + \frac{\partial^2 V}{\partial z^2} = 0 \quad .$$

It follows then that, for an electron having a spherically symmetric distribution about a nucleus,

$$\frac{\partial^2 V}{\partial x^2} = \frac{\partial^2 V}{\partial y^2} = \frac{\partial^2 V}{\partial z^2} = 0 \quad .$$

Hence an electron in an s-orbital, or electrons within a closed shell configuration cannot give rise to a quadrupole coupling interaction. A large electric field gradient at a nucleus is most often produced by electrons in a p-orbital.

F. Effect of Quadrupole Interaction upon
Nuclear Magnetic Resonance

Let us consider a magnetic nucleus with $I \geq \frac{1}{2}$ located within a single crystal where there is an electric field gradient q in the z direction. When a magnetic field H is applied in the z direction, the energy of the nucleus is given by the Zeeman interaction (14) and the quadrupole interaction (75):

$$W(M_I) = -\gamma\hbar H M_I + \frac{eqQ}{4I(2I-1)}[3M_I^2 - I(I+1)] \quad . \tag{76}$$

The splitting schemes given by Eq. (76) when $\gamma\hbar H \gg eqQ$ and when $q = 0$ are also illustrated in Fig. 16 for our example of $I = 3/2$. Thus the NMR transitions which are degenerate at ν_L when $q = 0$ become split into three lines at $\nu_0 = \nu_L$ and $\nu_\pm = \nu_L \pm eqQ/2$, respectively, when $q \neq 0$. The splitting depends upon the direction of the magnetic field relative to the direction of the electric field gradient. When they make an angle α and $eqQ \ll \gamma\hbar H$, the three resonance frequencies are given by $\nu_0 = \nu_L$ and $\nu_\pm = \nu_L \pm \frac{eqQ}{2}(\frac{3}{2}\cos^2\alpha - \frac{1}{2})$.

In a polycrystalline sample the random distribution of the angle α causes the NMR signals to be spread over the range defined by these equations. The lattice motion of a solid causes an additional broadening by producing fluctuation in the electric field gradient. When a nucleus possesses a quadrupole moment, this coupling to the lattice motion via the quadrupole interaction becomes the dominant spin-lattice relaxation mechanism for the nucleus.

V. FERROMAGNETIC RESONANCE

Electron spin resonance can also be observed in ferromagnetic materials. It is known as ferromagnetic resonance. It differs from the ordinary magnetic resonance in that one must consider the effect of the demagnetizing field H_d:

$$H_d = -NM \quad . \tag{77}$$

Here N and M are the demagnetization factor and the magnetization of a sample, respectively. Thus, in the presence of an external field H_0 applied in the z direction, the magnetization M experiences a torque exerted by the internal field, given by

$$H_x = -N_x M_y$$

$$H_y = -N_y M_x \tag{78}$$

$$H_z = H_0 - N_z M_z \quad .$$

The N's are the demagnetization factors in the direction indicated, and their values depend upon the shape of the sample. Substitution of these values into Eq. (23) yields the following equations:

$$\frac{dM_x}{dt} = \gamma[H_0 - M_z(N_z - N_y)]M_y$$

$$\frac{dM_y}{dt} = -\gamma[H_0 - M_z(N_z - N_x)]M_x \tag{79}$$

$$\frac{dM_z}{dt} = \gamma M_x M_y (N_x - N_y) \quad .$$

Since the external field is in the z direction, M_x and M_y are small. Hence, in a first-order approximation, we may assume that $dM_z/dt = 0$ and $M_z = M$. A solution to the above set of equations can then be shown to be

$$M_x = \alpha^{\frac{1}{2}} \sin[(\alpha\beta)^{\frac{1}{2}} t]$$

$$M_y = \beta^{\frac{1}{2}} \cos[(\alpha\beta)^{\frac{1}{2}} t] \tag{80}$$

where

$$\alpha = \gamma[H_0 - M(N_z - N_x)]$$

$$\beta = \gamma[H_0 - M(N_z - N_y)] \quad .$$

Equations (80) state that the magnetization M precesses about the z axis with the angular velocity ω_0:

$$\omega_0 = \gamma[H_0 - M(N_z - N_x)]^{\frac{1}{2}}[H_0 - M(N_z - N_y)]^{\frac{1}{2}} \quad . \tag{81}$$

For a ferromagnetic material with a spherical shape, $N_x = N_y = N_z$, so that $\omega_0 = \gamma H_0$. For a sample with a shape of a flat plate lying in the x-z plane, $N_z = N_x = 0$, and $N_y = 4\pi$. Therefore,

$$\omega_0 = \gamma[H_0(H_0 + 4\pi M)]^{\frac{1}{2}} \quad . \tag{82}$$

The resonance frequency of ferromagnetic resonance is thus dependent upon the shape of the sample. The spectroscopic splitting factors g determined through the relation $\gamma\hbar = g\beta$ for metallic Fe, Co, and Ni are all in the neighborhood of 2.2, indicating that a major fraction of the magnetism comes from the spin, rather than orbital motion, of the electrons.

Chapter 7

OPTICAL PROPERTIES OF SOLIDS

John D. Axe

Department of Physics
Brookhaven National Laboratory
Upton, New York

I. INTRODUCTION

This chapter is concerned with two closely related questions.
First of all, how does light interact with solid matter? Secondly, what
can we learn about solids from a study of their optical properties? In
line with the general concept of the book as a whole, the emphasis is
on solids, not optics. There is, however, no effort made to acquaint
the reader with anything past the barest essentials of the body of experimental facts that make up the subject under discussion. The emphasis instead is on concepts.

Qualitatively, at least, the facts are reasonably familiar. Metals
are opaque and highly reflecting, whereas some insulators are transparent to the eye and others are highly colored or opaque. Yet quantitatively, as we shall see, it is possible to uniquely define all of the
optical properties of a given system by a single function, a dielectric
response function, which relates a driven wave of polarization within
the solid to a driving electromagnetic wave. Furthermore, the form of
this dielectric response function for most solids is basically similar
and can be derived and understood by an analysis of a classical harmonic oscillator.

The great variety of optical phenomena found in solids arises
from the variation in the energy level structure which exists between
different materials. Just as spectroscopy has provided our most
detailed knowledge of the microscopic nature of atoms and molecules
by a study of their energy levels, so the study of the interaction of
light with solids provides direct information about the possible states
of excitation of solids. For the most part, the important excitations
of crystalline solids, rather than being localized about one particular
atom, have a wavelike nature, and each excitation extends throughout
the volume of the solid. The idea of excitation waves in solids and
the associated paraphernalia of reciprocal lattices and Brillouin zones
are very important and provide a unifying viewpoint from which to

discuss the optical properties of conduction electrons, excitons, phonons, and so on.

   The chapter is reasonably self-contained, even at the expense
of discussions of topics not specifically concerned with optical properties (Brillouin zones or density-of-states functions, for example),
which also appear elsewhere in this book.   It is felt that all topics in
this category are sufficiently important to be understood thoroughly and
from many points of view, and thus no apologies are made for such
duplication.

   An attempt is made to present and analyze simple models and to
extract and emphasize the generalities that can be inferred.   In this
way the mathematical complexities are kept to a minimum but what is
presented is in rather complete form.   A bibliography is included at
then end of the chapter to assist the interested student in the selection
of further reading.

II.   DIELECTRIC POLARIZATION IN SOLIDS

   We begin our discussion of the optical properties of solids by
considering how a solid body reacts upon application of a static electric field.   Even if the solid has no net charge (as is the usual case),
it is composed of positively and negatively charged entities which
experience oppositely directed forces due to the electric field.   As a
result, the body becomes polarized, that is, it develops a dipole moment.   The magnitude of the polarization developed depends upon the
strength of the applied electric field and upon the number and kinds of
charge carriers that are causing the charge separation.   There can be
contributions from conduction electrons, bound electron charge distributions which become distorted by the electric field, motion of charged
ions, and so forth.

If there are several kinds of mobile charge carriers and each moves
by an amount $\underline{\delta}_i$ under the action of the applied electric field, each
creates locally a dipole moment $q_i \underline{\delta}_i$. The dielectric polarization $\underline{P}$,
defined as the dipole moment per unit volume, is thus given by

$$\underline{P} = \sum_i N_i q_i \underline{\delta}_i \quad ,$$

where N_i is the number of charges of the ith type per unit volume. This
polarization may vary from point to point in the solid, but the polariza-
tion field $\underline{P}(\underline{r})$ at any given point $\underline{r}$ is found to be proportional to the
electric field $\underline{E}(\underline{r})$ at the same point, as long as $\underline{E}$ is not too large.
We express this observation by writing

$$\underline{P}(\underline{r}) = \chi^0 \underline{E}(\underline{r}) \quad , \tag{1}$$

where the constant of proportionality χ^0 is known as the dielectric
susceptibility. χ^0 is an intrinsic property of the material which char-
acterizes the way in which it behaves in an electric field. To be a
little more exact we can think of the right-hand side of Eq. (1) as being
the first term of a power series expansion in $\underline{E}$, the higher order terms
in $\underline{E}^2$, etc., being negligible in ordinary circumstances.

Mostly for historical reasons there is another more common way
to characterize the dielectric response of a medium than by the susceptibility χ^0. We define a new vector field $\underline{D}$ as the following linear
combination of $\underline{E}$ and $\underline{P}$

$$\underline{D} = \underline{E} + 4\pi \underline{P} \quad . \tag{2}$$

Since $\underline{P}$ is to a good approximation proportional to $\underline{E}$, so is $\underline{D}$, and we
can write

$$\underline{D}(\underline{r}) = \varepsilon^0 \underline{E}(\underline{r}) \quad , \tag{3}$$

where

$$\varepsilon^0 = 1 + 4\pi\chi^0 \quad . \tag{4}$$

TABLE 1

The Form of the Dielectric Susceptibility Tensor

for the Seven Crystal Systems

System	Characteristic symmetry	Number of independent coefficients	Form of tensor
Triclinic	A center of symmetry or no symmetry	6	$\begin{pmatrix} \chi_{xx} & \chi_{xy} & \chi_{xz} \\ \chi_{xy} & \chi_{yy} & \chi_{zy} \\ \chi_{xz} & \chi_{zy} & \chi_{zz} \end{pmatrix}$
Monoclinic	Twofold rotation about z axis	4	$\begin{pmatrix} \chi_{xx} & \chi_{xy} & 0 \\ \chi_{xy} & \chi_{yy} & 0 \\ 0 & 0 & \chi_{zz} \end{pmatrix}$
Orthorhombic	Twofold rotation about x, y, z axes	3	$\begin{pmatrix} \chi_{xx} & 0 & 0 \\ 0 & \chi_{yy} & 0 \\ 0 & 0 & \chi_{zz} \end{pmatrix}$
Tetragonal	Fourfold rotation about z axis		
Hexagonal	Sixfold rotation about z axis	2	$\begin{pmatrix} \chi_{xx} & 0 & 0 \\ 0 & \chi_{xx} & 0 \\ 0 & 0 & \chi_{zz} \end{pmatrix}$
Trigonal	Threefold rotation about z axis		
Cubic	Fourfold rotation about x, y, z axes	1	$\begin{pmatrix} \chi_{zz} & 0 & 0 \\ 0 & \chi_{zz} & 0 \\ 0 & 0 & \chi_{zz} \end{pmatrix}$

ε^0 is known as the (static) dielectric constant and it contains no more (or less) information than χ^0 does. (The definition of ε^0 does involve the introduction of a new field $\underline{D}$, which seems less "physical" than $\underline{P}$. About the only justification that can be given for the introduction of a $\underline{D}$ field is that it simplifies the appearance of certain of Maxwell's equations, as we shall see.)

We have made the further assumption in writing Eqs. (1)-(4) that the induced $\underline{P}$ and $\underline{D}$ are in the same direction as $\underline{E}$. Solids such as cubic crystals and amorphous substances that have this property are termed electrically (or optically) isotropic. In general, however, crystalline solids are anisotropic, and a relation such as Eq. (1) must be written for each component of $\underline{P}$ and $\underline{E}$:

$$P_x = \chi^0_{xx} E_x + \chi^0_{xy} E_y + \chi^0_{xz} E_z$$

$$P_y = \chi^0_{yx} E_x + \chi^0_{yy} E_y + \chi^0_{yz} E_z \qquad (5)$$

$$P_z = \chi^0_{zx} E_x + \chi^0_{zy} E_y + \chi^0_{zz} E_z \;.$$

Thus, generally a crystal has not a single susceptibility but a maximum of nine components making up the susceptibility tensor $\underset{=}{\chi^0}$. Not all components are independent, however. For example, all components χ^0_{ij} with $i \neq j$ obey the relation $\chi^0_{ij} = \chi^0_{ji}$. Furthermore, any rotations or reflections that take a solid into itself may impose certain relations between components. The form of the susceptibility tensor as determined by symmetry requirements for each of the seven crystal systems is shown in Table 1. Now the preceding arguments made for $\underset{=}{\chi^0}$ can be made for $\underset{=}{\varepsilon^0}$ as well, so that the susceptibility tensor and the dielectric constant tensor must be of the same form. In the next section we are going to investigate the susceptibility of crystals for time-varying fields, and they too must have the forms given in Table 1. Symmetry-determined relations, such as these, occur often in solid state properties and are very useful. They allow certain properties to be readily deduced from a knowledge of symmetry alone, and conversely the

symmetry of a material can often be deduced from experimentally established relations among the components of the dielectric constant, the
resistivity, or some other tensor quantity.

In most of the remainder of this chapter, we will treat only isotropic solids. Many important real systems do fall in this category and
all of the important physical ideas encountered in isotropic systems
occur in anisotropic ones as well. There are, however, important
optical effects (those associated with double refraction, for example)
which are peculiar to anisotropic materials but which are cumbersome
to treat mathematically.

III.   TIME-DEPENDENT FIELDS AND
COMPLEX SUSCEPTIBILITIES

Now we want to consider how the previous discussion must be
modified if the static applied field is replaced by a time-varying one.
A sinusoidally varying field, say $\underline{E}(\underline{r}, t) = \underline{E}(\underline{r}) \cos \omega t$, is the only
type of time dependence we need to consider, because after we find
the effect of any single frequency field, we can synthesize any time-
dependent field by appropriate combinations of harmonic ones of differing frequency (Fourier analysis).

For very slowly varying fields (very low frequencies), we should
expect that the polarization adjusts itself to the instantaneous value
of $\underline{E}$, that is, $\underline{P}(\underline{r}, t) = \chi \underline{E}(\underline{r}, t)$. But as the frequency becomes higher
it becomes increasingly impossible for all of the charge carriers to move
fast enough to stay in step with the driving field. Contributions involving nuclear motion would be expected to lag behind more quickly than
those involving motion only of the much lighter electrons. By studying
the polarization of a material at various frequencies, and noting at
what frequencies various components of the polarization begin to lag
behind the driving field, we can learn how much of the polarization is

due to lattice vibrations, how much due to conduction electrons, and so forth. Thus, it is important to have a convenient way to describe and measure both the magnitude of the induced polarization and the amount by which it is out of phase with the $\underline{E}$ field.

We can modify Eq. (1) to represent this tendency of $\underline{P}$ to lag behind the driving $\underline{E}$ field by introducing a phase angle $\varphi(\omega)$:

$$\underline{P}(\underline{r}, t) = \chi_r \underline{E}(\underline{r}) \cos(\omega t - \varphi)$$

$$= (\chi_r \cos \varphi) \underline{E}(\underline{r}) \cos \omega t + (\chi_r \sin \varphi) \underline{E}(\underline{r}) \sin \omega t \quad , \tag{6}$$

where, by comparison with Eq. (1), $\chi_r(\omega) \to \chi^0$ and $\varphi(\omega) \to 0$ as $\omega \to 0$. We expect χ_r as well as φ to change with the driving frequency ω. That is, a particular charge is going to respond more at some frequencies than at others. For example, at frequencies so high that the fastest moving carriers can no longer respond there will be no polarization whatever, so $\chi_r \to 0$ as $\omega \to \infty$.

Although introduction of a phase angle is a perfectly satisfactory way of describing the phase lag, there is a more convenient and conventional method of representation which involves the use of complex numbers. Recalling that

$$\exp(-i\omega t) = \cos \omega t - i \sin \omega t$$

we see that our sinusoidal $\underline{E}$ field can be represented as the real part of a complex field

$$\hat{\underline{E}}(\underline{r}, t) = \underline{E}(\underline{r}) \exp(-i\omega t)$$

(we indicate a complex quantity by a caret, their real and imaginary parts by primes; for example $\hat{a} = a' + ia''$)

$$\underline{E}(\underline{r}, t) = Re[\hat{\underline{E}}(\underline{r}, t)] = \underline{E}(\underline{r}) \cos \omega t \quad . \tag{7}$$

Now we further define a complex susceptibility

$$\hat{\chi}(\omega) = \chi'(\omega) + i\chi''(\omega)$$

and a complex polarization given by

$$\hat{\underline{P}}(\underline{r}, t) = \hat{\chi}(\omega)\hat{\underline{E}}(\underline{r}, t) \quad . \tag{8}$$

We cannot equate $\hat{\underline{P}}$ with the polarization of a medium any more than we can say that $\hat{\underline{E}}$ is the electric field. Both $\underline{P}$ and $\underline{E}$ are physically measurable quantities and must therefore be real. The imaginary parts of $\underline{P}$ and $\underline{E}$ have no physical meaning whatever. However, the definition

$$\underline{P}(\underline{r}, t) = \text{Re } \hat{\underline{P}}(\underline{r}, t) = \chi'(\omega)\underline{E}(\underline{r}) \cos \omega t + \chi''(\omega)\underline{E}(\underline{r}) \sin \omega t \tag{9}$$

is equivalent to the previous definition of $\underline{P}(\underline{r}, t)$ [Eq. (6)] with

$$\chi'(\omega) = \chi_r(\omega) \cos \varphi \quad \text{and} \quad \chi''(\omega) = \chi_r(\omega) \sin \varphi \quad . \tag{10}$$

Thus we can specify the polarization response by the two numbers χ_r, φ or equivalently by the number pair $\hat{\chi} = \chi' + i\chi''$, the relation between the two descriptions being given by Eq. (10). Both the real and imaginary parts of $\hat{\chi}$ (or $\hat{\epsilon}$) are meaningful because $\hat{\chi}$ represents a functional relationship between two real observables but is not itself a physical observable. Any "measurement" of the dielectric response really consists of measuring the in- and out-of-phase components of the polarization produced by a known driving field. Because the algebra of exponentials is easier than that of sines and cosines it is customary not only in optics but in many branches of physics and engineering to represent phase shifts (time lags) by this trick of replacing real quantities by complex ones which we invent. The algebra is performed on these complex quantities and only at the end of the calculation do we revert to the real part of the solution. Note that in Eq. (9) we have resolved a polarization that lags behind the driving field by some angle into two components. One component, proportional to χ', is in phase with $\underline{E}$, and the other, proportional to χ'', is out of phase by $90°$ (see Fig. 1).

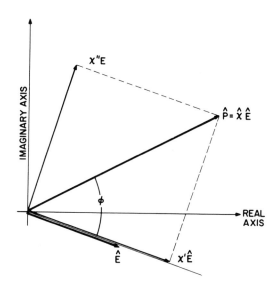

FIG. 1. Vector diagram showing components of the complex polarization. The complex $\hat{\underline{P}}$ field lags $\hat{\underline{E}}$ by the angle φ. The component of $\hat{\underline{P}}$ in phase with $\hat{\underline{E}}$ is $\chi'\hat{\underline{E}}$, the out-of-phase component is $\chi''\hat{\underline{E}}$. The real (physical) $\underline{P}$ and $\underline{E}$ fields are the projections of $\hat{\underline{P}}$ and $\hat{\underline{E}}$ upon the real axis.

Proceeding in exactly the same way, we can "invent" a complex $\hat{\underline{D}}$ field related to $\hat{\underline{E}}$ by a complex dielectric response function, $\hat{\varepsilon}(\omega) = \varepsilon'(\omega) + i\varepsilon''(\omega)$. We find that the modifications of Eqs. (1)-(4) necessary to describe sinusoidally varying fields are quite simply written down in our complex notation. They are

$$\hat{\underline{P}}(\underline{r}, t) = \hat{\chi}(\omega)\hat{\underline{E}}(\underline{r}, t) \tag{11}$$

$$\hat{\underline{D}}(\underline{r}, t) = \hat{\underline{E}}(\underline{r}, t) + 4\pi\hat{\underline{P}}(\underline{r}, t) \tag{12}$$

$$\hat{\underline{D}}(\underline{r}, t) = \hat{\varepsilon}(\omega)\hat{\underline{E}}(\underline{r}, t) \tag{13}$$

$$\hat{\varepsilon}(\omega) = 1 + 4\pi\hat{\chi}(\omega) \quad . \tag{14}$$

IV.   CURRENT FLOW AND ABSORPTION OF ENERGY

Now we want to examine some of the consequences of a time-varying polarization within a material medium. First of all, a time-varying polarization inevitably has an electrical current associated with it. The current associated with a charged particle is the product of its charge times its velocity. Each of the oscillating charges q_i thus contributes a current $q_i \, d\underline{\delta}_i/dt$, and the total current density $\underline{j}$ (current/unit volume) is $d(\Sigma_i N_i q_i \underline{\delta}_i)/dt$ or

$$\underline{j}(\underline{r},\, t) = d\underline{P}(\underline{r},\, t)/dt \quad . \tag{15}$$

In order to see how the current varies with the driving field $\underline{E}$, we substitute the expression given in Eq. (8) for $\underline{\hat{P}}(\underline{r},\, t)$ into Eq. (15)

$$\underline{j}(\underline{r},\, t) = \frac{d}{dt} \, \mathrm{Re}\, \underline{\hat{P}}(\underline{r},\, t) = \mathrm{Re}[-i\omega\hat{\chi}\,\underline{\hat{E}}(\underline{r})\, \exp(-i\omega t)]$$

$$= \omega \underline{E}(\underline{r})[\chi'' \cos \omega t - \chi' \sin \omega t] \quad . \tag{16}$$

We see that the current response also has components in- and out-of-phase with $\underline{E}$. By analogy with our previous treatment of the polarization response we can resort to complex notation and define a conductivity $\hat{\sigma}(\omega) = \sigma'(\omega) + i\sigma''(\omega)$ for harmonic fields by the relation

$$\underline{\hat{j}}(\underline{r},\, t) = \hat{\sigma}(\omega)\underline{\hat{E}}(\underline{r},\, t) \quad , \tag{17}$$

and by writing out the real and imaginary parts of Eq. (17) and comparing them with Eq. (16) we find that

$$\sigma'(\omega) = \omega\chi''(\omega) \quad \text{and} \quad \sigma''(\omega) = -\omega\chi''(\omega) \quad . \tag{18}$$

Notice that because the polarization and current are related by a time derivative, the imaginary part of the susceptibility determines the real conductivity and vice versa. You see that $\hat{\chi}$ or $\hat{\varepsilon}$ and $\hat{\sigma}$ are directly related, and as a result they are referred to almost interchangeably. Some people prefer to mix the two response functions, describing the polarization in phase with $\underline{E}$ by a dielectric response and that out-of-phase with a conductivity. One can write, for example,

$$\hat{\varepsilon}(\omega) = \varepsilon(\omega) + i\left(\frac{4\pi}{\omega}\right)\sigma(\omega) \quad ,$$

where ε and σ stand for the real parts of $\hat{\varepsilon}$ and $\hat{\sigma}$. It is largely a matter of personal preference, there being no firmly established conventions. It is important to understand that $d\underline{P}/dt$, even if it arises from the polarization of electrons firmly bound or localized in space, is a perfectly real current. Thus even "insulators" can have a quite high conductivity at high frequencies. The apparent inconsistency in this remark arises from the fact that we loosely term a material an "insulator" or "conductor" based upon its static or dc conductivity. You can easily see that bound electrons cannot give rise to a dc current, whereas unbound or metallic electrons do.

The rate (per unit volume) at which the driving field does work upon the medium is given by

$$\underline{E} \cdot \frac{d\underline{P}}{dt} = \underline{E} \cdot \underline{j}$$

which must be averaged over one complete cycle to obtain the net power per unit volume absorbed from the field and dissipated within the medium:

$$\text{absorbed power density} = \langle \underline{j} \cdot \underline{E} \rangle_{av} = \tfrac{1}{2}\omega\chi''(\omega)|E(\underline{r})|^2 \tag{19}$$

Only the out-of-phase component of $\underline{P}$ (or equivalently the in-phase current $\underline{j}$) continuously drains away energy from the field and appears as Joule heating of the material. The energy associated with the in-phase component of $\underline{P}$, on the other hand, is alternately stored in the material and released twice during each cycle of the driving field and thus makes no net contribution to the energy balance.

In addition to giving rise to a current through its time variation, an induced polarization can disturb the electrical neutrality that exists within a material. If the polarization $\underline{P}(\underline{r})$ is uniform in the region

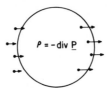

FIG. 2. If the polarization field $\underline{P}(\underline{r})$ is nonuniform, less charge can enter a small volume than leaves it, leading to a charge imbalance.

about $\underline{r}$, neutrality is preserved, for as much charge as leaves through one side of an infinitesimal volume element is replaced by charge coming in on the opposite side. However, if $\underline{P}(\underline{r})$ is not uniform, more (or less) charge enters than leaves a tiny volume element and the charge neutrality is locally disturbed. This is illustrated in Fig. 2. From the divergence theorem, this excess charge density $\hat{\rho}$ (charge per unit volume) is given by

$$\hat{\rho}(\underline{r}, t) = -\text{div}\,\underline{P}(\underline{r}, t) = -\left(\frac{\partial P_x}{\partial x} + \frac{\partial P_y}{\partial y} + \frac{\partial P_z}{\partial z}\right) \tag{20}$$

Thus any nonuniform polarization which causes a bunching of charges can itself act as a source of an electric field.

V. ELECTROMAGNETIC WAVES IN SOLIDS

A. Maxwell's Equations

Now that we have some idea of how the polarization response of a dielectric medium behaves with time, we can turn to another aspect of the problem. Electromagnetic disturbances propagate as waves, and waves vary simultaneously both in time and space. What, if anything, can we learn about the spatial dependence of the polarization response? It turns out that, knowing the time behavior, we can completely specify the spatial behavior, because the two are not independent but are closely linked by Maxwell's equations, which are the most fundamental relationships that exist between the electromagnetic $\underline{E}$, ρ, $\underline{j}$ and the magnetic field $\underline{B}$. Maxwell's equations can be written in the form

$$\text{div } \underline{E} = 4\pi\rho \tag{21}$$

$$\text{div } \underline{B} = 0 \tag{22}$$

$$\text{curl } \underline{E} = -\frac{1}{c}\frac{\partial \underline{B}}{\partial t} \tag{23}$$

$$\text{curl } \underline{B} = \frac{1}{c}\left(\frac{\partial \underline{E}}{\partial t} + 4\pi\underline{j}\right) \quad . \tag{24}$$

Since they are so fundamental, Maxwell's equations in this form apply
on an atomic scale. That is, the fields associated with individual
atomic particles obey Eqs. (21)-(24), so that in principle ρ can be
taken as the exact distribution of electronic and nuclear charges within
the atoms and $\underline{j}$ the associated currents. You can see that in a solid
these ρ's and $\underline{j}$'s act as sources for strong $\underline{E}$ and $\underline{B}$ fields which
furthermore vary tremendously from one atom to another (or for that
matter from one electronic shell to another even within one atom). In
practice, as you might imagine, such detailed calculations are extremely difficult to perform and fortunately such fine grained solutions
are not necessary in order to discuss optical properties. What we do
is interpret $\underline{E}$, $\underline{B}$, ρ, and $\underline{j}$ not as the atomic quantities just described
but instead as the average of these quantities over a volume that is
large enough to contain many atoms or molecules but small enough that
it can be associated with a point in the material on a macroscopic
scale. It is these spatially averaged or macroscopic fields that we
have really been discussing in the preceding sections, although we
did not bother to make such a distinction. It is the wavelike motion of
these average fields that we associate with the presence of "light" in
a medium, and which we will determine with the help of Maxwell's
equations.

Although Eqs. (21)-(24) involve the real fields $\underline{E}$, $\underline{B}$, ρ, and $\underline{j}$,
we can equally well substitute their complex equivalents without
destroying the equality. For example, the complex version of Eq. (21)
is

$\mathrm{div}\,(\underline{E}' + i\underline{E}'') = \mathrm{div}\,\underline{E}' + i\,\mathrm{div}\,\underline{E}'' = 4\pi(\rho' + i\rho'')$

which must be satisfied separately for both real and imaginary parts. Since we invented the imaginary field components anyway, we are free to ignore them. The important thing is that the real parts satisfy Eq. 21. [It is this same property of linearity, $f(x + y) = f(x) + f(y)$, which allows us to substitute averaged fields into Maxwell's equations in place of the microscopic ones.]

We want to consider the situation in which the time-dependent polarization of the preceding sections is the only source of charge and current present. This is justified as long as the medium has no net charge, and we neglect currents induced by the magnetic field $\underline{B}$. Substituting from Eq. (20) for ρ and Eq. (15) for $\underline{j}$ and passing to the spatially averaged complex fields, Maxwell's equations take the form

$$\mathrm{div}\,(\underline{\hat{E}} + 4\pi\underline{\hat{P}}) = \hat{\varepsilon}\,\mathrm{div}\,\underline{\hat{E}} = 0 \tag{25}$$

$$\mathrm{div}\,\underline{\hat{B}} = 0 \tag{26}$$

$$\mathrm{curl}\,\underline{\hat{E}} = -\frac{1}{c}\frac{\partial\underline{\hat{B}}}{\partial t} \tag{27}$$

$$\mathrm{curl}\,\underline{\hat{B}} = \frac{1}{c}\frac{\partial}{\partial t}(\underline{\hat{E}} + 4\pi\underline{\hat{P}}) = \frac{\hat{\varepsilon}}{c}\frac{\partial\underline{\hat{E}}}{\partial t} \tag{28}$$

B. Wavelike Solutions of Maxwell's Equations

Any electromagnetic disturbance within the material must now be a solution of Eqs. (25)-(28). We will continue to analyze for a single frequency component, that is, for fields with an $\exp(-i\omega t)$ time dependence. Having thus prescribed the time dependence, Eqs. (25)-(28) must then determine the spatial behavior of the fields. We anticipate wavelike behavior, characterized by a wavelength λ and a direction of propagation which can be specified by a unit vector $\underline{s}$. Both of these can be combined into a single propagation vector $\underline{k} = (2\pi/\lambda)\underline{s}$. Since we are considering an isotropic medium, we can simplify the vector algebra somewhat by picking one of the coordinate directions, the z axis, for example, to coincide with $\underline{s}$. Then we can look for solutions of the complex field quantities with time and space variations of the following form:

$$\left.\begin{array}{c} \hat{\underline{E}}(\underline{r},\, t) \\ \\ \hat{\underline{B}}(\underline{r},\, t) \end{array}\right\} = \left\{\begin{array}{c} \underline{E}(k,\, \omega) \\ \\ \underline{B}(k,\, \omega) \end{array}\right\} \exp\, i(kz - \omega t) \quad .$$

Solutions of this type are correct for samples with dimensions ℓ large enough to contain many waves, $k\ell \gg 1$, which is the usual case at optical frequencies.

Let us further restrict ourselves to waves with transverse electric fields, that is, with $\hat{\underline{E}}$ perpendicular to the propagation direction $\underline{s}$. With no loss of generality, we can pick the x coordinate axis to coincide with $\hat{\underline{E}}$, so that $\hat{\underline{E}}(k,\, \omega) = E_x \underline{i}$. (Since $\hat{\varepsilon}$ and $\hat{\chi}$ are scalar quantities, $\hat{\underline{P}}$ and $\hat{\underline{D}}$ are also directed along the x axis.) Note that

$$\mathrm{div}\, \hat{\underline{E}} = \frac{\partial}{\partial x}[E_x \exp\, i(kz - \omega t)] = 0 \quad ,$$

so that transverse electric waves automatically satisfy Eq. (25). Now since

$$\mathrm{curl}\, \hat{\underline{E}} = \underline{j}\frac{\partial}{\partial z}[E_x \exp\, i(kz - \omega t)] = \underline{j}[i\, k\, E_x \exp\, i(kz - \omega t)]$$

and

$$-\frac{1}{c}\frac{\partial B}{\partial t} = -i\frac{\omega}{c}\underline{B}(k,\, \omega)\, \exp\, i(kz - \omega t) \quad ,$$

Eq. (27) tells us that

$$\underline{B}(k,\, \omega) = B_y \underline{j} = (ck/\omega)\, E_x \underline{j} \quad . \tag{29}$$

Similarly, Eq. (28) becomes

$$\underline{B}(k,\, \omega) = B_y \underline{j} = (\hat{\varepsilon}\omega/ck)\, E_x \underline{j} \quad . \tag{30}$$

Note that, since $\underline{B}$ is also transverse, the remaining Maxwell relation $\mathrm{div}\, \underline{B} = 0$ is automatically satisfied. Equations (29) and (30) are simultaneously satisfied only if $(ck/\omega) = (\hat{\varepsilon}\omega/ck)$ or

$$k^2 = \hat{\varepsilon}(\omega/c)^2 \quad , \tag{31}$$

which is the fundamental dispersion relation, which relates the spatial
variation k to the time variation ω. Maxwell's equations, which seem
so formidable, boil down to this single simple result when applied to
transverse plane waves. Since $\hat{\varepsilon}$ is a complex quantity, k^2 is also
complex. It is customary in optics to eliminate k in favor of a complex
index of refraction defined by

$$\hat{k} = \hat{n}(\omega/c) = (n + i\kappa)(\omega/c) \quad . \tag{32}$$

Then the real and imaginary parts of Eq. (24) become simply

$$\varepsilon' = Re[(\hat{n})^2] = n^2 - \kappa^2 \tag{33}$$

and

$$\varepsilon'' = Im[(\hat{n})^2] = 2n\kappa \quad . \tag{34}$$

The meaning of a complex refractive index becomes clear upon writing

$$\exp ikz = [\exp in(\frac{\omega}{c})z] \exp -\kappa(\frac{\omega}{c})z \quad .$$

Re $\hat{n}$ = n determines the wavelength $\lambda = (2\pi c/\omega n)$ or equivalently the
phase velocity $v_p = Re(\omega/\hat{k}) = c/n$ of the wave. The imaginary component Im $\hat{n}$ = κ determines the rate at which the wave decays spatially
within the medium. The attenuation coefficient $\alpha = 2\kappa\omega/c$ gives the
rate at which the energy of the electromagnetic wave (proportional to
$|E|^2$) decays, $I(z) = I(0) \exp(-\alpha z)$.

We learned in Section IV [Eq. (19)] that ε'' (or χ'') > 0 is a necessary and sufficient condition for a material to absorb and dissipate
energy from the electromagnetic fields. Now we see from Eq. (34) that,
if $\varepsilon'' > 0$, then $\kappa > 0$, so that any absorption of power from the wave
is accompanied by the attenuation of the wave itself, which is all well
and good. The converse, however, is not true; it is possible to have
attenuation without absorption. Suppose $\varepsilon'' = 0$ so that there is no
absorption. If $\varepsilon' > 0$, the medium is perfectly transparent, n > 0,
$\kappa = 0$. But if $\varepsilon' < 0$ (we shall see in the next section how this can

come about) then we must have $n = 0$, $\kappa > 0$, which represents a "wave" which is purely exponential. Crudely speaking, what happens in this case is that the energy in the "wave" decays not through absorption, but through reradiation in the backward direction, i. e., the energy is reflected rather than transmitted.

Equation (31) shows, in a sense, how the polarization properties of a medium $\hat{\epsilon}(\omega)$ determine $\hat{n}(\omega)$, which is a property of light within the medium. Thus $\hat{\epsilon}$ can be thought of as the more fundamental of the two quantities. In the following sections we shall be able to deduce what the dielectric response of conduction electrons, or lattice vibrations, etc., should be like. If we want to know how the optical properties of a material with both conduction electrons and optical lattice vibrations will behave, we merely add the two contributions, $\hat{\epsilon} = \hat{\epsilon}_{el} + \hat{\epsilon}_{vib}$, because the total polarization is the sum of all contributions. It would make no sense to calculate individual contributions to $\hat{n}$, because they are not additive.

VI.   EXPERIMENTAL METHODS

Although the material response function $\hat{\epsilon}(\omega)$ can be obtained directly at low frequencies by measurement of the capacitance and dielectric loss of a slab of material, at optical frequencies, we must resort to measuring the complex refractive index $\hat{n}(\omega)$ and calculating $\hat{\epsilon}(\omega)$, using Eqs. (33) and (34). While there are many variants, most methods of measuring n and κ as a function of frequency involve intensity measurements of light either transmitted through or reflected from a sample of the material under investigation. Both have in common the need for either a source of light of relatively high spectral purity or a method of analyzing the spectral components of white or broad band light which has interacted with the sample in some way. In either case a grating or prism monochromator with modest resolution is usually

suitable. The choices among various monochromator designs, and of
detectors and sources (usually black bodies), are dictated by the spectral range in question.

A.   Transmission Spectra

By a study of the fraction of light energy transmitted through a
sample of known thickness, one can determine $\alpha(\omega)$. In order to avoid
corrections for reflection losses, samples of two different thicknesses
can be studied. If the absorption is relatively weak, of the type arising
from impurities in a transparent host, for example, $n(\omega)$ is essentially
that of the host and is nearly constant. Under these conditions peaks
in the observed $\alpha(\omega)$ correspond to peaks in $\varepsilon''(\omega)$, which are in turn
characteristic of the energy absorption of the (impurity) system under
study.

A direct and precise method of determining $n(\omega)$ is by measurement of the angular deviation of light passing through a prism of material. Internal reflections in samples with very parallel sides give rise
to interference effects, which cause periodic modulation of the light
transmission as a function of the wavelength of incident light. The
period of the modulation is determined by the "optical thickness" of
the sample, $n\ell$, providing another method for determining $n(\omega)$.

B.   Reflectance Spectra

The fundamental optical processes in a solid, those arising from
the bulk solid rather than some minor constituent, produce spectral
regions of very strong absorption. Materials can become so opaque in
these regions that light is rapidly attenuated within several tens of
lattice spacings. It becomes mechanically difficult to prepare samples
thin enough to give meaningful transmission spectra, although evapo-
rated films are useful if crystallographic orientation is not essential.
Direct measurement of $n(\omega)$, which is no longer a slowly varying function in regions of strong absorption, is no less difficult. In such

cases, one often resorts to measurements involving the fraction of light energy reflected from a flat surface of the solid. At a given angle of incidence, θ, the fraction of energy reflected is a known function of n and κ, $R(\theta) = f(n, \kappa)$. By measuring R at two different angles of incidence, n and κ can be separately deduced. There are various modifications of this basic technique, known collectively as polarimetric methods.

Another technique which has found wide acceptance involves measuring the reflectivity at only one angle of incidence (usually near-normal incidence). If $R(\omega)$ is measured over a wide range of frequencies (in principle, from $\omega = 0$ to $\omega = \infty$), it is possible to deduce $n(\omega)$ and $\kappa(\omega)$ via a rather complicated numerical integration procedure, which can be readily handled by digital computers.

Looked at in detail, reflection from a surface is not simply described by light somehow "bouncing off" a surface, but is a much more subtle process by which polarization induced in a thin surface layer of material radiates energy in the backward direction. The thickness of this backward radiating layer is of the order of one-half the wavelength of light in the material. Since it is only material within this thin layer that one samples in a reflectivity experiment, a certain amount of care in surface preparation is necessary to insure that the optical constants obtained are representative of the bulk sample rather than of a mechanically strained or chemically different surface layer. The problem is particularly severe in the short wavelength spectral regions.

VII. RESPONSE FUNCTION OF AN HARMONIC OSCILLATOR

In order to illustrate what is involved in understanding response functions, we are going to investigate a very simple model of an atom. The model was proposed by the Dutch physicist H. A. Lorentz in a (largely successful) attempt to understand the optical properties of atoms many years before quantum theory was developed. It may seem a little "unsophisticated," but in fact it contains, in its simplest form, the basis for an understanding of most of the optical properties of matter, and for this reason we shall study it in some detail. In the next section we will go back and try to understand why such an oversimplified model works so well.

A. Equations of Motion of a Driven Oscillator

Suppose we imagine each electron within an atom as being smeared out into a spherical shell with the nucleus at the center. A solid is just some dense collection of such atoms. Under the influence of an electric field each electron feels a force $-e\underline{E}$, and we will assume that as a result of this force the ith electron "cloud" moves in such a way that its center-of-charge is displaced from the position of the nucleus by an amount $\underline{\delta}_i$. If the atom is to maintain its stability, the displaced electron must experience a restoring force which we shall choose to be of the form $-K_i\underline{\delta}_i$, the restoring force being just proportional to the displacement. Finally we postulate a kind of frictional retarding force proportional to the time rate of change of $\underline{\delta}_i$. This damping force $\Gamma_i d\underline{\delta}_i/dt$ can be thought of as representing in a very crude way the interaction between electrons as one moves through the rest. Damping forces can usually be regarded as small compared with the driving and restoring forces, but they still play an important role, as we shall see. Combining these forces and equating them to the mass times acceleration, $m d^2\underline{\delta}_i/dt^2$, the resulting equation of motion for the electron shell in a time-dependent field is

$$m\frac{d^2\underline{\delta}_i}{dt^2} + m\gamma_i\frac{d\underline{\delta}_i}{dt} + m\Omega_i^2\underline{\delta}_i = -e\underline{E}(t) \quad , \tag{35}$$

where for convenience we have rewritten $\Gamma_i = m\gamma_i$ and $K_i = m\Omega_i^2$, m being, of course, the electron mass. Ω_i is identified with the natural or resonance frequency of the electron because, if initially disturbed, the electron can continue to oscillate at this frequency even after the disturbance is removed.

> Exercise. Undriven motion of a damped oscillator, if the driving field were suddenly removed, for example, satisfies the equation
>
> $$m\left(\frac{d^2\delta}{dt^2} + \gamma\frac{d\delta}{dt} + \Omega^2\delta\right) = 0 \quad .$$
>
> Investigate solutions of the form $\delta = \delta_0 \exp(-i\Lambda t)$, where $\Lambda = \Lambda' + i\Lambda''$ is now complex. Show that if $\Omega^2 > (\gamma/2)^2$, the motion consists of an oscillation of frequency $\Lambda' = [\Omega^2 - (\gamma/2)^2]^{\frac{1}{2}}$, which is additionally damped as $\exp[-(\gamma t/2)]$. If $(\gamma/2)^2 > \Omega^2$, show that the response is an entirely real exponential decay with no oscillatory components. We say in this case that the motion is overdamped.

Each electron in an atom should obey an equation of motion similar to (35) but with appropriate values of Ω_i and γ_i. Equation (35) is just the equation of motion for a driven damped harmonic oscillator, and the response function we derive will be exactly correct for this classical system. It is the application of this equation to an atomic system that is approximate and that we are trying to make plausible.

We are interested in the response to an harmonic field $\underline{E}(t) = \text{Re}\,\underline{\hat{E}}(t) = \text{Re}\,\underline{E}_0 \exp(-i\omega t)$, and we can easily obtain the solutions we want for $\underline{\delta}_i(t)$ using complex representations. It is necessary to know only those components of $\underline{\delta}_i$ that have the same frequency as the driving field in order to determine the response function, so we substitute $\underline{\hat{\delta}}_i = \underline{\delta}_i^0 \exp(-i\omega t)$ into Eq. (35) and obtain immediately

$$m(-\omega^2 - i\gamma_i\omega + \Omega_i^2)\underline{\hat{\delta}}_i = -e\underline{\hat{E}} \quad .$$

If there are N such electrons per unit volume, the polarization due to them is $\Sigma_i \text{Re}(-N e\underline{\hat{\delta}}_i)$, or we can write

$$\underline{\hat{P}}(t) = \chi(\omega)\underline{\hat{E}}(t) = \Sigma_i\left(\frac{Ne^2}{m}\right)\frac{1}{\Omega_i^2 - \omega^2 - i\gamma_i\omega}\underline{\hat{E}}(t) \tag{36}$$

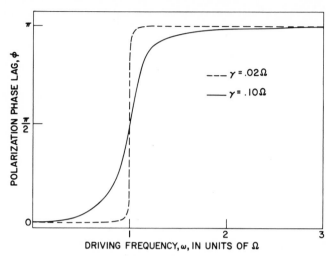

FIG. 3. Polarization phase lag develops abruptly as the driving frequency approaches the natural resonance frequency Ω if the damping is slight. The transition is less abrupt with greater damping.

Multiplying both the numerator and denominator of (36) by $\Omega_i^2 - \omega^2 + i\gamma_i\omega$, we can separate the real and imaginary parts of $\hat{\chi}(\omega)$,

$$\chi'(\omega) = \sum_i \left(\frac{Ne^2}{m}\right) \frac{(\Omega_i^2 - \omega^2)}{(\Omega_i^2 - \omega^2)^2 + \gamma_i^2\omega^2} \tag{37a}$$

$$\chi''(\omega) = \sum_i \left(\frac{Ne^2}{m}\right) \frac{\gamma_i\omega}{(\Omega_i^2 - \omega^2)^2 + \gamma_i^2\omega^2} \quad . \tag{37b}$$

B.   Discussion of Results

Equations (37) are an important result, and we have to understand their content. For simplicity we consider the effects of a single term of the sum, as would be the case if all electrons in the solid were equivalent. This is, of course, scarcely the case, but often the resonance frequencies of various electrons are well separated, and a single equivalent group of electrons dominates the response function in a particular frequency interval about its characteristic Ω_i.

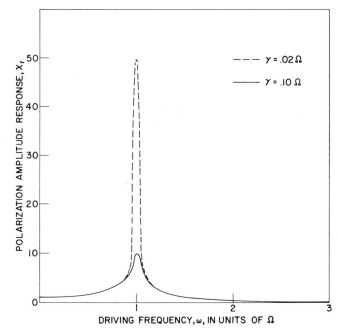

FIG. 4. The amplitude of the polarization response is sharply
peaked about the natural resonance frequency Ω for slight damping.
Increased damping reduces the response at all frequencies. The polarization amplitude, here normalized to unity at $\omega = 0$, goes to zero
rapidly above Ω.

First let us examine Eqs. (37) from the point of view of the polar-
ization response. The phase angle φ by which $\underline{P}$ lags the driving $\underline{E}$
field is, from Eq. (10), given by

$$\tan \varphi = \chi''/\chi' = \gamma\omega/(\Omega^2 - \omega^2) \quad .$$

The behavior of $\varphi(\omega)$ is shown in Fig. 3. As we expected $\varphi(\omega) \to 0$ as
$\omega \to 0$, but the phase changes by $180°$ in the vicinity of Ω, more or less
abruptly depending upon the value of γ. So much for the phase of the
induced polarization. How about the amplitude? According to Eq. (6)
the amplitude of the polarization response is proportional to χ_r, which
by Eq. (10) is equal to $[(\chi')^2 + (\chi'')^2]^{\frac{1}{2}}$ or

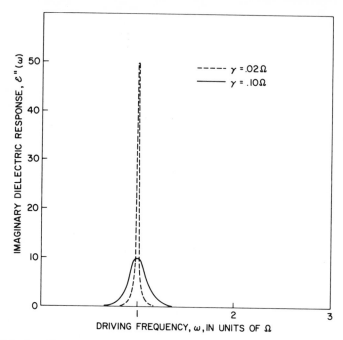

FIG. 5a. The imaginary dielectric response is peaked about the resonance frequency Ω but is broadened by increased damping of the oscillator. Note the differences between $\varepsilon''(\omega)$ shown in this figure and the polarization amplitude response χ_r shown in Fig. 4.

$$\chi_r = \left(\frac{Ne^2}{m}\right)\frac{1}{[(\Omega^2 - \omega^2)^2 + \gamma^2\omega^2]^{\frac{1}{2}}} \tag{38}$$

Figure 4 shows what is intuitively obvious, that the system responds most at the frequency at which it naturally oscillates. Imagine two systems, one with a greater damping constant γ than the other, but otherwise identical. (Often in real systems γ depends rather strongly upon temperature, whereas the resonance frequency does not, so you can imagine comparing the behavior of such a material at two different temperatures.) As you would expect, the system with the greater damping shows a smaller polarization response. This is true regardless of the driving frequency, but it is particularly pronounced near the natural resonance frequency.

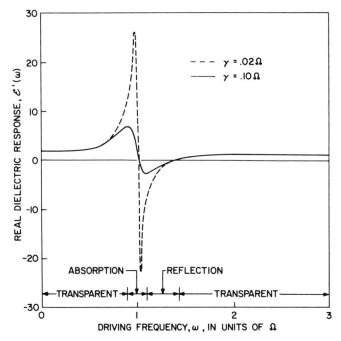

FIG. 5b. The real dielectric response is a positive increasing
function of ω well below the resonance frequency (normal dispersion)
and a (usually negative) increasing function well above Ω. If the
strength of the resonance is insufficient to cause $\varepsilon'(\omega)$ to become
negative the reflection region disappears.

How would we expect light to behave in a medium whose dielectric
properties were described by a single term of the form of Eq. (37)? Here
it is most convenient to discuss not χ_r and φ but the complex $\hat{\varepsilon}$ response function, whose real and imaginary parts vary with frequency as
shown in Fig. 5. At frequencies much smaller than the resonance fre-
quency Ω, $\varepsilon'' \approx 0$ while $\varepsilon' > 0$, so that, by Eqs. (33) and (34), the
complex index of refraction is $n \approx \varepsilon'^{\frac{1}{2}}$ and $\kappa \approx 0$. In this frequency
region, the material is nearly completely transparent and $n(\omega)$ increases
with frequency. $\varepsilon'' \approx 0$ and $\varepsilon' > 0$ at very high frequencies also,
giving rise to another transparent region. $n(\omega)$ decreases with frequency
in this region, and this behavior is sometimes called "anomalous" dispersion in contrast to the behavior of $n(\omega)$ below Ω, which historically

was studied first. Now, of course, we understand this behavior so it is no longer really "anomalous."

Remembering that the power absorbed from the radiation is proportional to $\omega\varepsilon''$ [Eq. (19)], we recognize a region centered near Ω and roughly of width γ which is the region of strongest absorption. Only in this limited region is the polarization response both large and significantly out of phase with the driving field.

Finally, there may exist a frequency range with a third qualitatively different behavior with regard to light propagation. As we have drawn Fig. 5, there is a region just above the absorption region in which $\varepsilon'' \approx 0$ while $\varepsilon' < 0$. As we pointed out in a previous section, under these conditions we find that $n \approx 0$ and $\kappa > 0$, so that incident radiation in this frequency band is rapidly attenuated and cannot propagate in the material, it is reflected rather than absorbed. Whether or not such a reflection region occurs depends upon the magnitude of the dielectric response, i. e., the size of the quantity Ne^2/m, as well as upon γ. If the former is too small, or the latter too large, ε', at its minimum, will still be greater than zero and the absorption region will pass directly into the high frequency transmission region. At the concentrations at which they are usually studied, impurities do not normally disturb the dielectric response of the host sufficiently to give rise to a reflection region, whereas many intrinsic optical processes do. We must point out that in many solids the transitions between regions occur gradually rather than having a sharp dividing line as Fig. 5 suggests. There is, for example, always some absorption in the transparent and reflecting regions.

VIII.   DIELECTRIC RESPONSE OF QUANTUM SYSTEMS

The analysis of the dielectric response of the classical harmonic oscillator furnishes a simple introduction to the kind of physics involved in calculating how a system reacts to an electric field. However, you may not be convinced that the results of the preceding section have much direct application to the optical properties of solids, and with some justification. After all, we are interested in the dielectric response of electrons and nuclei which obey quantum rather than Newtonian equations of motion. Furthermore, the forces exerted upon an electron by its neighbors are by no means harmonic. In truth, however, as sometimes happens in physics, the form of Eq. (38) has a far greater range validity than the simple method of derivation leads one to assume. The fact is that all quantum systems show basically the same type of dielectric response, namely, that of a classical oscillator. In applying this idea to the various optical phenomena in solids, you will see what kind of hedging is necessary and implied by insertion of the word "basically." But the important point for now is that quantum systems and harmonic oscillators do behave similarly insofar as their interactions with light are concerned, and that this fact makes possible a unified and systematic treatment of many of the diverse optical phenomena that solids exhibit.

A. Time-Dependent Perturbations

Since we will repeatedly use the above result, it is worthwhile to see how it comes about, particularly because it can be derived rather simply in an approximate way. Suppose we have a quantum system characterized by a ground-state energy $\mathscr{E}_0$ and an associated wave function of definite energy

$$\Psi_0(\underline{r}, t) = \phi_0(\underline{r}) \exp[-i(\mathscr{E}_0/\hbar)t] \quad .$$

Let there be any number of excited states with distinctly different energies $\mathscr{E}_j$ and wave functions

$$\Psi_j(\underline{r}, t) = \phi_j(\underline{r}) \exp[-i(\mathscr{E}_j/\hbar)t] \quad .$$

Furthermore, let $\mathcal{E}_j - \mathcal{E}_0 \gg kT$, so that only the ground state is occupied. We subject this system to a time-dependent electric field $\underline{E}(t)$. The wave function of the perturbed ground state is no longer Ψ_0 but is given by Ψ_0', which must satisfy the time-dependent Schrödinger equation:

$$i\hbar \frac{\partial \Psi'}{\partial t} = [\mathcal{H} - \underline{\mathcal{P}} \cdot \underline{E}(t)] \Psi' \quad . \tag{39}$$

$\mathcal{H}$ is the hamiltonian for the unperturbed system (we assume that it does not depend explicitly upon t) and $-\underline{\mathcal{P}} \cdot \underline{E}$ takes care of additional interaction energy in the electric field. $\underline{\mathcal{P}}$ is the dipole moment operator for the system (just $\Sigma_i \, q_i \underline{r}_i$ for a system of particles with charges q_i).

If we assume that $|\underline{\mathcal{P}} \cdot \underline{E}| \ll |\mathcal{E}_j - \mathcal{E}_0|$, then we can look for a ground-state solution of Eq. (39) of the form

$$\Psi_0' = \Psi_0 + \sum_j c_j(t) \Psi_j \quad ,$$

and we can expect all $|c_j(t)| \ll 1$. Inserting this expression into Eq. (39) produces the following relation among the quantities:

$$i\hbar \sum_j \frac{\partial c_j}{\partial t} \phi_j \exp\left[-i\left(\frac{\mathcal{E}_j t}{\hbar}\right)\right] = -(\underline{\mathcal{P}} \cdot \underline{E})\left\{\phi_0 \exp\left[-i\left(\frac{\mathcal{E}_0 t}{\hbar}\right)\right] + \sum_j c_j \phi_j \exp\left[-i\left(\frac{\mathcal{E}_j t}{\hbar}\right)\right]\right\}$$

By multiplying by any particular excited state wave function, say ϕ_k^*, integrating over all the spatial variables (which we will denote as v), and making use of the orthonormality of the ϕ_j (i.e., $\int \phi_k^* \phi_j \, dv = \delta_{kj}$), we arrive at an explicit differential equation for c_k:

$$i\hbar \frac{\partial c_k}{\partial t} = -[\underline{\mathcal{P}}_{ko} \exp(i\Omega_{ko} t) + \sum_j c_j \underline{\mathcal{P}}_{kj} \exp(i\Omega_{kj} t)] \cdot \underline{E}(t) \tag{40}$$

where

$$\underline{\mathcal{P}}_{ko} = \int \phi_k^* \underline{\mathcal{P}} \phi_0 \, dv \quad \text{and} \quad \Omega_{ko} = (\mathcal{E}_k - \mathcal{E}_0)/\hbar \quad .$$

Since all $c_j \ll 1$, we can drop all but the first term in Eq. (40). But we also want to add something analogous to the damping force for the classical oscillator. We know that, if $\underline{E}(t)$ were suddenly turned off, the system would decay back to its original equilibrium condition, i. e., $c_k = 0$. We can assure this behavior by adding a term linear in c_k to c_k Eq. (40):

$$i\hbar \frac{\partial c_k}{\partial t} = -\underline{\mathscr{P}}_{ko} \cdot \underline{E}(t) \exp(i\Omega_{ko}t) - i\hbar \frac{\gamma_k}{2} c_k \quad , \tag{41}$$

where $\gamma_k > 0$, so that $c_k \approx \exp(-\gamma_k t/2)$ in the absence of $\underline{E}(t)$. It is possible to justify an added term of this form somewhat more rigorously but not in an elementary way. A detailed quantum description of damping involves the exchange of energy of the system under discussion with other systems (e. g., electrons with other electrons or with phonons in a "heat bath"), which must be handled by statistical mechanics because there are so many ways in which a given deexcitation can occur.

As always, we are interested in a sinusoidal driving field, which we write as

$$\underline{E}(t) = \underline{E}_0 [\exp(i\omega t) + \exp(-i\omega t)] \quad .$$

Since Eq. (41) is linear, we can find separately the solutions $c_k(+)$ and $c_k(-)$ for the driving fields $\underline{E}_0 \exp(i\omega t)$ and $\underline{E}_0 \exp(-i\omega t)$ and add them together. The solution for $\underline{E}_0 \exp(i\omega t)$ is readily found upon substituting $c_k(+) = c_k^0 \exp(i\Lambda t)$. Upon differentiating and collecting terms, we find that $\Lambda = (\Omega_{ko} + \omega)$ and

$$c_k(+) = \left(\frac{\underline{\mathscr{P}}_{ko} \cdot \underline{E}_0}{\hbar}\right) \frac{\exp[i(\Omega_{ko} + \omega)t]}{\Omega_{ko} + \omega - i\gamma_k/2} \quad .$$

$c_k(-)$ is obtained by replacing ω by $-\omega$ in the above expression, so that

$$c_k(t) = \left(\frac{\mathscr{P}_{ko} \cdot E_0}{\hbar} \right) \left[\frac{\exp\left[i(\Omega_{ko} + \omega)t\right]}{\Omega_{ko} + \omega - i\gamma_k/2} + \frac{\exp\left[i(\Omega_{ko} - \omega)t\right]}{\Omega_{ko} - \omega - i\gamma_k/2} \right] . \tag{42}$$

It is now a simple matter to find the time-dependent polarization of the ground state:

$$\mathscr{P}(t) = \int \Psi'(\underline{r}, t)^* \underline{\mathscr{P}} \Psi'(\underline{r}, t) \, dv$$

$$= \underline{\mathscr{P}}_0 + \sum_j [c_j \underline{\mathscr{P}}_{oj} \exp(-i\Omega_{jo} t) + c_j^* \underline{\mathscr{P}}_{jo} \exp(i\Omega_{jo} t)] ,$$

where we have dropped terms of order c_j^2. $\underline{\mathscr{P}}_0 = \int \phi_0^* \underline{\mathscr{P}} \phi_0 \, dv$ is the (time-independent) dipole moment of the unperturbed ground state, which vanishes in a great many cases. The remaining terms give the induced moment:

$$\mathscr{P}(t) - \underline{\mathscr{P}}_0 = \sum_j \frac{\underline{\mathscr{P}}_{oj} \underline{\mathscr{P}}_{jo}}{\hbar} \left[\frac{1}{\Omega_{jo} + \omega + i\gamma_j/2} + \frac{1}{\Omega_{jo} - \omega - i\gamma_j/2} \right] E_0 \exp(-i\omega t) +$$

$$\sum_j \frac{\underline{\mathscr{P}}_{oj} \underline{\mathscr{P}}_{jo}}{\hbar} \left[\frac{1}{\Omega_{jo} + \omega - i\gamma_j/2} + \frac{1}{\Omega_{jo} - \omega + i\gamma_j/2} \right] E_0 \exp(i\omega t) .$$

We can obtain the polarization by multiplying the above by N, and since there is a term in $\underline{P}(t)$ proportional to $\underline{E}_0 \exp(i\omega t)$ and $\underline{E}_0 \exp(-i\omega t)$ separately, we can use the definition of $\chi(\omega)$ given in Eq. (8) to obtain the desired result:

$$\chi(\omega) = N \sum_j \frac{\underline{\mathscr{P}}_{oj} \underline{\mathscr{P}}_{jo}}{\hbar} \left[\frac{1}{\Omega_{jo} + \omega + i\gamma_j/2} + \frac{1}{\Omega_{jo} - \omega - i\gamma_j/2} \right] . \tag{43}$$

In order to emphasize the similarity between Eq. (43) and the classical result of Eq. (36) for the harmonic oscillator, one sometimes introduces a quantity f_{oj}, know as the <u>oscillator strength</u>:

$$f_{oj} = (2m\Omega_{oj}/e^2) |\underline{\mathscr{P}}_{oj}|^2/\hbar . \tag{44}$$

Then Eq. (43) can be written with a little rearrangement as

$$\chi(\omega) = \frac{Ne^2}{m} \sum_j \frac{f_{oj}}{\Omega^2_{oj} - \omega^2 - i\gamma_j\omega} \quad , \tag{45}$$

where we have also neglected a term $(\gamma_j/2)^2$ in the denominator since in most real cases it is small compared to Ω^2_{oj}.

Equation (45) proves our assertion of the essential similarity of classical and quantum systems. For despite its similarity to the classical oscillator result of Eq. (36), it is a general quantum mechanical expression for the susceptibility. The natural resonance frequency Ω_j is replaced by the frequencies corresponding to the allowed energy levels: $\mathscr{E}_j - \mathscr{E}_0 = \hbar\Omega_{jo}$. The strength of the resonance, that is, the numerator of Eq. (45), is modified by the factor f_{oj}, which is a measure of the quantum mechanical transition probability between the ground and excited states of the system.

Exercise. We want to investigate the nature of the dielectric response in the limit of no damping.
   (a) In the limit as $\gamma \to 0$, the χ' for a single level $\Omega_{jo} = \Omega$ becomes

$$\chi'(\gamma \to 0) = \frac{Ne^2}{m} \frac{1}{\Omega^2 - \omega^2} .$$

One must be more careful with χ''. First show that

$$\left[\frac{1}{\Omega^2 - \omega^2 - i\gamma\omega}\right] = \frac{1}{2\Omega}\left[\frac{1}{\Omega + \omega + i\eta} + \frac{1}{\Omega - \omega - i\eta}\right]$$

if $\eta \equiv (\gamma/2) \to 0$, and then that

$$\chi''(\gamma \to 0) = \frac{Ne^2}{m} \frac{1}{2\Omega} \left[\frac{-\eta}{(\Omega+\omega)^2 + \eta^2} + \frac{\eta}{(\Omega - \omega)^2 + \eta^2}\right] .$$

   (b) Suppose we define a new function

$$\delta(x) = \frac{1}{\pi}\left[\frac{\eta}{x^2 + \eta^2}\right]_{\eta \to 0} .$$

In terms of this function,

$$\chi''(\gamma \to 0) = \frac{Ne^2}{m} \frac{\pi}{2\Omega} \left[\delta(\Omega - \omega) - \delta(\Omega + \omega)\right] .$$

The function $\delta(x)$, known as the Dirac delta function, has the interesting property of being zero everywhere except at the point $x = 0$, where it is infinitely large.

There is, however, a finite area under the curve. Show that

$$\int_{-\infty}^{\infty} \delta(x)\, dx = 1 \quad .$$

The interpretation of the preceding exercise is of interest. It shows that only in the limit of no damping, $\gamma = 0$, are the energy levels of a quantum system precisely defined. Then the imaginary response χ'' is nonzero only when the radiation has exactly the energy necessary to cause energy-conserving transitions between two levels. (The term involving $\delta(\Omega - \omega)$ represents upward transitions from the ground state to the excited state. The remaining term describes downward transitions involving the emission of radiation.)

The ability of light to exchange energy with a resonant quantum system and giving rise to an imaginary susceptibility component is quite familiar. But note carefully that this mechanism cannot explain the existence of the real part of the susceptibility, which is present even at frequencies well removed from resonance. The real susceptibility arises because the field causes a kind of reversible mixing of ground and excited state wave functions that causes no energy to be gained or lost in the field and is sometimes described as resulting from "virtual" transitions between quantum states.

B.   Local Field Problem

There is one further point that we have thus far been careless about. By its definition, $\hat{\chi}$ is the constant of proportionality between the polarization and the average of the electric field over a volume containing many atoms. An individual electron, if it is localized at a particular point $\underline{r}$, feels not this averaged field $\underline{E}_{av}$, but the actual local field $\underline{E}_{loc}(\underline{r})$. Thus our formulas for the susceptibility should be multiplied by the factor $(\underline{E}_{loc}/\underline{E}_{av})$. It is not difficult to see why $\underline{E}_{loc}$ and $\underline{E}_{av}$ may be quite different. There is an electric field associated with each of the elementary dipoles themselves, which is strongest along the dipole direction and decreases as the inverse cube of the distance. Each dipole thus responds to a field made up of the applied field plus the dipolar fields of the other dipoles, and this latter contribution may be large and can vary rapidly, even over distances small compared with atomic dimensions. To give an idea of how large these "local field" corrections can be, for dipoles situated at the lattice points of a cubic lattice, the local field correction is

$$\frac{E_{loc}}{E_{av}} = \frac{1}{1 + 4\pi\chi/3} \quad .$$

The actual situation is complicated by the fact that even strongly bound electrons are not localized at a point and thus sample a varying E_{loc}. This so-called local field problem is probably the most difficult one to handle, in a practical way, that we will encounter in discussing the optical properties of solids. The only really simple situation occurs for nonlocalized or free electrons, which respond to E_{av} rather than E_{loc} and thus require no local field correction.

IX. EXCITATION WAVES IN CRYSTALS

Since, in quantum mechanical language, the optical properties of a material result from transitions induced by light between allowed energy states of the material, it is very helpful to know something of the nature of these energy states. As it turns out, although there are many possible kinds of allowed excitations in crystals, many have certain common characteristics, which follow more or less directly from the fact that crystals are built up by repetition in three dimensions of a single unit cell. There is an elegant branch of mathematics known as group theory which one can use to deduce these properties, which are the result of symmetry alone, but it may be more physically enlightening to see what we can discover using elementary quantum mechanics.

A. Excitations in a One-Dimensional Lattice

To make things as easy as possible, we start with a one-dimensional "crystal" made up of an infinite number of identical cells extending along a line. We label the cells by an integer ℓ. See Fig. 6. There exists some state of this system which is lower than any other and which we will denote as Ψ_0. How do we go about constructing an excited state?

If interactions between cells were negligible, so that each behaved as a separate entity, then we could excite just one of the cells, the ℓth one, say, in some way, and the wave function for the resulting state of the crystal could be written

$$\Psi_\ell(x, t) = \phi_\ell(x) \exp[-i(\bar{\mathscr{E}}t/\hbar)] \quad .$$

The time-dependent factor $\exp(-i\bar{\mathscr{E}}t/\hbar)$ tells us that Ψ_ℓ is an allowed energy state with energy $\bar{\mathscr{E}}$. Another way of saying this is that if the hamiltonian function for this "noninteracting" crystal is called $\mathscr{H}_0$, then Schrödinger's equation tells us that

$$\mathscr{H}_0\,\Psi_\ell = i\hbar\,\frac{\partial\Psi_\ell}{\partial t} = i\hbar\left(-\frac{i\bar{\mathscr{E}}}{\hbar}\right)\Psi_\ell = \bar{\mathscr{E}}\,\Psi_\ell \quad .$$

Because we are aiming at a general discussion, we need not say anything specific about the nature of the excitation associated with Ψ_ℓ. But to give you something concrete to think about, Table 2 is a list of some of the more important excitations we have in mind. In the left column we describe a possible type of localized excitation to which Ψ_ℓ might refer. From each of these localized excitations one can construct

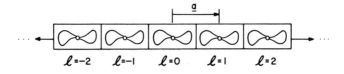

$$\ell\text{=-2} \qquad \ell\text{=-1} \qquad \ell\text{=0} \qquad \ell\text{=1} \qquad \ell\text{=2}$$

FIG. 6. A one-dimensional lattice with an arbitrary basis. The lattice point associated with the ℓth cell is specified by $\underline{x}_\ell = \ell\underline{a}$.

TABLE 2

A List of Important Excitations in Solids

Localized excitation in noninteracting crystals	Appropriate generalization in real crystals
1. Vibrational oscillations of nuclei within unit cell	Phonon
2. Extra electron (in addition to number required for charge neutrality) added to cell	Conduction electron in insulator or semiconductor
3. Absence of electron necessary for normal valency and charge requirements from cell	Conduction "hole" in insulator or semiconductor
4. Electron promoted from ground state to excited state still bound by atom or molecule in unit cell	Exciton
5. Reversal of magnetic moment in one cell of an otherwise magnetically oriented (ferromagnetic, antiferromagnetic) crystal	Magnon or spin wave

nonlocalized excitations appropriate to interacting solids, a matter which we will immediately discuss. The names commonly associated with these nonlocalized excitations are given in the right column. As we have just implied, the difficulty with our description thus far is that a "noninteracting" crystal does not make much sense physically. In fact, a crystalline arrangement is precisely the result of such inter-action. We can represent the sum of these interactions between cells by an additional term $\mathcal{H}'$ in the hamiltonian function, the correct total hamiltonian being $\mathcal{H} = \mathcal{H}_0 + \mathcal{H}'$. Ψ_ℓ, an allowed energy state of the "noninteracting" crystal, i.e., of $\mathcal{H}_0$, is no longer an allowed state of the real crystal, i.e., of $\mathcal{H}$. We will describe a method of finding the

allowed states known as the tight binding approximation, which works
well if $\mathcal{H}_0 \gg \mathcal{H}'$.

Physically, the effect of the interaction term $\mathcal{H}'$ is easily described. Instead of the ℓth cell, we could have singled out any other
cell ℓ' upon which to localize the excitation. That is, insofar as the
crystal as a whole is concerned, Ψ_ℓ is but one of an infinite number of
identical states $\Psi_{\ell'}$. $\mathcal{H}'$ provides a mechanism, lacking in $\mathcal{H}_0$, to
accomplish the transfer of excitation from state Ψ_ℓ to $\Psi_{\ell'}$, i.e., from
cell to cell. Again keeping things as simple as possible, suppose that
the forces between cells extend only to adjacent neighbors. As we
shall verify, in order for $\mathcal{H}'$ to correctly express this ability to transfer
excitation from cell ℓ to ℓ', it must have the following property:

$$\mathcal{H}'\Psi_\ell = -\Gamma(\Psi_{\ell+1} + \Psi_{\ell-1}) \quad , \tag{46}$$

$\Psi_{\ell+1} = \phi_{\ell+1}(x) \exp(-i\bar{\varepsilon}t/\hbar)$, of course, represents the system when the
excitation is localized in cell $\ell + 1$, and Γ is a constant which speci-
fies the magnitude of the intercell forces.

The situation here is rather similar to that encountered in a hydrogen molecule where an electron can lower its energy by becoming simultaneously associated with both protons. To pursue this analogy further,
we can investigate the properties of a composite wave function:

$$\Psi'(x, t) = \sum_{\ell=1}^{\infty} c_\ell(t) \Psi_\ell(x, t)$$

which allows the excitation to extend spatially over the entire system.
If Ψ' is to satisfy Schrödinger's equation

$$i\hbar \frac{\partial \Psi'}{\partial t} = (\mathcal{H}_0 + \mathcal{H}')\Psi' \quad ,$$

we find, upon substituting into Eq. (46) and knowing how $\mathcal{H}_0$ and $\mathcal{H}'$
act separately on the Ψ_ℓ, that

$$i\hbar \sum_\ell \frac{\partial c_\ell}{\partial t} \Psi_\ell = -\Gamma \sum_\ell c_\ell(\Psi_{\ell+1} + \Psi_{\ell-1}) \quad .$$

(In performing $\partial\Psi'/\partial t$, note that both the c_ℓ and the Ψ_ℓ are time dependent.) Following a procedure similar to that adopted in the preceding section, we multiply by $\Psi_{\ell'}^*$, and take advantage of the orthonormal property of $\Psi_{\ell'}^*, \Psi_\ell$ to obtain an equation for $c_\ell(t)$:

$$i\hbar\, \partial c_\ell/\partial t = -\Gamma(c_{\ell+1} + c_{\ell-1}) \quad . \tag{47}$$

This equation looks simple, but keep in mind that there is one such equation for each cell, so that (47) really stands for an infinite set of coupled differential equations. A particularly easy situation to discuss, at least for short times, is the following one. Initially, at $t = 0$, the excitation is localized in the ℓth cell, so that $c_\ell(0) = 1$, $c_{\ell'}(0) = 0$ ($\ell' \neq \ell$). Equation (47) tells us that, after a small time interval δt, $c_{\ell+1}$ and $c_{\ell-1}$ will no longer be zero but will be given by

$$c_{\ell+1}(\delta t) = c_{\ell-1}(\delta t) = -(\Gamma/i\hbar)\,\delta t \quad .$$

Probability amplitude is transferred to adjacent cells at a rate $|\Gamma/\hbar|$, and as you now see this is the direct result of having set $\mathcal{H}'\Psi_\ell = -\Gamma(\Psi_{\ell+1} + \Psi_{\ell-1})$.

The solutions of the simultaneous equations, represented by (47), that we are most interested in, however, are the steady-state ones, in which all of the $c_\ell(t)$ show the same time dependence $\exp(-i\mathcal{E}'t/\hbar)$. These represent stationary states of allowed energy $\mathcal{E} = \bar{\mathcal{E}} + \mathcal{E}'$. (Because of the factor $\exp(-i\bar{\mathcal{E}}t/\hbar)$ occurring in each of the Ψ_ℓ, the total time dependence of Ψ' is $\exp[i(\bar{\mathcal{E}} + \mathcal{E}')t/\hbar]$.) Rather than try to solve (47) directly, we will instead investigate trial solutions of the form

$$c_\ell(t) = c_0 \exp[i(K\ell a - \mathcal{E}'t/\hbar)] \quad . \tag{48}$$

Inserting (48) into (47), performing the indicated operations, and canceling a common factor $c_0 \exp[i(K\ell a - \mathcal{E}'t/\hbar)]$ on both sides, we find

$$(\mathcal{E}_K - \bar{\mathcal{E}}) = -\Gamma(e^{iKa} + e^{-iKa}) = -2\Gamma\cos Ka \quad , \tag{49}$$

which tells us that no matter what value of K we choose, there is a

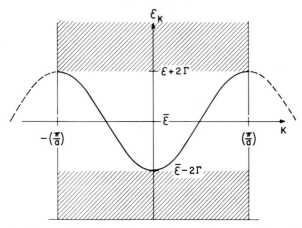

FIG. 7. Energy levels of excitations in a one-dimensional lattice are in an allowed "band" of width 4Γ. The region of K-space bounded by $\pm(\pi/a)$ defines the first Brillouin zone.

solution of the form (48) with the energy $\mathscr{E}_K$. The solutions of Eq. (49) are plotted as a function of the parameter K in Fig. 7, which illustrates that the single excitation energy of the "noninteracting" crystal $\bar{\mathscr{E}}$ is broadened into a band of allowed energies of width 4Γ. The width of the band is determined directly by the strength of the interaction be-tween cells.

B. Brillouin Zone

In Fig. 7, we have shown K values between only (π/a) and $-(\pi/a)$, but this is all that is necessary. Any K value outside this region can be written in the form $K'' = (2\pi/a)n + K'$, where n is an integer ± 1, ± 2, etc., and $-(\pi/a) < K' < (\pi/a)$. We see immediately from the form of (49) that the energy is periodic in K, so that $\mathscr{E}_{K''} = \mathscr{E}_{K'}$. The wave function

$$\Psi'_{K''}(x, t) = \exp(2\pi n i) \, \Psi'_{K'}(x, t) = \Psi'_{K'}(x, t)$$

has the same property. Thus solutions outside the region of K-space bounded by $-(\pi/a) < K < (\pi/a)$, which is known as the first Brillouin zone (B. Z.) of a linear lattice of spacing a, are simply repetitions of those solutions within the B. Z. and can be ignored.

It is a fairly simple matter to generalize our results to three dimensions, for example, a simple cubic crystal with lattice spacing a. Each cell is now specified by a triplet of integers $\underline{\ell} = (\ell_x, \ell_y, \ell_z)$ giving its location $\underline{r}_\ell = \underline{\ell}a$ in the lattice. If this problem is worked through in the same way, expansion coefficients of the form

$$c_K(t) = c_0 \exp\{i[(K_x\ell_x + K_y\ell_y + K_z\ell_z)a - \mathcal{E}'t/\hbar]\}$$

$$= c_0 \exp[i(\underline{K}\cdot\underline{\ell}a - \mathcal{E}'t/\hbar)]$$

(50)

provide solutions with energies $\mathcal{E}_K$ given by

$$\mathcal{E}' = \mathcal{E}_K - \bar{\mathcal{E}} = -2\Gamma(\cos K_x a + \cos K_y a + \cos K_z a) \quad .$$

(51)

The first B. Z. within which all allowed solutions lie is now a cubic volume in $\underline{K} = K_x, K_y, K_z$ space with $-(\pi/a) < K_i < (\pi/a)$. The band of allowed values of $\mathcal{E}_K$ can be represented as a surface in a four-dimensional $\mathcal{E}_K, K_x, K_y, K_z$ space. If we slice the B. Z. along the plane $K_z = 0$, we can represent the energy surface as contours mapped on the K_x, K_y plane, as shown in Fig. 8.

> Exercise. Show how the equation of motion [Eq. (47)] must be modified for a three-dimensional cubic crystal and verify that solutions of the form (50) lead to the energies given in (51).

Next we might allow for interactions of longer range than between adjacent cells. This is also relatively simple to work through. A more troublesome problem is that, if the interaction term $\mathcal{H}'$ is comparable to or larger than $\mathcal{H}_0$, we can no longer get an adequate representation of the wave function Ψ' using only localized states Ψ_ℓ of a single energy $\bar{\mathcal{E}}$. We must then interpret the sum in Eq. (46) as including all energy states of the localized excitation. In practice these improvements lead to so many algebraic complications that some different

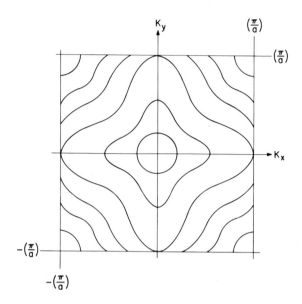

FIG. 8. A contour mapping of the $\mathcal{E}_K$, $\underline{K}$ surface for excitations in
a three-dimensional lattice. This represents a cross section of the
surface in the K_x, K_y plane.

method of "guessing" the form of Ψ' is usually employed if $\mathcal{H}' \geq \mathcal{H}_0$,
but, in principle at least, it can be done by modifying the tight binding
approximation in the way we have indicated. Regardless of the method
used in solving the problem, the same parameter $\underline{K}$ figures prominently
in the solutions. For example, the allowed energy states can always
be written in the form of a <u>Bloch wave</u>

$$\Psi'_{\underline{K}}(\underline{r},\ t)\ =\ u_{\underline{K}}(\underline{r})\ \exp[i(\underline{K}\cdot\underline{r}\ -\ \mathcal{E}t/\hbar)]\quad,\tag{52}$$

where the $u_{\underline{K}}(\underline{r})$ depends upon the details of the problem but must have
the periodicity of the lattice itself, i. e.,

$$u_{\underline{K}}(\underline{r}\ +\ \ell a)\ =\ u_{\underline{K}}(\underline{r})\quad.$$

The allowed $\Psi'_{\underline{K}}$ are nothing more than plane waves, with wavelength
$\lambda\ =\ 2\pi/K$ and frequency $\omega\ =\ \mathcal{E}/\hbar$, modulating a rapidly varying function
$u_{\underline{K}}$, as shown in Fig. 9. For every value of $\underline{K}$ within the first B. Z.,

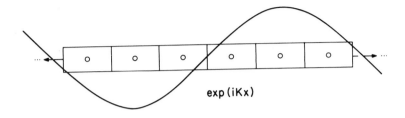

exp (iKx)

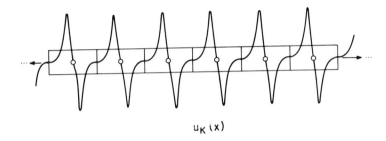

$u_K (x)$

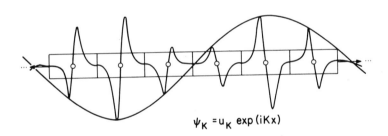

$\psi_K = u_K \, exp \, (iKx)$

FIG. 9.   The composition of a Bloch wave.

there is an allowed energy $\mathscr{E}_K$, and again $\underline{K}$ values outside the first
B.Z. give nothing new. It is thus always possible to represent the
band of allowed energies as a surface in $\mathscr{E}, \underline{K}$ space.

Exercise. In the one-dimensional tight binding approximation, we can allow for interaction between distant
cells by setting

$$\mathscr{H}' \Psi_\ell = - \sum_q \Gamma_q (\Psi_{\ell+q} + \Psi_{\ell-q}) \quad,$$

where Γ_q specifies the strength of the interactions

between cells which are qth neighbors. Show that this leads to the following energy bands:

$$\mathscr{E}_K - \bar{\mathscr{E}} = -\sum_q \Gamma_q [\exp(iKqa) + \exp(-iKqa)] \quad .$$

Thus, within the tight binding approximation, $\mathscr{E}_K$ vs $\underline{K}$ is a Fourier cosine transform of the intercell "forces." The corresponding result for three dimensions is useful for estimating intercell forces from experimentally determined $\mathscr{E}_{\underline{K}}$, $\underline{K}$ surfaces.

Exercise. Show that the tight binding result [Eq. (50)] can be written in Bloch wave form with

$$u_{\underline{K}}(\underline{r}) = \sum_\ell c_0 \, \phi_\ell \, \exp[i\underline{K} \cdot (\underline{r} - \underline{\ell}a)]$$

and verify that

$$u_{\underline{K}}(\underline{r}) = u_{\underline{K}}(\underline{r} + \underline{\ell}a) \quad .$$

C. Crystal Momentum

We began by asking if we could label the allowed excitation states of a crystal by specifying a particular lattice point $\underline{\ell}$ where the excitations resided. What we found was, no, that will not work, but every excitation does "reside" at a particular point $\underline{K}$ in reciprocal space, and we can use the value of $\underline{K}$ to classify our states. We sometimes express the fact that we can label our excitation states by their value of $\underline{K}$ by saying that $\underline{K}$ is a "good" quantum number. An exactly analogous phenomenon occurs in free atoms where electron states are labeled by angular momentum quantum "numbers" s, p, d, etc., by which we mean that a certain state has not only a definite energy but also a definite amount of angular momentum 0, $\hbar$, $2\hbar$, or whatever. To exploit this analogy, which is really deeper than these superficial arguments convey, we can "invent" a new quantity called crystal momentum, defined by

crystal momentum = $\hbar\underline{K}$.

On the basis of this definition, we can summarize the principal result of this section by stating that excitations in crystals have a definite energy and a definite value of crystal momentum. In the following sections the usefulness of the crystal momentum concept will become clearer.

X. PHONONS

The atoms in a crystal are not static but are in constant motion about their average position. By the arguments of the preceding section the vibrational energy associated with any given atomic displacement cannot be isolated in one cell, and the resulting lattice vibration wave is called a phonon.

The most concise way to discuss the characteristics of these excitations is by reference to their $\mathcal{E}_{\underline{K}}$, $\underline{K}$ surfaces. For any value of $\underline{K}$

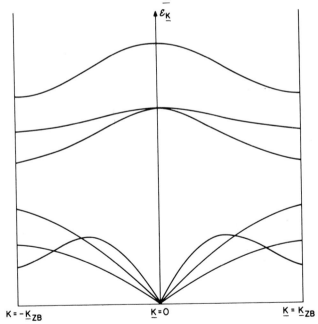

FIG. 10. Possible $\mathcal{E}_{\underline{K}}$ vs $\underline{K}$ diagram for phonons in a diatomic lattice. There are three optic mode branches and three acoustic mode branches.

there are 3n distinctly different vibrational excitations possible (equal
in number to the 3n independent vector components necessary to specify
the displacements of every atom from its average position). Every $\mathcal{E}_{\underline{K}}$, $\underline{K}$
diagram, therefore, has 3n branches. Such a diagram for a crystal
with a diatomic unit cell (e.g., NaCl) might look like Fig. 10 for some
particular direction of $\underline{K}$. Note that two or more branches may cross at
some point (or even coalesce into a single line along a particular direction of $\underline{K}$) but in such cases the states are distinct, that is, the vibrations involved are different, only the excitation energies associated
with them happen to be the same. A general feature of all phonon energy
surfaces is that, regardless of the number of atoms per unit cell, there
are always three branches with $\mathcal{E}_{\underline{K}}$ proportional to $|\underline{K}|$ for small $\underline{K}$.
Thus there is no cost in energy to excite at $\underline{K} = 0$. These three branches
are known as the acoustic branches because these excitations provide
the mechanism for the transmission of sound waves through a crystal.
The remaining 3n - 3 branches describe what are called the optic phonon
modes. This nomenclature is somewhat unfortunate, for as we shall see,
most of the optic modes contribute very little to the optical properties of
the crystal.

A. Interaction Diagrams and Conservation Rules

There are several different processes by which phonons interact
with electromagnetic radiation, and it is sometimes useful to represent
the various processes graphically, such as those shown in Fig. 11.
Figure 11(a) represents the simplest, strongest, and, therefore, most
important interaction, the so-called one-phonon process, where, in
quantum mechanical terms, one photon is taken out of the radiation
field and in its place there appears a single phonon. Then there is a
two-phonon process [Fig. 11(b)] in which one photon disappears (is
destroyed) and two phonons appear (are created) in its place. Three
and higher order phonon processes also exist, in principle, but become
increasingly improbable. There are also multiple-photon processes

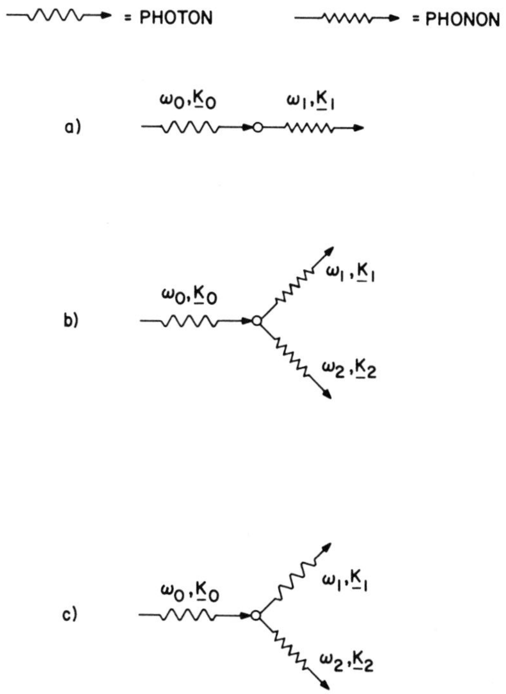

FIG. 11.   Photon-phonon interaction diagrams. (a) One phonon or direct interaction. (b) Two-phonon process. (c) Scattering of light by a phonon (Raman scattering).   More complicated interactions are possible but are weaker and therefore less probable.

involving two or more photons and a single phonon. Figure 11(c) is an example of a photon scattering process in which an incident photon creates a phonon plus another photon of different (lower) energy and direction. This process, which is of some importance in studying lattice vibrations, is known as Raman scattering (or Mandelshtam scattering, depending upon whom you believe should get credit for the discovery of the effect).

No matter how complicated they become, interaction diagrams of the type shown in Fig. 11 have the following properties. The sum of the frequencies of the waves going into a vertex must equal the sum of the frequencies of the waves going out. A similar statement holds true

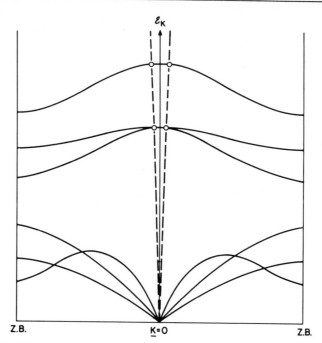

FIG. 12.   Energy and momentum are conserved in direct phonon-
photon interaction at the intersection of their $\mathcal{E}_K$, K surfaces.   For a
given energy a photon has much less momentum than a typical phonon,
so that the intersection is nearly at $\underline{K} = 0$.

for the sum of the wave vectors of the waves associated with a vertex.
To see how this comes about, consider the one-phonon process
[Fig. 11(a)] involving the interaction of two waves.   The interaction
energy $-\underline{P} \cdot \underline{E}$ is proportional to the product of the two wave fields,
since the polarization associated with a phonon is proportional to the
displacements that make up the phonon Bloch wave.   The time-dependent
part of the interaction is the product of two factors:

$$[\exp(i\omega_0 t) + \exp(-i\omega_0 t)][\exp(i\omega_1 t) + \exp(-i\omega_1 t)] = \cos(\omega_0 t)\cos(\omega_1 t) \quad ,$$

where we have written each factor as a sum of a positive and negative
exponential to insure that the result is real.   Assuming the two cosine
terms are in phase at $t = 0$, and that $\omega_0 \neq \omega_1$, as time goes on they
will be out of phase, then in phase again, and so on, so that the time

averaged interaction is zero. This does not happen if $\omega_0 = \omega_1$, where two waves initially in phase stay in phase and thus have time to interact. For interactions involving more than two waves, the interaction energy always involves the product of all the waves involved, and the above argument is easily generalized. If we write our rule in the form

$$\hbar(\omega_1^{in} + \omega_2^{in} + \ldots) = \hbar(\omega_1^{out} + \omega_2^{out} + \ldots) \quad , \tag{53}$$

we see that it is just the condition for energy conservation among quantum mechanical "particle-waves," each of which has an energy of $\hbar\omega_i$. The conservation of wave vectors occurs for similar reasons, since even if two waves have the same frequency, their product averaged over the interacting volume vanishes unless the waves have the same wavelength and are traveling in the same direction. The resulting general relation among interacting wave vectors,

$$\hbar(\underline{K}_1^{in} + \underline{K}_2^{in} + \ldots) = \hbar(\underline{K}_1^{out} + \underline{K}_2^{out} + \ldots) \quad , \tag{54}$$

is an expression of conservation of the quantity we defined in the last section as crystal momentum.

These two conservation rules create some very serious restrictions on the way in which light can interact with phonons. Let us again take the one-phonon process as an example. The conservation rules tell us that, if we superimpose the $\mathcal{E}_K$ vs $\underline{K}$ surface for a photon upon the $\mathcal{E}_K$ vs $\underline{K}$ surface for the phonons, only where the surfaces intersect is direct photon-phonon interaction possible. The cross section along a particular direction for such a set of intersecting surfaces might appear as in Fig. 12. The curve for the photon, for which $\mathcal{E}_K = \hbar cK$, is a straight line with slope equal to $c = 3 \times 10^{10}$ cm/sec. The photon curve appears very steeply rising, and we can see that this should be the case by noting that the highest phonon frequencies are of the order of $\omega_{max} \approx 3 \times 10^4$ sec^{-1}, which corresponds to a photon wave vector of $\omega_{max}/c \approx 10^4$ cm^{-1}. But the wave vector associated with the B.Z. edge is of the order $\pi/a \approx 10^8$ cm^{-1}. Thus the photon curve intersects

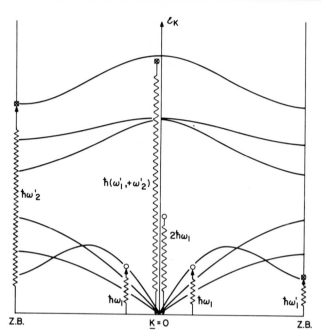

FIG. 13.   Two-phonon interactions involving two phonons from the same branch (O) or one acoustic and one optic phonon (⊗). Similar diagrams involving photon energies over a broad band can be drawn by shifting K_{phonon}.

that of the phonons at a point of the order of only $1/10^4$ of the distance to the edge of the B. Z. For most purposes we can ignore the differences in the properties of the phonons at the actual intersection point as compared with the B. Z. center ($\underline{K}$ = 0) and simply say that electromagnetic radiation interacts directly with only $\underline{K}$ = 0 phonons. In this regard optical effects are quite different from such properties as heat capacity and thermal conductivity, to which all phonons (i. e., phonons of all wavelengths) make roughly equal contributions.

Matching of the frequencies and wave vectors of the photon and phonon waves is a necessary condition for interaction but it is not sufficient. The phonon displacements must give rise to a polarization field parallel to the electric field of the electromagnetic wave, and thus perpendicular or transverse to the mutual wave vector of the two waves in order that the interaction term $-\underline{P} \cdot \underline{E}$ not vanish. Therefore, $\underline{K} = 0$ optic modes which give rise to either longitudinal polarization or no polarization at all are optically inactive. Examples of the latter behavior occur in Si and Ge, which are cubic structures with two atoms per unit cell. The $\underline{K} = 0$ optic modes consist of equal displacements of the two atoms in opposite directions, so that any polarization produced by one atom is neutralized by the other. In such materials, where there is no strong direct phonon absorption to mask it, it is relatively easy to observe two-phonon absorption [Fig. 11(b)]. Crystal momentum conservation is much less stringent in this case and can usually be satisfied over a range of frequencies. So long as the two phonons created have equal and opposite values of $\underline{K}$, they can come in general from any combination of branches throughout the B.Z. Figure 13 shows two possibilities. In contrast to absorption due to the one-phonon process, which is very intense and strongly localized about a single frequency, absorption due to two- (or, more generally, multiple-) phonon processes are weak and distributed over a broad band of frequencies.

B. One-Phonon Interaction

For a more detailed discussion of the most important process, involving one phonon and one photon at $\underline{K} = 0$, we will consider a cubic lattice with two dissimilar atoms per unit cell, as found, for example, in the alkali halides. Think of such a lattice as composed of two interpenetrating sublattices, one made up of all of the positive ions, the other of the negative ions. The special property of $\underline{K} = 0$ phonons is that all like ions on one sublattice are displaced by equal amounts, and we can, therefore, describe these vibrations as rigid displacements of entire sublattices relative to one another. Since we know in advance that the dielectric response for a cubic crystal is going to be isotropic, i.e., independent of the orientation of the crystal, we can pick a coordinate axis (the z axis, say) to coincide with the direction of both the electric field of the radiation and the two sublattice displacements,

which we will call u_+ and u_-, respectively. The mutual wave vector
of the two interacting waves is nearly zero and confined to the x, y
plane, so that both waves are transverse.

If the ions making up the crystal could be thought of as carrying
point charges Z_+ and Z_-, the polarization due to sublattice displacements would be $P = N(Z_+u_+ + Z_-u_-)$, where N stands for the number
of cells per unit volume. There are several complications which make
the point charge assumption questionable in real crystals, so it is better
to proceed more generally. Since the lattice dielectric response is concerned only with that part of P that depends on u_+ and u_-, rather than
trying to work out P from first principles we can expand P in terms of
u_+ and u_-, stopping after the lowest order terms:

$$P_{lattice} = (\partial P/\partial u_+)u_+ + (\partial P/\partial u_-)u_-$$

$$= N(Q_+u_+ + Q_-u_-)$$

$$= NQ(u_+ - u_-) \quad .$$

We will call

$$Q = \frac{1}{N} \frac{\partial P}{\partial u_+} = -\frac{1}{N} \frac{\partial P}{\partial u_-}$$

the apparent charge of the ions. The relation $(\partial P/\partial u_+) = -(\partial P/\partial u_-)$
follows from the observation that a uniform translation of both sublat-
tices, $u_+ = u_-$, cannot give rise to a dipole moment.

We have a choice at this point of considering the lattice as a
classical or a quantum system. We arrive at the correct susceptibility
either way. Classically, we proceed by writing down an equation of
motion for the ions of each sublattice. For small displacements we
assume that each sublattice experiences a restoring force proportional
to the relative sublattice displacements, $u_+ - u_-$. This linear or harmonic approximation is not exactly correct but is very nearly true for
most crystals well below their melting points. There is an additional
driving force of $\pm QE$ due to the radiation field. Thus the equations of
motion for the two sublattices are

$$m_+ d^2 u_+/dt^2 = -F_t(u_+ - u_-) + QE \tag{55a}$$

$$m_- d^2 u_-/dt^2 = -F_t(u_- - u_+) - QE \quad , \tag{55b}$$

where F_t is the harmonic coupling constant between the two sublattices. (The subscript "t" is to remind us that the displacements we are considering are transverse with respect to the nonzero, but small, wave vector.) The simultaneous solution of these two coupled equations corresponding to the optic mode is obtained by multiplying the first by m_-, the second by m_+, and subtracting. The result, divided by $(m_+ + m_-)$ can be written as

$$\mu \frac{d^2}{dt^2}(u_+ - u_-) + F_t(u_+ - u_-) = QE \quad , \tag{56}$$

where

$$\mu = m_+ m_-/(m_+ + m_-)$$

is the <u>reduced mass</u> of the lattice. If we add in the usual way, and with the usual justification, a damping term $\Gamma\, d(u_+ - u_-)/dt$ to Eq. (56), then we have the familiar equation of motion for a driven damped oscillator. That is, Eq. (56) becomes identical with Eq. (35) if we make the following substitutions:

$$(u_+ - u_-) \to \delta_i \quad , \qquad \mu \to m$$

$$Q \to -e \quad , \qquad (F_t/\mu) \equiv \Omega_t^2 \to \Omega_i^2$$

$$\gamma \equiv (\Gamma/\mu) \to \gamma_i \quad .$$

Since the lattice behaves in all respects like an harmonic oscillator, it must in particular exhibit the dielectric response of an harmonic oscillator [Eq. (36) with the appropriate substitutions]:

$$\chi(\omega)_{lattice} = \left(\frac{NQ^2}{\mu}\right) \frac{1}{\Omega_t^2 - \omega^2 - i\gamma\omega} \quad . \tag{57}$$

$\Omega_t^2 = F_t/\mu$ is, of course, just the vibrational frequency of the transverse $K = 0$ optic phonon.

Exercise. It is interesting to use the general quantum mechanical susceptibility we derived in Section VIII to calculate χ_{lattice} for the diatomic lattice. One arrives at Eq. (57) by a rather different series of steps.

(a) Write the hamiltonian function $\mathscr{H}$ = kinetic energy + potential energy and the dipole moment operator $\mathscr{P}$ for a system of two sublattices in terms of the displacements u_+ and u_-.

(b) Show that the substitutions

$$u_+ = (\mu/m_+)\xi \quad \text{and} \quad u_- = (\mu/m_-)\xi$$

transform the hamiltonian into the harmonic oscillator form:

$$\mathscr{H} = \tfrac{1}{2}\mu(d^2\xi/dt^2) + \tfrac{1}{2}F_t\xi^2$$

and the dipole moment operator into

$$\mathscr{P} = Q\xi \quad ;$$

ξ is known as the optic mode normal coordinate.

(c) The normalized ground state and first excited state wave functions for the harmonic hamiltonian above are

$$\phi_0(\xi) = (\alpha/\pi)^{\frac{1}{4}} \exp(-\tfrac{1}{2}\alpha\xi^2)$$

$$\phi_1(\xi) = (4/\pi)^{\frac{1}{4}} \xi \exp(-\tfrac{1}{2}\alpha\xi^2)$$

where $\alpha \equiv (\mu\Omega_t/\hbar)$. Show that the dipole moment integral between the two states is

$$\mathscr{P}_{01} = \int \phi_0^*(Q\xi)\phi_1 \, d\xi = (\hbar/2\mu\Omega_t)Q$$

and verify that this result, used in Eq. (45), leads to a susceptibility identical to the classical result calculated above.

(NOTE: It is a well-known property of the harmonic oscillator that the integral $\mathscr{P}_{oj} = 0$ for all other excited states.)

Some of these results are easily generalized to more complicated lattices. If there are n atoms per unit cell, the vibrations can be described in terms of relative displacements of the n sublattices. In a cubic crystal the modes can always be chosen as purely transverse or longitudinal, regardless of the direction of $\underline{K}$. Each $\underline{K} = 0$ transverse optic (t. o.) mode will in general contribute a term of the form of Eq. (57) to χ_{lattice}. Each mode has a characteristic frequency, apparent charge, and mass, which can be worked out, but for which no simple closed expressions in general exist. Any modes that are nonpolar will be found to have $Q = 0$ and can be ignored, as we mentioned before.

There are two kinds of information available from a study of the optical properties of a crystal at lattice vibrational frequencies:

(1) the forces coupling the sublattices together, determined from the resonant frequencies,

(2) the apparent charges of the sublattices, determined from the resonance strengths [i.e., the numerator of Eq. (57)].

Quite often, as we pointed out in Section VI, the reflectivity is the the quantity measured experimentally. One way to obtain the resonance frequencies and strengths is to calculate a "theoretical" reflectivity, using Eq. (57) and treating Ω, Q, and γ as parameters which are adjusted to produce as good a fit as possible to the measured reflectivity. How successful this fitting procedure is in a favorable case is illustrated by Fig. 14. The "flat" appearance at the top of the reflection band is the

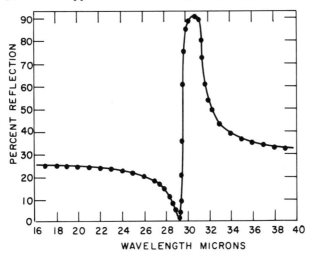

FIG. 14. The reflectivity of AlSb vs wavelength in the restrahlen region. The solid curve is calculated using Eq. (57), while the points are the experimental data. Reprinted from W. J. Turner and W. E. Reese, Phys. Rev. 127, 126 (1962) by courtesy of the American Institute of Physics.

result of the extended reflection region at energies just above the resonance frequency, as discussed in Section VII. This property finds
practical application in isolating radiation of a desired band of frequencies from a "white" source. Optical elements of this kind are called
reflection filters.

C. Apparent Ionic Charges

In Table 3 the apparent charges Q obtained by an analysis of
infrared optical data for several alkali halides are listed. Notice that
in all cases Q is larger than the nominal static charge $|Z_+| = |Z_-| =$
1.0 $|e|$, assuming ionic bonding. Any covalency, of course, would be
expected to reduce the static charges. One major reason for the difference between the apparent charge Q and the static charge Z of a lattice
is that the ions themselves possess polarizable electrons. When a
sublattice is displaced, the resultant polarization sets up local electric
fields at the ions, which are just proportional to the displacements.
These local fields induce a further polarization of the electrons in the
ions themselves, also proportional to the sublattice displacements.
But since Q, by its definition, includes all sources of polarization proportional to sublattice displacement, it is enhanced over Z by this
mechanism. The actual enhancement factor for a cubic crystal (assuming
that the polarizable electrons are affected by the local field at the nuclei
of the ions) is $(\varepsilon_\infty + 2)/3$, where ε_∞ is the dielectric response of the
electrons alone. (In practice, ε_∞ is the dielectric response measured
at frequencies too high for the lattice to respond to, but still well below
any electronic resonances, so that the electrons respond perfectly.) The
quantity Q_s defined by

$$Q = Q_s(\varepsilon_\infty + 2)/3$$

and known as the Szigeti effective charge is seen to be the static charge
an ion must have to produce the observed apparent charge after the above
polarization effects are accounted for. The effective charges Q_s cor-

TABLE 3

Comparison of Q and Q_s for Several Alkali Halides[a,b]

		Li	Na	K
F	Q	1.13	1.15	1.03
	Q_s	0.87	0.93	0.88
Cl	Q	1.14	1.05	1.11
	Q_s	0.73	0.74	0.81
Br	Q	1.18	1.07	1.09
	Q_s	0.68	0.70	0.76

[a]After E. Burstein, in Phonons and Phonon Interactions, Benjamin, New York, 1964, p. 296.

[b]The upper number is the apparent charge Q, while the lower number Q_s is corrected for local field effects.

rected in this manner are also listed in Table 3. The fact that the Q_s charges are now somewhat smaller than $1.0|e|$ indicates that there are further corrections to the apparent charge that we have not considered. It appears that much of this discrepancy can be removed by calculating the additional distortion of the ions due to short-range repulsive forces, rather than covalency per se.

Local electrostatic field effects also make a contribution to the forces coupling the sublattices. One consequence is that the restoring forces for longitudinal displacements are larger than for transverse displacements. (We are still talking about displacements parallel or perpendicular to a nonzero but very small wave vector.) For a diatomic cubic lattice the force constants differ by

$$F_\ell = F_t + 4\pi N Q^2 / \varepsilon_\infty \quad . \tag{58}$$

This is entirely the result of the different local electric fields acting on

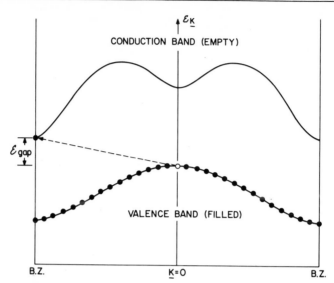

FIG. 15a. Filled electron states are schematically represented by filled circles, unfilled states by unfilled circles. In order to change its momentum an electron is forced to acquire a minimum additional energy of $\mathcal{E}_{gap}$. This leads to an insulator or semiconductor depending upon the magnitude of $\mathcal{E}_{gap}$.

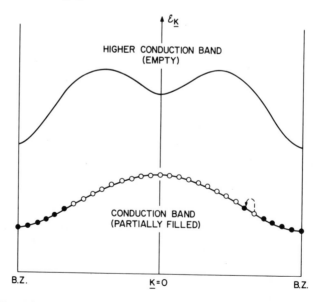

FIG. 15b. Electrons in partially filled bands can change momentum with a small energy expenditure leading to metallic conduction.

the ions in the two cases. If, as in Si and Ge, there is no polarization associated with the $K = 0$ phonons, the transverse and longitudinal force constants and vibrational frequencies are the same.

Exercise. As a result of the increased electrostatic forces just discussed for longitudinal displacements, the frequencies of longitudinal optical vibrations, Ω_ℓ, are higher than those for corresponding transverse vibrations, Ω_t. For a crystal of the NaCl type, use Eq. (58) to show that the ratio (Ω_ℓ/Ω_t) is given by

$$(\Omega_\ell/\Omega_t)^2 = \left(\varepsilon_\infty + \frac{4\pi NQ^2}{\mu\Omega_t^2}\right)\Big/\varepsilon_\infty .$$

Notice that

$$(4\pi NQ^2/\mu\Omega_t^2) = 4\pi\chi_{lattice}(\omega = 0) = \varepsilon_{lattice}(\omega = 0) .$$

This leads immediately to the Lyddane-Sachs-Teller relation

$$(\Omega_\ell/\Omega_t)^2 = \varepsilon(0)/\varepsilon_\infty ,$$

where $\varepsilon(0) = \varepsilon_{lattice}(0) + \varepsilon_\infty$ is the total dielectric response of the crystal at zero frequency.

XI. INTERBAND AND EXCITON TRANSITIONS

A. Electronic Energy Bands

We learned in Section IX that wavelike excitation states exist in solids, which can be characterized as having a well-defined crystal momentum $\hbar K$ and energies falling within certain permitted bands. In this section we will study the properties of optical transitions between electron states which can be described in this way.

We now describe the band model of a solid. All individual electron states in a solid are Bloch states [Eq. (52)]. The ground electronic state of the total solid is the one in which the states for the individual electrons are filled up successively, starting with states of lowest energy and working up. If, when the band-filling procedure is complete, the available states of one or more of the lower bands are just used up, then to create an excited state an electron must be taken from a filled band and put into a higher unfilled one, a process which takes a rather

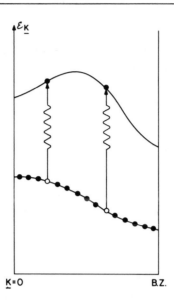

FIG. 16. In a direct (or vertical) transition an electron is transferred vertically to a higher band, the difference in energy being sup-
plied by the photon absorbed. For simplicity only the right half of the
$\mathcal{E}_K$, K diagram is shown.

large amount of energy. See Fig. 15a. The electrons, somewhat trapped
in their lowest energy configuration, cannot respond and adjust to small
external fields, and the material is, therefore, an insulator.

If, on the other hand, one or more bands are incompletely filled
after all of the electrons are accounted for, a weak electric field is
capable of accelerating electrons from the uppermost occupied states
into adjacent states within the band having nearly the same energy.
See Fig. 15b. The material then exhibits metallic conductivity. Unfilled
or incompletely filled bands in which electrons can transport charge in
an external field are known as conduction bands. Filled bands, or at
least the filled band of highest energy, is known as the valence band.

Only interband electronic transitions, that is, where an electron
is taken from one band to another, are possible in insulators. In metals
both interband and intraband transitions are possible. We will consider
optical properties associated with interband transitions in this section,
and those arising from intraband transitions in the next section.

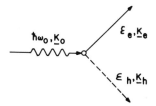

FIG. 17. Interaction diagram for direct electron-photon interaction.

In analyzing interband transitions caused by photons, we can again make use of energy and crystal-momentum conservation. As before, we find that the wave vector of a photon at optical frequencies, $K_0 = \omega_{opt}/c$, is negligible on the scale of the B.Z., so that light-induced transitions are essentially "vertical" when represented on an electron $\mathcal{E}_K$, $\underline{K}$ surface. The simplest and most strongly allowed interband transitions are those shown in Fig. 16. The electron, which is transferred to the upper band, leaves behind an unoccupied state in the lower band, which we call an electron "hole."

An interaction diagram representing these direct vertical transitions is shown in Fig. 17. (The energy $\mathcal{E}_h$ and momentum $\hbar K_h$ of a hole are defined as the negative of the energy and momentum of an electron if it occupied the same state. With these definitions, energy and momentum are conserved at the vertex of Fig. 17.) The diagram can be summarized by saying that a photon is annihilated to create an electron-hole pair. Note that, like the photon – two-phonon process we studied in the last section, because there are a pair of particles produced, transitions are not limited to the center of the B.Z. nor to a single well-defined energy. Absorption due to interband transitions is, as you can see, roughly as broad as the width of the bands themselves. There is no process analogous to the one-phonon process, which dominates the phonon dielectric response. Although phonons and photons can be created and annihilated singly, electronlike excitations appear or disappear only in pairs in interaction with neutral "particles." There is another fundamental conservation law operating here, conservation of charge.

B.   Direct Transitions

It is a straightforward problem to work out the direct interband contribution to the dielectric response.   Every possible vertical transition, of the type shown in Fig. 16, makes a contribution, which we can calculate by application of the general quantum mechanical result, Eq. (45).   It is only necessary to sum the contributions from all of the occupied states of the lower band.   Thus

$$\chi(\omega) \; = \; 2\left(\frac{1}{V}\right) \frac{e^2}{m} \sum_{K} \frac{f_{vc}(\underline{K})}{\Omega^2_{vc}(\underline{K}) - \omega^2 - i\omega\gamma_{vc}(\underline{K})} \quad , \tag{59}$$

where

$$\hbar\Omega_{vc}(\underline{K}) \; = \; \mathcal{E}_c(\underline{K}) - \mathcal{E}_v(\underline{K}) \quad .$$

There are, per band, two electron states for every allowed $\underline{K}$ value, one with intrinsic electron spin $+\frac{1}{2}$ and one with spin of $-\frac{1}{2}$, so that the number of states (with a specified $\underline{K}$) per unit volume of the solid is just $2(1/V)$ where V is the total volume of the sample.   Equation (59) is a complete and exact result, and if one knows in advance $\mathcal{E}_K$ as a function of $\underline{K}$ and has explicit expressions for the wave functions in order to evaluate f_{vc}, Eq. (59) is quite convenient.

Often, however, this information is not available.   In this case it is convenient to reexpress Eq. (59) not as a sum over $\underline{K}$ but instead as an equivalent integral over energy:

$$\chi(\omega) \; = \; \left(\frac{2}{V}\right)\left(\frac{e^2}{m}\right) \int \frac{f_{vc}\rho(\Omega_{vc}) \, d\Omega_{vc}}{\Omega^2_{vc} - \omega^2 - i\omega\gamma_{vc}} \quad . \tag{60}$$

The conversion from sum over $\underline{K}$ to integral over Ω_{vc} is made possible formally by introducing a joint density of states function $\rho(\Omega_{vc})$ defined as follows.   The number of initial-final $\underline{K}$-state pairs (disregarding electron spin which has already been accounted for) whose energy separation is between Ω_{vc} and $\Omega_{vc} + d\Omega_{vc}$ is $\rho(\Omega_{vc}) \, d\Omega_{vc}$.   As you can prove by doing the following Exercise, it is possible to approximately carry out the integration over $d\Omega_{vc}$, at least for the imaginary part of Eq. (60), and to obtain the expression given in the Exercise for the absorption coefficient, showing that the absorption coefficient for light of a given energy is proportional to the density of energy-state pairs with that energy separation.

$$\alpha(\omega) = \left(\frac{4\pi^2}{V}\right)\left(\frac{e^2}{mc}\right)\frac{1}{n} f_{vc}(\omega)\, \rho(\omega) \quad . \tag{61}$$

Exercise. In most cases of physical interest, the damping term γ_{vc} in Eq. (60) is considerably smaller than Ω_{vc}, and under these conditions the integral can be simplified considerably.
(a) Using the results of the Exercise in Section VIII, show that the imaginary part of the susceptibility can be written as

$$\chi''(\omega) = \left(\frac{2}{V}\right)\left(\frac{e^2}{m}\right) \int\left(\frac{\pi}{2\Omega_{vc}}\right) f(\Omega_{vc})\rho(\Omega_{vc})[\delta(\Omega_{vc} - \omega) - \delta(\Omega_{vc} + \omega)]d\Omega_{vc}$$

The terms involving $\delta(\Omega_{vc} + \omega)$ describing downward transitions can be ignored if the upper band is not occupied.
(b) Integrals involving δ-functions are easily performed because the integrand is everywhere zero except at a single point, thus

$$\int g(x)\, \delta(x - a)\, dx = g(a) \quad .$$

Use this result to derive Eq. (61).

C. Density of States

The concept of a density-of-states function is an important one, which recurs in many connections, and is thus worth pursuing further. Qualitatively, the more nearly an $\mathscr{E}_K$, $\underline{K}$ surface approaches the horizontal (i. e., $\partial\mathscr{E}_K/\partial\underline{K} = 0$), the more allowed $\underline{K}$-states there are with nearly the same energy, and thus the higher the density-of-states function for the band. The joint density-of-states function, which is the joint property of two bands, becomes large when the slopes of the two energy surfaces are nearly the same. Then there are a large number of $\underline{K}$ values for which the energy separation of the two bands is nearly constant.

We can attack the whole problem of counting of states more quantitatively as follows. Given that the states within a band can be

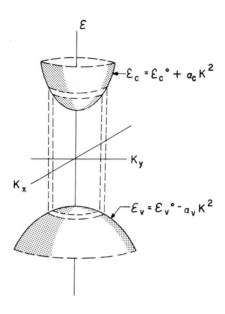

FIG. 18. A sketch of parabolic valence and conduction bands used to calculate the joint density-of-states function.

labeled by their crystal momentum $\hbar\underline{K}$, how many allowed values of $\underline{K}$ are there within one band? Or, asking the same question in a slightly different way, how closely are allowed $\underline{K}$-states packed within $\underline{K}$-space? You will recall that, in the tight binding approximation, the $\underline{K}$-states were formed from linear combinations of wave functions localized within each of the cells,

$$\Psi'(\underline{K}) = \sum_{\ell}^{N} c_\ell \Psi_\ell \quad ,$$

N being the total number of cells in the crystal. The number of $\underline{K}$-states in a band is just equal to the number of possible independent combinations of the Ψ_ℓ, which is, of course, N. <u>The total number of allowed $\underline{K}$-states located on one $\mathscr{E}_{\underline{K}}$, $\underline{K}$ surface equals the total number of cells in the sample.</u> To find the density of states in $\underline{K}$-space, we must divide the number of states by the volume of the first B.Z. within which all of the states must lie. The first B.Z. of a cubic lattice is also a cube $(2\pi/a)$ on a side, giving (volume of the first B. Z.) $= (2\pi)^3/V_c$ (62)

where V_c is the volume of the unit cell in real space. This result is true for noncubic, as well as cubic, lattices. Therefore,

(density of allowed K-states in K-space) $= \dfrac{N}{(2\pi)^3/V_c} = \dfrac{V}{(2\pi)^3}$.   (63)

The point to notice is that the size of the B. Z. is determined by V_c, an intrinsic property of the substance, but the total number of allowed states, and thus how tightly these K-states are packed into the B.Z., varies with the size of the sample.

In order to translate the density of states in K-space into a density of states within a given energy interval, we need only know how the energies of the allowed transitions vary with K, which, of course, depends upon the detailed form of the $\mathcal{E}_K$, K surface. We will consider the following specific case, which often arises at the edge of an energy gap. We assume that the minimum separation between bands occurs at K = 0 and that, at least near K = 0, both bands are parabolic, i.e., the energy varies linearly with $K^2 = K_x^2 + K_y^2 + K_z^2$, as shown schematically in Fig. 18. The energy associated with a vertical transition at any wave vector K is given by

$$\mathcal{E}_{vc}(K) = (\mathcal{E}_c^0 - \mathcal{E}_v^0) + (\alpha_c + \alpha_v)K^2$$

or

$$\hbar\Omega_{vc} = \hbar(\Omega_{gap} + \alpha K^2) \quad . \qquad (64)$$

Now from Eq. (63) we know that, within a spherical shell of radius $|K|$ and thickness dK in K-space, there are

$$V/(2\pi^3) \cdot 4\pi K^2 \, dK$$

allowed K-states. Using Eq. (64) and $d\Omega_{vc} = 2\alpha K \, dK$ to eliminate K from the above expression, we find that the joint density of states is

$$\rho(\Omega_{vc}) \, d\Omega_{vc} = \dfrac{V}{4\pi^2 \alpha^{\frac{3}{2}}} (\Omega_{vc} - \Omega_{gap})^{\frac{1}{2}} \, d\Omega_{vc} \qquad (65)$$

for direct transitions between parabolic bands. Near the edge of an
energy band gap it is usually true that the density-of-states function
changes with frequency more rapidly than either the refractive index
$n(\omega)$ or the oscillator strength f_{vc}, so from Eq. (61) the frequency
dependence of absorption just above an energy gap $\mathcal{E}_{gap}$ is often determined by the frequency dependence of the density of states, for the
band shapes we are considering.

D. Indirect Transitions

There are many substances which have indirect band gaps, i.e.,
the minima of the conduction band and the maxima of the valence band
do not occur at the same value of $\underline{K}$ as, for example, in Fig. 15. In
such cases the energy marking the onset of direct (vertical) transitions
does not correspond to the energy gap, which is always defined as the
energy difference between the top of the lower band and the bottom of
the upper one. It is, however, often possible to observe indirect optical transitions connecting bands at different points in the B.Z. The
difference in crystal momentum is taken up by a phonon. These indirect
or phonon-assisted transitions can be imagined as occurring in two
steps, as shown in Fig. 19. First a photon of energy $\hbar\omega_0$ is absorbed
causing a vertical transition into a state of the conduction band. But
this state is not a stable energy state of the system because it appears
within the forbidden energy gap and must, therefore, quickly decay.
(In fact, by the uncertainty principle, the lifetime of such a state is of
order $\hbar/\Delta\mathcal{E}$, $\Delta\mathcal{E}$ being in this case the discrepancy between $\hbar\omega_0$ and
the actual vertical separation of the bands.) The most probable decay
path is vertically downward, again reemitting a photon at the same
energy as the one absorbed. (This "virtual" absorption and reemission
of photons is a way of understanding the real part of the dielectric
response of a quantum system, as we mentioned in Section VIII.)

Occasionally, however, before this unstable state within the gap
decays back to the initial state, it may undergo a "collision" with a

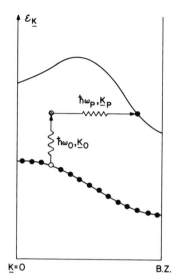

FIG. 19. An indirect or phonon-assisted electronic transition. Absorption of the phonon causes a vertical transition to a "virtual" state from which the electron may be scattered "horizontally" (since the phonon energy is often nearly negligible) by a phonon to a stable state on the $\mathcal{E}_K$, K surface.

phonon, annihilate it, and take on its energy and momentum. The energy of the phonon is usually practically negligible by comparison with the photon energy, but the additional momentum can cause a "horizontal" transition, which may result in the electron finding itself in a stable state on the conduction band energy surface. The net result is the creation of a hole with momentum K_h, a conduction electron with momentum $K_p - K_h$, and the disappearance of a photon with momentum $K_0 \approx 0$, and a phonon with momentum K_p. The energy of the photon absorbed is less than the difference in the energies of the initial and final electron states by $\hbar\omega_p$, the amount of energy donated by the destroyed phonon. At finite temperatures, thermally excited phonons are always present, so that no external "supply" of phonons is necessary for this process. However, as you might suppose, absorption due to these phonon-assisted transitions is generally much weaker than that due to direct transitions and is, therefore, of importance only at energies where direct transitions do not occur. In particular, indirect transitions can be used to deduce the energies of indirect band gaps.

E.   Excitons

In discussing the energy states of electrons in terms of Bloch waves, we have thus far assumed that, once formed, a conduction electron and a hole are independent noninteracting entities. But an electron and a hole are oppositely charged and thus exert some electrostatic attraction upon one another. One can also understand that in a metal, for example, other conduction electrons act to neutralize or screen out this attraction, making the previous treatment nearly correct. But, on the other hand, in insulators it may be energetically favorable for the electron and hole to remain more or less tightly bound to one another and to move through the crystal as an electrically neutral pair, with many of the attributes of a single "particle." Such a bound electron-hole pair is known as an exciton.

It is possible to think of excitons rather simply in two extreme limits. In the first limit, that of the tightly bound exciton, the attraction between the electron and hole is so great that they remain localized about the same atom. An equally appropriate description of such a state is as an excitation of a single atom or molecule, which is, however, free to hop to adjacent cells. This situation can be understood quite adequately in terms of the tight binding model we discussed in Section IX and gives rise to bands of exciton states with wave vectors spanning the B.Z. Examples of tightly bound excitons are to be found in alkali halides, many organic crystals, and inorganic salts containing transition metal or rare earth ions with well-shielded inner d- or f-electron states.

At the other extreme an electron and hole can be so weakly bound that the excited electron leaves its atom but remains in the immediate vicinity, forming a kind of hydrogenlike "atom" with the hole it left behind. The simplest treatment considers the electron-hole potential

$V(\underline{r})$ to be screened by the electronic polarization of intervening cells:

$$V(\underline{r}) = -e^2/\varepsilon_\infty r \quad .$$

(The nuclei cannot respond quickly enough to screen the fields produced by the faster moving electron-hole pair, which accounts for the appearance of the electronic response function ε_∞ rather than the static value ε_0.) If we measure the exciton energy relative to the bottom of the conduction band, we can use the well-known result for the energy levels of a hydrogen atom:

$$\mathcal{E}_n = -\frac{m^*e^4}{2\varepsilon_\infty^2 \hbar^2}\frac{1}{n^2}$$

$$= -13.6(m^*/m\varepsilon_\infty^2)(1/n^2) \quad eV \quad ,$$

(66)

where for the time being we will consider the effective reduced mass m^* simply as an adjustable parameter. As in the hydrogen atom, one thus expects a progression of bound states becoming successively closer together as one approaches the "ionization limit," which is in this case just the energy of the conduction band.

Regardless of whether the excitons are strongly or weakly bound, they behave as a single particle in the sense that the electron-hole pair has a well-defined crystal momentum $h\underline{K}$ although neither the electron nor hole wave function can individually be written any longer in the simple Bloch form. The pair, considered as a single "particle," can be created or destroyed singly, so the selection rules are the same ones we encountered in Section X for phonons. Referring back to Fig. 11, exciton processes involving only $\underline{K} = 0$ excitons produce the strongest absorption, and weaker processes involving two excitons (or one exciton and one phonon) with equal but opposite values of $\underline{K}$ are also possible.

Figure 20 is the experimentally determined exciton spectrum of the semiconductor Cu_2O showing transitions from the top of the valence band to exciton levels with successively higher values of the principal quantum number n. The sequence of levels fits Eq. (66) quite well

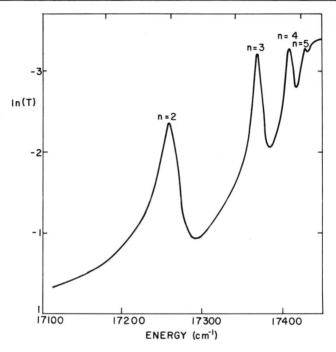

FIG. 20. The absorption spectrum of Cu_2O at 77 °K showing the exciton lines corresponding to several values of the quantum number n. Reprinted from P. W. Baumeister, Phys. Rev., **121**, 359 (1961), by courtesy of the American Institute of Physics.

with $\varepsilon_\infty = 10$ and $m^* = 0.7\,m$. The radius of an electron orbit scales as $(m\varepsilon_\infty/m^*)$, so that the radius is about thirteen times larger in Cu_2O than in the hydrogen atom. For many semiconductors (few of which, incidentally, have as pretty an exciton spectrum as Cu_2O), the effective mass m^* is much smaller than m and the electron radius is correspondingly further increased. We will discuss the meaning of effective masses of electrons and holes more fully in the following section.

XII. OPTICAL PROPERTIES OF METALS

A. Free Electron Theory

Many features characteristic of metals depend in a direct way upon the presence of conduction electrons, and this is particularly true of the optical properties. This is because electrons in an unfilled band are more or less unrestricted in their ability to move in an electric field and thereby dominate the dielectric response of a metal almost completely. We will begin by considering a free electron theory, in which the electrons are considered to have only kinetic energy; any potential energy arising from interactions of the electrons with the positive ion cores or with themselves is neglected. It turns out that the analysis is simple and closely related to our previous treatment of the classical harmonic oscillator. We will discuss, then, how the free electron theory is modified in light of the more realistic quantum mechanical energy band picture.

Since the lattice plays no part in the free electron theory, we can think of an electron "gas" as occupying the volume of the solid, where the electrons are moving about (due to thermal agitation) and colliding occasionally with one another. Because of the random nature of the thermal motion, the average or mean velocity $\langle \underline{v} \rangle$ of the electrons is zero, as is the mean force acting on an electron. When a uniform electric field $\underline{E}(t)$ is applied, there is a mean force $-e\underline{E}$ to which the electron system responds with a mean velocity given by

$$m \, d\langle \underline{v} \rangle /dt = -e\underline{E} \quad .$$

Of course, each individual electron is undergoing a much more complicated random motion and the mean velocity $\langle \underline{v} \rangle$ is simply superimposed upon it. As we have done in previous sections, we postulate an additional "frictional" force $-m\langle \underline{v} \rangle /\tau$ to account for collisions between electrons and phonons or impurities and which allows $\langle \underline{v} \rangle$ to relax to zero as $\exp[-(t-t_0)/\tau]$ if $\underline{E}$ is suddenly turned off at time t_0.

Thus the equation for the mean electron velocity becomes

$$m\left(\frac{d}{dt} + \frac{1}{\tau}\right) \langle \underline{v} \rangle = -e\underline{E}(t) \quad ,$$

which we can put into more familiar form upon setting $\langle \underline{v} \rangle = d\langle \underline{\delta} \rangle/dt$, where $\langle \underline{\delta} \rangle$ is the mean electron displacement resulting from the applied field $\underline{E}$. The equation of motion of $\langle \underline{\delta} \rangle$ is then

$$m\frac{d^2\langle \underline{\delta} \rangle}{dt^2} + \frac{1}{\tau}\frac{d\langle \underline{\delta} \rangle}{dt} = -e\underline{E}(t) \quad , \tag{67}$$

which is identical to the driven oscillator equation of motion, Eq. (35), if we set $\Omega_i = 0$ and replace δ_i by $\langle \underline{\delta} \rangle$ and γ_i by $1/\tau$. Because of this and the fact that the polarization due to the conduction electrons is $\underline{P}_{cond} = -Ne\langle \underline{\delta} \rangle$, the free conduction electron susceptibility is given by an equation similar in form to that of Eq. (36), with $\Omega_i = 0$ and $\gamma_i = 1/\tau$:

$$\hat{\chi}_{cond}(\omega) = -\left(\frac{Ne^2}{m}\right)\frac{1}{\omega^2 + i(\omega/\tau)} \tag{68a}$$

or separating the real and imaginary parts,

$$\chi'_{cond} = -\left(\frac{Ne^2}{m}\right)\frac{1}{\omega^2 + (1/\tau)^2} \tag{68b}$$

$$\chi''_{cond} = \left(\frac{Ne^2}{m}\right)\frac{(1/\omega\tau)}{\omega^2 + (1/\tau)^2} \tag{68c}$$

It is sometimes useful to have the equivalent expression for the frequency-dependent complex conductivity. Using Eq. (18), this is given by

$$\hat{\sigma}(\omega) = \omega(\chi''_{cond} - i\chi'_{cond}) = \sigma_0/(1 - i\omega\tau) \quad , \tag{69}$$

where we have recombined real and imaginary parts and defined

$$\sigma_0 = Ne^2\tau/m \quad , \tag{70}$$

which is, thus, the static or dc conductivity of the electron gas. Equation (70) provides an excellent way to evaluate τ, which until now has been an arbitrary parameter in our theory, in terms of measurable quantities. The values of τ derived in this way for good metals (the alkali metals or Cu, for example) are of the order of 10^{-14} sec at room temperature. (This corresponds to electron mean free paths of ~ 500Å.)

Of course, all metals have interband, as well as intraband, contributions to the dielectric response, but in many cases the most important interband gaps are at higher-than-visible light frequencies, and their effect upon the dielectric response may be approximated by a single, frequency-independent, real term. Thus we can write, approximately,

$$\hat{\varepsilon}(\omega) = (1 + 4\pi\hat{\chi}_{interband}) + 4\pi\hat{\chi}_{cond} = \varepsilon_{\infty} + 4\pi\hat{\chi}_{cond} , \tag{71}$$

and upon substituting Eq. (68) for $\hat{\chi}_{cond}$ we obtain, with a little rearranging,

$$\varepsilon'(\omega) = \varepsilon_{\infty}\left(1 - \frac{\omega_p^2}{\omega^2 + (1/\tau)^2}\right) \tag{72a}$$

$$\varepsilon''(\omega) = \left(\frac{\varepsilon_{\infty}}{\omega\tau}\right)\frac{\omega_p^2}{\omega^2 + (1/\tau)^2} \tag{72b}$$

where we have defined a new quantity

$$\omega_p^2 = 4\pi Ne^2/m\varepsilon_{\infty} . \tag{73}$$

ω_p is known as the electron plasma frequency and is a parameter which recurs in the discussion of all the collective motions of conduction electrons. Electron densities typical of metals ($N \sim 10^{22}$ cm^{-3}) produce plasma frequencies of $\sim 5 \times 10^{15}$ sec^{-1} (which corresponds to a light wavelength of ~ 3000Å). Therefore, the condition $\omega_p \gg (1/\tau)$ generally is true for metals.

What optical properties do we expect from a dielectric response given by Eq. (71)? There are three different limiting frequency regions:

(1) $\omega \ll (1/\tau) \ll \omega_p$. In this low frequency region, which typically extends to the near infrared, $\varepsilon' \approx -\varepsilon_{\infty}\omega_p^2/\tau$ is large and negative but is negligible compared to the still larger

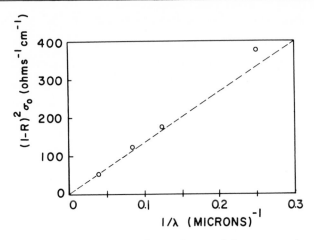

FIG. 21. A comparison of the observed frequency dependence of the reflectivity of constantan compared with the prediction of the Hagen-Rubens relation, Eq. (74). After E. Hagen and H. Rubens, Ann. Phys., 14, 936 (1904).

$$\varepsilon'' \approx \varepsilon_\infty \omega_p^2 \tau/\omega = 4\pi\sigma_0/\omega \quad .$$

The optical constants are approximately given by $n \approx \kappa \approx (\varepsilon''/2)^{\frac{1}{2}}$, which is large compared to unity. The reflectivity at normal incidence,

$$R \approx 1 - (2/n) = 1 - (2\omega/\pi\sigma_0)^{\frac{1}{2}} \quad , \tag{74}$$

is nearly perfect. If we use Eq. (74) along with measured dc conductivities to calculate $(1 - R)$ for real metals, we find that it gives the observed frequency dependence and in many cases about the correct magnitude, as shown in Fig. 21. This constitutes the first test of our free electron model.

Since $\kappa \gg 1$, waves within the metal are quickly attenuated. In metals it is customary to talk not about the absorption coefficient α but instead about a skin depth $\delta = (2/\alpha)$ which is the distance the $\underline{E}$ field penetrates before attenuating by $1/e$. In the low frequency region,

$$\delta = c/(2\pi\sigma_0\omega)^{\frac{1}{2}} \tag{75}$$

In very pure metals at low temperatures the electron mean free path may become longer than skin depth δ. Since the electric field $\underline{E}$ varies over the distance an electron moves between collisions, the ordinary theory of conductivity we have presented is not valid, and all of the equations we have derived are in error. Deviations of this sort are referred to as an "anomalous skin effect."

(2) $(1/\tau) \ll \omega \ll \omega_p$. In this region $\varepsilon'' \approx \varepsilon_\infty \omega_p^2/\omega^3 \tau$ is negligible compared to $\varepsilon' \approx -\varepsilon_\infty \omega_p^2/\omega^2$, which is large and negative. This gives rise to a region of nearly lossless reflection of the type we have already encountered above the resonance frequency of a harmonic oscillator. Since the free electrons behave like oscillators with zero resonance frequencies, this should not come as much of a surprise. Most metals do have a region in which the frequency dependence of $\hat{\varepsilon}$ is as predicted; however, the agreement in magnitude is generally less good than in the low frequency region. The discrepancies can be partly traced to the approximate way we have treated electron scattering. We should more correctly allow the scattering time to vary with the total velocity (both magnitude and direction) of the electrons. This, of course, complicates the model considerably.

(3) $(1/\tau) \ll \omega_p \ll \omega$. We see from Eq. (72a) that $\varepsilon'(\omega) \approx \varepsilon_\infty(1 - \omega_p^2/\omega^2)$, which changes from a negative to a positive quantity near the frequency $\omega = \omega_p$. In Section V we showed that the transition from $\varepsilon' < 0$ to $\varepsilon' > 0$ is one in which we pass from lossless reflection to transparency. The discussion in Section V is relevant here also. These equations thus make the surprising prediction that above the frequency ω_p a metal should abruptly become transparent. Just such an effect has been discovered for simple metals in the ultraviolet spectral region, and Table 4 gives a comparison of the observed wavelength of the onset of transparency with the plasma frequency calculated from the known values of the electron density N, for the alkali metals. The good agreement constitutes one of the most striking successes of the free electron theory of metals.

TABLE 4

Ultraviolet Transmission Edge of Alkali Metals

	Li	Na	K	Rb	Cs
λ_p (calculated, assuming one free electron of mass m per atom), Å	1550	2090	2870	3220	3620
λ_p (observed transparency edge), Å	1550[a]	2100[b]	3400[c]	3950[d]	4350[d]

[a]R. W. Wood, Phys. Rev., 44, 353 (1933).

[b]H. E. Ives and H. B. Briggs, J. Opt. Soc. Am., 27, 181 (1936).

[c]H. E. Ives and H. B. Briggs, J. Opt. Soc. Am., 26, 238 (1936).

[d]H. E. Ives and H. B. Briggs, J. Opt. Soc. Am., 27, 395 (1937).

B. Effective Mass of Electrons

Now we have to say a few words about the validity of the method we have chosen to perform these calculations. You should have been somewhat troubled by the fact that we used classical equations of motion and considered the electrons as localized point charges in deriving the free electron dielectric response. We have already established that the correct quantum mechanical description is (seemingly) quite different. An electron in a crystal is described by a Bloch wave extending over the entire volume of the crystal. For such an electron "wave" the classical concept of velocity is replaced by the group velocity $\underline{v}_g$:

$$\underline{v}_g = \partial\omega/\partial\underline{K} = (1/\hbar)\, \partial\mathcal{E}_{\underline{K}}/\partial\underline{K} \quad . \tag{76}$$

We want to find the quantum analog of Eq. (67), so we differentiate Eq. (76) with respect to time

$$\frac{\partial\underline{v}_g}{\partial t} = \left(\frac{1}{\hbar}\right)\frac{\partial}{\partial t}\left(\frac{\partial\mathcal{E}_{\underline{K}}}{\partial\underline{K}}\right) = \left(\frac{1}{\hbar}\right)\frac{\partial}{\partial\underline{K}}\left(\frac{\partial\mathcal{E}_{\underline{K}}}{\partial t}\right) \quad . \tag{77}$$

Now $\partial \mathcal{E}_K / \partial t$, which is the rate at which an electron in the $\underline{K}$th state changes energy, is zero in the absence of an external field, and, there-fore, $\underline{v}_g$ does not change with time. In an external field, $\partial \mathcal{E}_K / \partial t$ must be equal to the rate at which the field does work on the electron, which classically is just $-\underline{v} \cdot e\underline{E}$, so we set

$$\frac{\partial \mathcal{E}_K}{\partial t} = -\underline{v}_g \cdot e\underline{E} = -\frac{1}{\hbar} \left(\frac{\partial \mathcal{E}_K}{\partial t} \right) e\underline{E} \quad .$$

Inserting this value of $\partial \mathcal{E}_K / \partial t$ into Eq. (68), we find

$$\frac{\partial \underline{v}_g}{\partial t} = -\left(\frac{1}{\hbar} \right)^2 \left(\frac{\partial^2 \mathcal{E}_K}{\partial K^2} \right) e\underline{E} \quad , \tag{78}$$

which is the correct quantum mechanical equation of motion for a con-duction electron with crystal momentum $\hbar \underline{K}$. Comparing this result to its classical equivalent

$$m\underline{v} = -e\underline{E} \quad ,$$

we come to the important conclusion that an electron in a crystal reacts to an external field as if it had an effective mass m^*, given by

$$\frac{1}{m^*} = \frac{1}{\hbar^2} \left(\frac{\partial^2 \mathcal{E}_K}{\partial K^2} \right) \quad . \tag{79}$$

The important thing to notice in connection with this result is that we have nowhere made any restrictions on the magnitude of the potential of the positive ion cores. This is very nice because it means that, even if the electrons are not free (in the sense of having no potential energy), we can still expect our classical response function to be correct if we replace the true electron mass by the effective mass m^*.

Actually, it turns out that, for most simple metals (which includes all the elemental metals but the transition metals and the noble metals Cu, Ag, and Au), the kinetic energy of the conduction electrons really is much greater than the electron-core interaction, so a nearly free electron theory is a good approximation, and $m^* \approx m$ over much of the B.Z. This comes about because the strong electrical attraction of the core is opposed by a quantum mechanical repulsion between the conduction electrons and the inner electrons of the core. Providing the core electron energy levels are quite deep, the net effective potential, or pseudopotential as it is called, felt by the conduction electron is quite small.

In other metals and most semiconductors, the cancellation of the coulombic attraction is not as good, and the electrons by no means appear nearly free. It is not a trivial matter to calculate $\mathcal{E}_K$ vs $\underline{K}$, and thus m^*, reliably for a given material. It is often more fruitful to invert the process and infer the curvature of the energy bands from the values of m^* derived, for example, from studies of the optical properties.

The effective mass of an electron has rather "queer" properties. To begin with, $(1/m^*)$ is really a tensor with components like $(1/m^*)_{xy} = (1/\hbar)^2(\partial^2\mathcal{E}/\partial K_x \partial K_y)$, etc. For simplicity we have been assuming spherical energy band surfaces, for that is the only case in which m^* reduces to a scalar number. Near the bottom of a band, the curvature is necessarily positive, so the components of m^* are positive, but the values can be quite different from the true mass. For example, for the lowest conduction band of the semiconducting alloy $Cd_xHg_{1-x}Te$ ($x = 0.14$), $m^* \leq 0.0004\,m$. More odd still is the property of negative effective mass found for electrons near the top of an energy band. In this case, if the band is nearly full, we again focus attention upon the vacant hole states which are "trapped" into moving along with the electrons in an applied field. Equation (78) correctly describes the motion of these "holes" if we give them a positive charge $+|e|$ and a positive inverse mass tensor $(1/m^*) = -(1/\hbar)^2(\partial^2\mathcal{E}/\partial K^2)$.

XIII. OPTICAL PROPERTIES OF POINT IMPERFECTIONS

In the preceding several sections we have discussed the dielectric response of an ideal crystal, i.e., one free of all defects and impurities. The deviations from this ideal, which inevitably occur in real crystals, even nominally "pure" ones, do not normally produce major changes in their overall dielectric and optical behavior. Absorption associated with fundamental electronic or lattice absorption is very great, so great that light can be nearly completely attenuated within 10^{-6} cm or less, a few tens of atomic layers. Perturbations caused by imperfections are minor on this scale. They can, however, still give rise to readily observable and even striking effects, providing that absorption is introduced into an otherwise transparent region in the host crystal, or to pick another example, if a few electrons are introduced into an otherwise empty conduction band. Furthermore, the special property changes brought about by deliberate and controlled introduction of impurities into solids are of great practical importance, and optical techniques are among the most direct and powerful means available for studying such impurity states.

It is more difficult to give a general account of impurity spectra than it is for the intrinsic processes we have studied previously. This is mainly because we cannot fall back on such generalizations as Bloch wave functions and crystal momentum. The validity of these concepts rests upon lattice periodicity, which is disrupted in an essential way by the imperfections.

A. Localized States in a One-Dimensional Lattice

To see how localized states come about when impurities are present, we can go back to our linear lattice in the tight binding approximation (Section IX). Recall that we introduced an initial set of wave functions

$$\Psi_\ell(x, t) = \phi_\ell(x) \exp\left[-i(\bar{\mathscr{E}}t/\hbar)\right]$$

representing excitations entirely localized in each of the cells, but we found that, as a result of interactions between cells, these did not represent true energy states of the system. The allowed energy states could however be constructed by taking the appropriate linear combinations of the above functions:

$$\Psi'_K (x, t) = \sum_\ell \exp[i(K\ell a - \mathscr{E}'t/\hbar)]\Psi_\ell(x, t) \quad .$$

Now suppose we put an imperfection in the cell at $\ell = 0$. It is clear that even in the absence of interactions between cells the impurity will have a somewhat different energy than its neighbors, so we write

$$\Psi_{\ell = 0}(x, t) = \phi_{\ell = 0} \exp\{-i[(\bar{\mathscr{E}} + \Delta)t/\hbar]\} \quad ,$$

so that

$$\mathscr{H}_0 \Psi_{\ell = 0} = (\bar{\mathscr{E}} + \Delta)\Psi_{\ell = 0} \quad ,$$

whereas $\mathscr{H}_0 \Psi_\ell = \bar{\mathscr{E}} \Psi_\ell$ for all cells where $\ell \neq 0$. To complete the problem we consider only the nearest cell interactions, specified as before by

$$\mathscr{H} \Psi_\ell = -\Gamma(\Psi_{\ell+1} + \Psi_{\ell-1}) \quad .$$

(Of course we should really allow the $\ell = 0$ cell to interact differently with its neighbors, but this is an extra algebraic complication which is unnecessary to demonstrate the basic features of localized excitations.)

Now suppose we proceed as in Section IX, make up a new trial wave function

$$\Psi'(x, t) = \sum_\ell c_\ell(t) \Psi_\ell(x, t) \quad ,$$

and require that Ψ' be an allowed energy state of the total hamiltonian, i.e.,

$$i\hbar \, \partial\Psi'/\partial t = (\mathscr{H}_0 + \mathscr{H}')\Psi' = \mathscr{E}\Psi' \quad .$$

Now, however, rather than looking for a sinusoidal spatial variation as in Eq. (48), we look for something which falls off exponentially as we go away from $\ell = 0$:

$$c_\ell = c^0 \exp(-K'\ell a) \exp(-i\mathscr{E}'t/\hbar) \quad , \qquad \ell > 0$$

$$c_\ell = c^0 \exp(K'\ell a) \exp(-i\mathscr{E}'t/\hbar) \quad , \qquad \ell < 0 \qquad (80)$$

$$c_\ell = c^0 \exp[-i(\mathscr{E}' - \Delta)t/\hbar] \qquad , \qquad \ell = 0 \quad .$$

We leave it as an exercise for the reader to work out the equations governing the c_ℓ's [most, but not all, of them obey Eq. (47)] and to substitute the exponential forms given above into them. You will find that, if Eqs. (80) are to represent valid solutions, two relations must be satisfied. First of all,

$$\exp(K'a) - \exp(-K'a) = -\Delta/\Gamma \quad , \tag{81}$$

which shows that the ratio Δ/Γ, which is fixed in any particular problem, uniquely determines the single allowed value of K'. (Note that this differs from the energy band case, where a large number of K-values are possible.) In fact if Δ and Γ are of the same sign it is impossible to satisfy Eq. (81) at all since $\exp(K'a) - \exp(-K'a)$ is always positive. In this case no truly localized impurity state can be formed. If Δ and Γ are of opposite sign one can always find a value of K' which satisfies Eq. (81), and the larger the ratio $|\Delta/\Gamma|$ the larger is K' and thus the more quickly the excitation decays once away from the cell containing the impurity. In no case, however, is the excitation entirely localized within a single cell. So long as there is interaction with neighboring cells there is a finite probability that the excitation will leak away and be found elsewhere.

The second equation which must be satisfied just gives the energy $\bar{\mathscr{E}}$ of the localized state relative to the band center:

$$\mathscr{E}'_{loc} = \pm(4\Gamma^2 + \Delta^2)^{\frac{1}{2}}$$

in which one must take the plus sign if Δ is positive and the negative sign if Δ is negative. Recalling that the band in the defect-free lattice extends from $-2\Gamma \leq \mathscr{E}'(K) \leq 2\Gamma$, we see that, if a local mode exists, it lies in the forbidden gap either above or below the band. The greater the ratio $|\Delta/\Gamma|$, the more tightly is the state localized in space and the further does its energy pull away from the edge of the energy band.

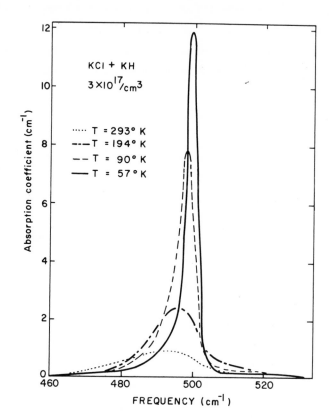

FIG. 22. Infrared spectrum of KCl showing absorption due to the U-center localized mode. Reprinted from G. Shaefer, J. Phys. Chem. Solids, 12, 233 (1960), by courtesy of Pergamon Press, Inc.

B. Localized Vibrational Modes

Many of the features of the simple model we have just presented have their counterparts in real physical systems. One of the easiest such systems to discuss is a crystal in which one or a relatively few light impurity atoms are substituted randomly throughout the lattice for normal atoms. Because the frequency at which an atom would like to vibrate varies inversely as the square root of its mass, an impurity atom, if it could vibrate independently (our initial assumption in the tight binding approximation), would do so at a higher frequency and thus a higher energy than its heavier normal neighbors. We have, for the sake of simplicity, additionally assumed that the forces holding the impurity to its equilibrium position are not sufficiently weakened to counteract the effect of the change in mass.

The analogy with our simple one-dimensional model is immediate. We should expect a localized vibrational mode to be formed above the band of energies at which the lattice phonons vibrate, and since in general there will be a dipole moment associated with the vibrating impurity, the localized excitation should absorb light of the appropriate frequency. Figure 22 shows an example of such impurity-induced absorption caused by U-centers in KCl. The U-center is an H^- ion that substitutes for a halide ion. In this case the mass difference is quite extreme, so that the localized mode frequency is considerably above the natural frequencies of the lattice, and the spatial localization of the vibration is correspondingly nearly complete.

In addition to localized states which exist in energy gaps in a host lattice, there also exist situations in which allowed energy states of the lattice-plus-impurity lie within the energy bands of the perfect lattice. These states are similar to the truly localized states, but the excitation amplitude, although it drops off considerably away from the impurity, takes on the character of ordinary wavelike band states. The amplitudes are most enhanced around the impurity position for states whose energies favor impurity ion participation. These quasilocalized or resonance states within the energy band are less well defined energetically than localized states because of the ease with which they interact and leak away into other bandlike states. As you might expect, it is possible to create quasilocalized vibrational excitations of this sort by introduction of a heavy mass defect into a light host lattice. Figure 23 shows a resonance due to Ag^+ ions substituted for K^+ in KBr and detected by absorption of far-infrared radiation. The impurity absorption in this case is at a considerably lower frequency than that due

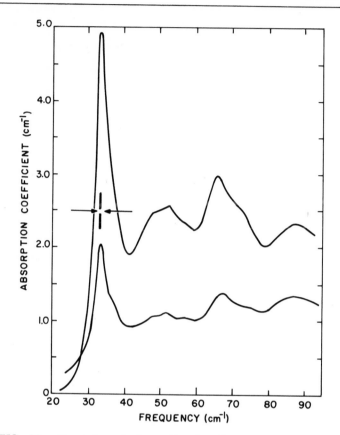

FIG. 23.   Far-infrared absorption in silver-activated potassium bromide.   The silver concentration is 7×10^{18} and 1.5×10^{18} atoms/cm^3 for the upper and lower curves, respectively.   The spectral slit width of the monochromator is indicated by the arrows.   The temperature is 2 °K. Reprinted from A.J. Sievers, Phys. Rev. Lett., 13, 310 (1964), by courtesy of the American Institute of Physics.

to the optic phonons and is situated energetically in the acoustic phonon band. (Refer to Fig. 10.) Although the mode appears quite well defined, its increased interaction with the lattice as compared to a mode within an energy gap is evidenced by the extremely low temperatures at which successful measurements must be performed.

C.   Shallow Donor and Acceptor Levels in Semiconductors

When a Group V element such as As substitutes for a host atom in a Si or Ge lattice, four of its valence electrons are used in the same way as those of the Si or Ge atom which it replaces in filling the valence band of the material. The fifth valence electron, if it were free, would be forced to begin filling the next higher conduction band. But the impurity atom, deprived of five electrons, attracts and can weakly bind this "extra" electron.

The tight binding scheme is not very appropriate to discuss this situation quantitatively. The state of affairs is really more similar to that of the weakly bound exciton which we discussed in Section XI. In fact, Eq. (66) which we derived for the energy of the bound exciton levels can also be used to estimate the binding of electrons to donor impurities. The effective mass of the donor ion may be taken as infinite since, unlike a hole, it is pinned in the lattice and cannot move. The reduced mass then becomes $\mu^* = m^*$, where m^* is the effective mass of the electron in the conduction band. Then, for example, assuming fairly typical values of $\epsilon_\infty = 10$ and $m^* = 0.1 m_e$, we should predict donors to have hydrogenlike spectra reduced in energy by a factor 10^3, giving a binding energy in the lowest ($n = 1$) level of $\sim 13 \times 10^{-3}$ eV. Figure 24 shows the results of a more careful calculation of the energy levels of donor states in Si, which takes into account the anisotropic effective mass of the conduction electrons. These results are compared with energy levels determined experimentally by observing infrared absorption corresponding to transitions between the $n = 1$ level and higher levels of the bound donors. The agreement is good for the excited states, but there are quite large individual deviations in the ground-state energies. The breakdown of this simple "effective mass" theory in the ground state is thought to occur because in the ground state the localized electron has a large probability of being within the donor cell, in which case the screened potential $V(r) = - e^2/\epsilon_\infty r$ is no longer appropriate. For the same reason, this simple treatment is completely inadequate for certain other non-Group V donors (e. g., Au) which have small orbital radii and consequently high binding energies (> 0.1 eV).

A model parallel to the one just given can be made for acceptor states, i. e., electronic states about an impurity which can accept an

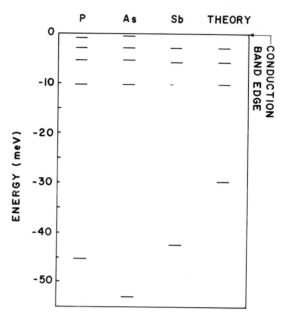

FIG. 24. Comparison of theoretically predicted energy levels of donor states in silicon with experimental observation. Reprinted from W. Kohn, in Solid State Physics, Vol. 5, p. 257, by courtesy of Academic Press, Inc.

electron from the valence band leaving a valence band "hole." Once again, estimates of the energies of acceptor states can be obtained from Eq. (66), using for μ^* the hole effective mass m_h^*, with about equally good results.

Of course, the occurrence of impurity donor and acceptor states is not limited to the Group IV semiconductors. Donor and acceptor levels can be introduced into the band gaps of all insulating materials. If the binding energy of the electron or hole to the impurity, $\mathcal{E}_b$, does not greatly surpass the available thermal energy kT, one observes thermally activated conduction. Direct absorption of light with an energy $\hbar\omega$ greater than $\mathcal{E}_b$ by a bound donor or acceptor is another important mechanism for introducing electrons or holes into the conduction band. This photoconductive effect is the basis of an important class of infrared light detectors.

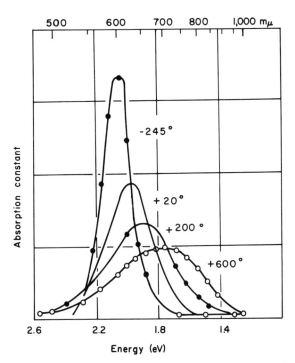

FIG. 25. The absorption spectrum of the F-center band in KBr at several temperatures. Reprinted from R. W. Pohl, Proc. Phys. Soc. (London), 49 (supplement), 3 (1937), by courtesy of the Institute of Physics.

D. Defect Color Centers

There are several ways by which coloration (i. e., optical absorption bands) can be introduced into an otherwise transparent spectral region of a crystal, in addition to the introduction of foreign atom impurities. For example, coloration results in many crystals upon exposure to ionizing radiation, x-rays, gamma rays, energetic electrons, or other particles. Other crystals can be colored by either chemical or electrolytic oxidation or reduction. In most of these instances the imperfection responsible is not an impurity atom but a vacancy or other intrinsic lattice defect. The best known and studied example of a defect color center is the F-center in the alkali halides.

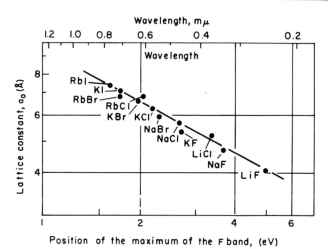

FIG. 26. Dependence of the F-band absorption maximum on the lattice constant of the host crystal. After H. Pick, Nuovo Cimento, VII, Series X, No. 2, 498 (1958), by courtesy of the Italian Physical Society.

An F-center is an electron bound at a negative ion vacancy. The optical coloration, which is caused by transitions to higher bound states, varies depending upon the alkali halide. The F-center absorption band produced in KBr, shown in Fig. 25, is in the red, so the crystal appears blue in transmitted light. The wavelength at which the peak of the rather broad absorption occurs can be correlated with the lattice constant a_0 of the crystal, as shown in Fig. 26. This can be approximately understood in the following simple way. The negative ion vacancy in the NaCl structure is bounded by an octahedron of positive ions, opposite vertices being separated by the distance a_0. If one approximates this positive ion "cage" by a cubic "potential box" with the potential going abruptly to infinity at $x, y, z = \pm a_0/2$, the sequence of energy levels is given by

$$\mathcal{E}(n_x, n_y, n_z) = \frac{\pi^2 \hbar^2}{2 m a_0^2} (n_x^2 + n_y^2 + n_z^2), \quad n_x, n_y, n_z = 1, 2, \ldots .$$

The excitation energy from the ground state to the first excited states is proportional to a_0^{-2}, which is quite close to what is actually observed.

At slightly elevated temperatures, F-centers are somewhat mobile and tend to aggregate in various ways to form more complex centers, each with its own characteristic spectrum. Thus an M-center consists of two adjacent F-centers, and R-centers consist of a cluster of three F-centers in a [111]-plane.

BIBLIOGRAPHY

Section I. General References

N. F. Mott and H. Jones, The Theory of the Properties of Metals and Alloys, Clarendon Press, Oxford, 1936.

F. Seitz, Modern Theory of Solids, McGraw-Hill, New York, 1940.

J. M. Ziman, Principles of the Theory of Solids, Cambridge Univ. Press, London and New York, 1964.

C. Kittel, Introduction to Solid State Physics, 3rd ed., Wiley, New York, 1967.

Sections II-V. Polarization Waves and Maxwell's Equations

L. D. Landau and E. M. Lifshitz, Electrodynamics of Continuous Media, Addison-Wesley, Reading, Massachusetts, 1960.

F. Stern, in Solid State Physics (F. Seitz and D. Turnbull, eds.), Vol. 15, Academic, New York, 1963.

R. P. Feynmann, Lectures in Physics, Addison-Wesley, Reading, Massachusetts, 1963.

Section VI. Experimental Methods

T. S. Moss, The Optical Properties of Semiconductors, Butterworth, London, 1959.

D. L. Greenaway and G. Harbeke, Optical Properties and Band Structure of Semiconductors, Pergamon, London, 1968.

Sections IX-XII. Excitation Waves in Solids

L. Brillouin, Wave Propagation in Periodic Structures, McGraw-Hill, New York, 1946.

M. Born and K. Huang, Dynamical Theory of Crystal Lattices, Clarendon Press, Oxford, 1954.

J. C. Slater, "Interaction of Waves in Crystals," Rev. Mod. Phys., 30, 197 (1958).

M. P. Givens, "Optical Properties of Metals," in Solid State Physics (F. Seitz and D. Turnbull, eds.), Vol. 6, Academic, New York, 1958.

H. R. Phillipp and H. Ehrenreich, in Semiconductors and Semimetals (R. K. Willardson and A. C. Beer, eds.), Vol. 3, Academic, New York, 1967.

Section XIII. Point Imperfections

W. Kohn, "Shallow Impurity States in Silicon and Germanium," in Solid State Physics (F. Seitz and D. Turnbull, eds.), Vol. 5, Academic, New York, 1957.

J. H. Schulman and W. D. Compton, Color Centers in Solids, Pergamon, London, 1962.

W. Beall Fowler, ed., Physics of Color Centers, Academic, New York, 1968.

RETURN TO: CHEMISTRY LIBRARY

100 Hildebrand Hall • 510-642-3753

LOAN PERIOD	1	2	1-MONTH USE	3
4		5		6

ALL BOOKS MAY BE RECALLED AFTER 7 DAYS.

Renewals may be requested by phone or, using GLADIS, type inv followed by your patron ID number.

DUE AS STAMPED BELOW.

FORM NO. DD 10
3M 7-08

UNIVERSITY OF CALIFORNIA, BERKELEY
Berkeley, California 94720–6000